Environmental Change in Siberia

ADVANCES IN GLOBAL CHANGE RESEARCH

VOLUME 40

Editor-in-Chief

Martin Beniston, *University of Geneva, Switzerland*

Editorial Advisory Board

B. Allen-Diaz, *Department ESPM-Ecosystem Sciences, University of California, Berkeley, CA, USA.*
R.S. Bradley, *Department of Geosciences, University of Massachusetts, Amherst, MA, USA.*
W. Cramer, *Earth System Analysis, Potsdam Institute for Climate Impact Research, Potsdam, Germany.*
H.F. Diaz, *Climate Diagnostics Center, Oceanic and Atmospheric Research, NOAA, Boulder, CO, USA.*
S. Erkman, *Institute for communication and Analysis of Science and Technology–ICAST, Geneva, Switzerland.*
R. Garcia Herrera, *Faculated de Fisicas, Universidad Complutense, Madrid, Spain.*
M. Lal, *Center for Atmospheric Sciences, Indian Institute of Technology, New Delhi, India.*
U. Luterbacher, *The Graduate Institute of International Studies, University of Geneva, Geneva, Switzerland.*
I. Noble, *CRC for Greenhouse Accounting and Research School of Biological Science, Australian National University, Canberra, Australia.*
L. Tessier, *Institut Mediterranéen d'Ecologie et Paléoécologie, Marseille, France.*
F. Toth, *International Institute for Environment and Sustainability, Ec Joint Research Centre, Ispra (VA), Italy.*
M.M. Verstraete, *Institute for Environment and Sustainability, Ec Joint Research Centre, Ispra (VA), Italy.*

For other titles published in this series, go to
www.springer.com/series/5588

Heiko Balzter
Editor

Environmental Change in Siberia

Earth Observation, Field Studies and Modelling

Editor
Heiko Balzter
Department of Geography
University of Leicester, Centre for Environmental Research
University Road
Leicester
United Kingdom
hb91@le.ac.uk

ISBN 978-90-481-8640-2   e-ISBN 978-90-481-8641-9
DOI 10.1007/978-90-481-8641-9
Springer Dordrecht Heidelberg London New York

Library of Congress Control Number: 2010927687

© Springer Science+Business Media B.V. 2010
No part of this work may be reproduced, stored in a retrieval system, or transmitted in any form or by any means, electronic, mechanical, photocopying, microfilming, recording or otherwise, without written permission from the Publisher, with the exception of any material supplied specifically for the purpose of being entered and executed on a computer system, for exclusive use by the purchaser of the work.

Cover illustration: Main photo: The foreground features subalpine meadows surrounding by Pinus Sibirica dominated woodlands, the background an alpine ridge in the Ergaki mountains called "Sleeping Sayan", photo by D.M. Ismailova. Top photo: A Larix sibirica above a landscape leading towards the South Altai Mountains, photo by V.I. Kharuk.

Printed on acid-free paper

Springer is part of Springer Science+Business Media (www.springer.com)

To Judith, Dominik and Julian...

Preface

The Siberian environment is a unique region of the world that is both very strongly affected by global climate change and at the same time particularly vulnerable to its consequences. The news about the melting of sea ice in the Arctic Ocean and the prospect of an ice-free shipping passage from Scandinavia to Alaska along the Russian north coast has sparked an international debate about natural resource exploitation, national boundaries and the impacts of the rapid changes on people, animals and plants. Over the last decades Siberia has also witnessed severe forest fires to an extent that is hard to imagine in other parts of the world where the population density is higher, the fire-prone ecosystems cover much smaller areas and the systems of fire control are better resourced. The acceleration of the fire regime poses the question of the future of the boreal forest in the taiga region. Vegetation models have already predicted a shift of vegetation zones to the north under scenarios of global climate change. The implications of a large-scale expansion of the grassland steppe ecosystems in the south of Siberia and a retreat of the taiga forest into the tundra systems that expand towards the Arctic Ocean would be very significant for the local population and the economy.

I have studied Russian forests from remote sensing and modelling for about 11 years now and still find it a fascinating subject to investigate. Over this time period Russia has undergone substantial social, political and economic changes and developed excellent remote sensing centres that now enjoy a world wide reputation. From 1998 to 2000 the European funded project SIBERIA, in which I started my post-doctoral research career and which was led by Professor Chris Schmullius from Jena, produced the first Synthetic Aperture Radar (SAR) map of forest growing stock over an area of 1 million square kilometers. At the time, the German Aerospace Agency (DLR) had to move a mobile receiving station to Lake Baikal to be able to record the first SAR images of the region. The forest map used over 600 images from three radar sensors, and led to the insight that the remaining forest cover in Siberia is much less than previous global change studies assumed. In the follow-on project SIBERIA-II we examined a much wider concept of using a whole range of biophysical data products from a multitude of satellites in a full greenhouse gas account over a region of 3 million square kilometers. This study was the first such attempt to incorporate many variables that would now be called Essential Climate Variables by the Global Climate Observing System (GCOS) into a real greenhouse gas account.

When I took up the Chair in Physical Geography at the University of Leicester in 2006 I invited a number of eminent researchers with interests in environmental change in Siberia to visit Leicester for a Symposium on Environmental Change in Siberia. We enjoyed 2 days packed with exciting presentations and full of inspiring conversations over coffee, tea and dinner. This book is primarily the outcome of this Symposium with a few additions from authors who I invited to contribute. I am particularly grateful to the University of Leicester for its financial support for the Symposium and to all participants for their contributions to this book. I also want to thank Alex Szumski who was a crucial helper in getting the book manuscript to the printing stage.

The structure of this book covers environmental change processes in the biosphere, hydrosphere and atmosphere and concludes with two contributions on environmental information systems that are being developed to safeguard data that are vital to further advance our understanding of Siberian ecosystems.

Leicester, September 2009   Prof. Heiko Balzter

Contents

Part I Biosphere

1 **Forest Disturbance Assessment Using Satellite Data of Moderate and Low Resolution** 3
M.A. Korets, V.A. Ryzhkova, I.V. Danilova, A.I. Sukhinin, and S.A. Bartalev

2 **Fire/Climate Interactions in Siberia** 21
H. Balzter, K. Tansey, J. Kaduk, C. George, F. Gerard, M. Cuevas Gonzalez, A. Sukhinin, and E. Ponomarev

3 **Long-Term Dynamics of Mixed Fir-Aspen Forests in West Sayan (Altai-Sayan Ecoregion)** 37
D.M. Ismailova and D.I. Nazimova

4 **Evidence of Evergreen Conifers Invasion into Larch Dominated Forests During Recent Decades** 53
V.I. Kharuk, K.J. Ranson, and M.L. Dvinskaya

5 **Potential Climate-Induced Vegetation Change in Siberia in the Twenty-First Century** 67
N.M. Tchebakova, E.I. Parfenova, and A.J. Soja

6 **Wildfire Dynamics in Mid-Siberian Larch Dominated Forests** 83
V.I. Kharuk, K.J. Ranson, and M.L. Dvinskaya

7 **Dendroclimatological Evidence of Climate Changes Across Siberia** 101
V.V. Shishov and E.A. Vaganov

8 Siberian Pine and Larch Response to Climate Warming
   in the Southern Siberian Mountain Forest: Tundra Ecotone 115
   V.I. Kharuk, K.J. Ranson, M.L. Dvinskaya, and S.T. Im

Part II Hydrosphere

9 Remote Sensing of Spring Snowmelt in Siberia 135
   A. Bartsch, W. Wagner, and R. Kidd

10 Response of River Runoff in the Cryolithic Zone
   of Eastern Siberia (Lena River Basin) to Future
   Climate Warming 157
   A.G. Georgiadi, I.P. Milyukova, and E.A. Kashutina

Part III Atmosphere

11 Investigating Regional Scale Processes Using Remotely
   Sensed Atmospheric CO_2 Column Concentrations
   from SCIAMACHY 173
   M.P. Barkley, A.J. Hewitt, and P.S. Monks

12 Climatic and Geographic Patterns of Spatial Distribution
   of Precipitation in Siberia 193
   A. Onuchin and T. Burenina

Part IV Information Systems

13 Interoperability, Data Discovery and Access:
   The e-Infrastructures for Earth Sciences Resources 213
   S. Nativi, C. Schmullius, L. Bigagli,
   and R. Gerlach

14 Development of a Web-Based Information-Computational
   Infrastructure for the Siberia Integrated Regional Study 233
   E.P. Gordov, A.Z. Fazliev, V.N. Lykosov, I.G. Okladnikov,
   and A.G. Titov

15 Conclusions 253
   H. Balzter

Appendix 255

Index 279

Contributors

Heiko Balzter
Department of Geography, University of Leicester, Centre for Environmental Research, University Road, Leicester LE1 7RH, UK
hb91@le.ac.uk

M.P. Barkley
School of GeoSciences, University of Edinburgh, Crew Building,
The King's Buildings, West Mains Road, Edinburgh EH9 3JN, UK
Michael.Barkley@ed.ac.uk

S.A. Bartalev
Space Research Institute (IKI), 117997, 84/32 Profsoyuznaya str.,
Moscow, Russia
bartalev@smis.iki.rssi.ru

A. Bartsch
Institute of Photogrammetry and Remote Sensing,
Vienna University of Technology, Gusshausstraße 27–29, 1040 Vienna, Austria
ab@ipf.tuwien.ac.at

Lorenzo Bigagli
Friedrich-Schiller-University, Institute for Geography,
Earth Observation, Grietgasse 6, 07743 Jena, Germany
lorenzo.bigagli@pin.unifi.it

T. Burenina
V.N. Sukachev Institute of Forest, SB RAS, 660036, Krasnoyarsk,
Akademgorodok, 50, Russia
burenina@ksc.krasn.ru

Maria Cuevas Gonzalez
Centre for Ecology and Hydrology, Maclean Building, Benson Lane,
Crowmarsh Gifford, Wallingford, Oxfordshire, OX10 8BB, UK
cuevasgonzalez@gmail.com

I.V. Danilova
Sukachev Institute of Forest (SIF), 660036, 50/28,
Akademgorodok str., Krasnoyarsk, Russia
tiv80@ksc.krasn.ru

M.L. Dvinskaya
V.N. Sukachev Institute of Forest, SB RAS, 660036,
Krasnoyarsk, Academgorodok, 50, Russia
mary_dvi@ksc.krasn.ru

A.Z. Fazliev
Institute of Atmospheric Optics SB RAS, 634055, Tomsk,
Akademicheski ave., 1, Russia
faz@iao.ru

Charles George
Centre for Ecology and Hydrology, Maclean Building, Benson Lane,
Crowmarsh Gifford, Wallingford, Oxfordshire, OX10 8BB, UK
ctg@ceh.ac.uk

A.G. Georgiadi
Institute of Geography, Russian Academy of Sciences,
Staromonetny per., 29, 119017 Moscow, Russia
galex50@gmail.com

France Gerard
Centre for Ecology and Hydrology, Maclean Building, Benson Lane,
Crowmarsh Gifford, Wallingford, Oxfordshire, OX10 8BB, UK
ffg@ceh.ac.uk

Roman Gerlach
Friedrich-Schiller-University, Institute for Geography, Earth Observation,
Grietgasse 6, 07743 Jena, Germany
roman.gerlach@uni-jena.de

E.P. Gordov
Siberian Center for Environmental research and Training
and Institute of Monitoring of Climatic and Ecological Systems SB RAS,
634055, Tomsk, Akademicheski ave., 10/3, Russia
gordov@scert.ru

A.J. Hewitt
Earth Observation Science group, Departments of Physics and Chemistry,
University of Leicester, University Road, Leicester, LE1 7RH, UK
ajh67@le.ac.uk

S.T. Im
V.N. Sukachev Institute of Forest, SB RAS, 660036, Krasnoyarsk,
Akademgorodok, 50, Russia
stim@ksc.krasn.ru

D.M. Ismailova
V.N. Sukachev Institute of Forest, SB RAS, 660036,
Krasnoyarsk, Akademgorodok, 50, Russia
dismailova@mail.ru

Jörg Kaduk
Centre for Environmental Research, Department of Geography
University of Leicester, University Road, Leicester LE1 7RH, UK
j.kaduk@leicester.ac.uk

E.A. Kashutina
Institute of Geography, Russian Academy of Sciences,
Staromonetny per., 29, 119017 Moscow, Russia
kategeo@mail.ru

V.I. Kharuk
V.N. Sukachev Institute of Forest, SB RAS, 660036, Krasnoyarsk,
Academgorodok, 50, Russia
kharuk@ksc.krasn.ru

R. Kidd
Institute of Photogrammetry and Remote Sensing, Vienna University of
Technology, Gusshausstraße, 27–29, 1040 Vienna, Austria and now at Spatial
Information & Mapping Centre, Banda Aceh, Indonesia
richard.a.kidd@gmail.com

M.A. Korets
Sukachev Institute of Forest (SIF), 50/28, Akademgorodok street,
660036, Krasnoyarsk, Russia
mik@ksc.krasn.ru

V.N. Lykosov
Institute for Numerical Mathematics RAS, Moscow, Russia
lykossov@inm.ras.ru

I.P. Milyukova
Institute of Geography, Russian Academy of Sciences,
Staromonetny per., 29, 119017 Moscow, Russia
mil-ira@list.ru

P.S. Monks
Earth Observation Science group, Departments of Physics and Chemistry,
University of Leicester, University Road, Leicester, LE1 7RH, UK
p.s.monks@le.ac.uk

Stefano Nativi
Italian National Research Council – IMAA and University of Florence at Prato
nativi@imaa.cnr.it

D.I. Nazimova
V.N. Sukachev Institute of Forest, SB RAS, 660036, Krasnoyarsk,
Akademgorodok, 50, Russia
inpol@mail.ru

I.G. Okladnikov
Siberian Center for Environmental research and Training and Institute of
Monitoring of Climatic and Ecological Systems SB RAS, 634055, Tomsk,
Akademicheski ave., 10/3, Russia
onuchin@ksc.krasn.ru

A. Onuchin
V.N. Sukachev Institute of Forest, SB RAS, 660036,
Krasnoyarsk, Akademgorodok, 50, Russia
onuchin@ksc.krasn.ru

E.I. Parfenova
V.N. Sukachev Institute of Forest, SB RAS, 660036,
Krasnoyarsk, Akademgorodok, 50, Russia
02611@rambler.ru

Evgeni Ponomarev
Sukachev Institute of Forest, Siberian branch of Russian Academy of Sciences,
660036, Krasnoyarsk, Academgorogok, Russia
evg@ksc.krasn.ru

K.J. Ranson
NASA Goddard Space Flight Center, Greenbelt, MD 20771, USA
jon.ranson@nasa.gov

V.A. Ryzhkova
Sukachev Institute of Forest (SIF), 50/28, Akademgorodok Street,
660036, Krasnoyarsk, Russia
vera@ksc.krasn.ru

Christiana Schmullius
Friedrich-Schiller-University, Institute for Geography, Earth Observation,
Grietgasse 6, 07743 Jena, Germany
c.schmullius@uni-jena.de

Vladimir V. Shishov
IT and Math. Modelling Department, Krasnoyarsk State Trade-Economical
Institute, L. Prushinskoi St., Krasnoyarsk, 660075, Russia
shishov@forest.akadem.ru
And Dendroecology Department, Sukachev Institute of Forest,
Siberian Branch of Russian Academy of Sciences,
Akademgorodok St., Krasnoyarsk, 660036, Russia

A.J. Soja
National Institute of Aerospace, Resident at NASA Langley Research Center 21
Langley Boulevard, Mail Stop 420, Hampton, VA 23681-2199, USA
Amber.J.Soja@nasa.gov

A.I. Sukhinin
Sukachev Institute of Forest (SIF), 660036, 50/28,
Akademgorodok str., Krasnoyarsk, Russia
boss@ksc.krasn.ru

Kevin Tansey
Centre for Environmental Research, Department of Geography
University of Leicester, University Road, Leicester LE1 7RH, UK
kjt7@le.ac.uk

N.M. Tchebakova
V.N. Sukachev Institute of Forest, SB RAS, 660036, Krasnoyarsk,
Akademgorodok, 50, Russia
ncheby@ksc.krasn.ru

A.G. Titov
Siberian Center for Environmental research and Training and Institute of
Monitoring of Climatic and Ecological Systems SB RAS, 634055, Tomsk,
Akademicheski ave., 10/3, Russia
titov@scert.ru

Eugene A. Vaganov
Siberian Federal University, 79 Svobodnji Ave, Krasnoyarsk 660041, Russia
institute@forest.akadem.ru

W. Wagner
Institute of Photogrammetry and Remote Sensing,
Vienna University of Technology, Gusshausstraße, 27–29, 1040 Vienna, Austria
ww@ipf.tuwien.ac.at

Part I
Biosphere

Chapter 1
Forest Disturbance Assessment Using Satellite Data of Moderate and Low Resolution

M.A. Korets, V.A. Ryzhkova, I.V. Danilova, A.I. Sukhinin, and S.A. Bartalev

Abstract Envisat-MERIS and SPOT Vegetation satellite data were tested for estimation of vegetation cover disturbances caused by fire and industrial pollution in central and northern Siberian test sites, respectively. MERIS data were used to assess forest disturbance levels on burned sites in Angara region. Chlorophyll indexes (REP and MTCI) were found to allow identifying up to five forest disturbance levels due to high space-borne sensor resolution and sensitivity to chlorophyll content of vegetation. A comparison of these chlorophyll indexes revealed that MTCI to show chlorophyll contents fairly precisely and to be useful for quantifying and mapping forest damage levels on burns. The current vegetation condition was assessed using MTCI index in the northern (Norilsk) test region. The lowest index values calculated for the most severely disturbed vegetation near Norilsk were found to correlate with sulphur concentrations in larch and spruce needles. Another approach to estimating spatial and temporal trends of vegetation condition used the 1998–2005 SPOT-Vegetation satellite data. The relationships obtained between MTCI, NDVI values, and forest mortality were based upon to map the1998–2005 forest degradation zone dynamics in the northern test site.

Keywords Chlorophyll indexes • Envisat-MERIS • SPOT vegetation • Vegetation condition assessment

M.A. Korets (✉), V.A. Ryzhkova, A.I. Sukhinin, and I.V. Danilova
Sukachev Institute of Forest (SIF), 50/28, Akademgorodok street, 660036, Krasnoyarsk, Russia
e-mail: mik@ksc.krasn.ru; vera@ksc.krasn.ru; boss@ksc.krasn.ru; tiv80@kgs.krasn.ru

S.A. Bartalev
Space Research Institute (IKI), 84/32 Profsoyuznaya street, 117997, Moscow, Russia
e-mail: bartalev@smis.iki.rssi.ru

1.1 Introduction

The need for real-time monitoring of terrestrial ecosystems in vast, remote areas of Siberia enhances the use of medium-to-low (250–1,000 m) resolution satellite data provided, with a sufficient frequency, by instruments having a wide field of view. Satellite data obtained in visible and near-IR spectral bands have been used efficiently for estimating terrestrial ecosystem characteristics and levels of disturbance by biotic and abiotic factors (Curran et al. 1997). NOAA AVHRR, SPOT Vegetation (1 km resolution), and TERRA/AQUA MODIS (250–500 m resolution) data have enjoyed an active application in detecting and assessing vegetation cover disturbances, such as logging and fire scars, insect outbreaks, and industrial pollution.

ENVISAT, one of the most current Earth observation spacecraft of the European Space Agency (ESA), was launched in 2002. Among ten sophisticated instruments, it carries MERIS spectrometer. MERIS (Medium Resolution Imaging Spectrometer) has fifteen programmable channels for investigating backscattered solar radiation in visible (seven channels) and near-IR (eight channels) bands with 300 m resolution (Curran and Steele 2005). With its 1,150 km viewing field, this instrument requires as few as 3 days to provide global coverage highly needed for atmospheric and ocean research, as well as for forest cover monitoring.

Regarding vegetation, MERIS IR channels 7 through 13 centred in 665–865 nm band appear to be most suitable, since this is where the so-called red edge position (REP), or the red boundary of the chlorophyll absorption zone, is found (Clevers et al. 2002). Absolute chlorophyll content and its abundance in the photosynthetically active green plant parts, which can be estimated from REP (Clevers et al. 2002), is considered to be a key indicator of plant health. Increasing chlorophyll content is manifested by REP movement (shift) towards longer waves. The REP can be calculated from vegetation reflectances in red and infrared satellite channels using so-called zonal ratios, or vegetation indexes (Vinogradov 1994). These forest cover state indicators remain invariant for a wide range of environmental factors. While many vegetation indexes are available nowadays, the normalized difference vegetation index (NDVI) ($NDVI = (p_{IR} - p_R)/(p_{IR} + p_R)$, where p_{IR} and p_R are image pixel reflectances in the near-IR and red spectral bands, respectively) enjoys the widest use.

REP can be determined as a point of maximum vegetation reflectance change in the band between 670 and 780 nm wavelengths. Various interpolation methodologies, such as three point Lagrangian interpolation (Jeffrey 1985) and linear interpolation (Dawson and Curran 1998), can help quantify REP. The latter can be calculated for MERIS channels 7, 9, 10, and 12 centred at 665, 708.75, 753.75, and 778.75 nm, respectively:

$$REP(MERIS) = 708.75 + 45\frac{(R_i(MERIS) - R_{Band9})}{(R_{Band10} - R_{Band9})} = 708.75 + 45\frac{(R_i(MERIS) - R_{708.75})}{(R_{753.75} - R_{708.75})},$$

where $R_i(MERIS) = \dfrac{(R_{Band7} + R_{Band12})}{2} = \dfrac{(R_{665} + R_{778.75})}{2}$ is reflectance cusp, with reflectance curved derived from MERIS data; and *REP(MERIS)* is REP, nm.

Another method of MERIS data-based chlorophyll content assessment has been recently proposed (Dash and Curran 2004). This is actually a new index called MERIS Terrestrial Chlorophyll Index (MTCI), which is a ratio of reflectance difference between MERIS channels 10 and 9 to that between MERIS channels 9 and 8:

$$MTCI = \frac{R_{Band10} - R_{Band9}}{R_{Band9} - R_{Band8}} = \frac{R_{753.75} - R_{708.75}}{R_{708.75} - R_{681.25}}$$, where $R_{753.75}$, $R_{708.75}$ and $R_{681.25}$ are reflectances in the respective MERIS channels.

The purpose of this study was to assess fire- and industrial emission-caused forest ecosystem disturbance level and spatial and temporal patterns. This included the following tasks: obtain satellite imagery for the study area; select data processing methodologies providing forest disturbance estimates; conduct thematic satellite imagery processing; and build thematic forest disturbance maps.

1.2 Study Area

Our study was carried out in two sites (Fig. 1.1): a Central Siberia site (57°–60° N; 95°–100° E) disturbed by fire in Angara region and Northern Siberia site (67°–71° N; 85°–95° E) experiencing industrial pollution, near Norilsk.

The Angara region is the southernmost central Siberian province, where a slightly continental climate of western Siberia and highly continental climate of Lena river catchment and north-eastern Siberia meet. We investigated the Chuna–Angara (these are rivers) forest vegetation sub-province with fairly smooth topography and highly continental climate. For the major conifer woody species, the forest is dominated by southern taiga Scots pine, mixed Scots pine/larch, larch, and larch/Scots pine stands, with secondary mixed birch/aspen stands of fire origin being also common, as fire is the main forest disturbance here.

The second study site was selected in the area that experiences direct pollutions from Norilsk industrial complex. Since this area is situated at the boundary between western and central Siberia, it is markedly diverse in terms of natural zones ranging from plain bogs to mountain tundra. The area is generally represented by plain (lowland tundra-forest and raised forest-tundra plains) and low-mountain (low mountains occupied by open woodland-tundra and taiga-open woodland) landscape types.

1.3 Vegetation in Burned Sites

MERIS FR (Full Resolution Geophysical Product) images with 300 m resolution[1] and the field (ground) data for the 1996–2004 burns[2] were used to estimate vegetation condition on burned sites in Angara region. The field data collected during

---

[1] ENVISAT MERIS images were provided by FEMINE project (Forest Ecosystem Monitoring in Northern Eurasia) ESA-IAF, 2004.
[2] Field data were provided by FireBear project (NASA 04-05-476).

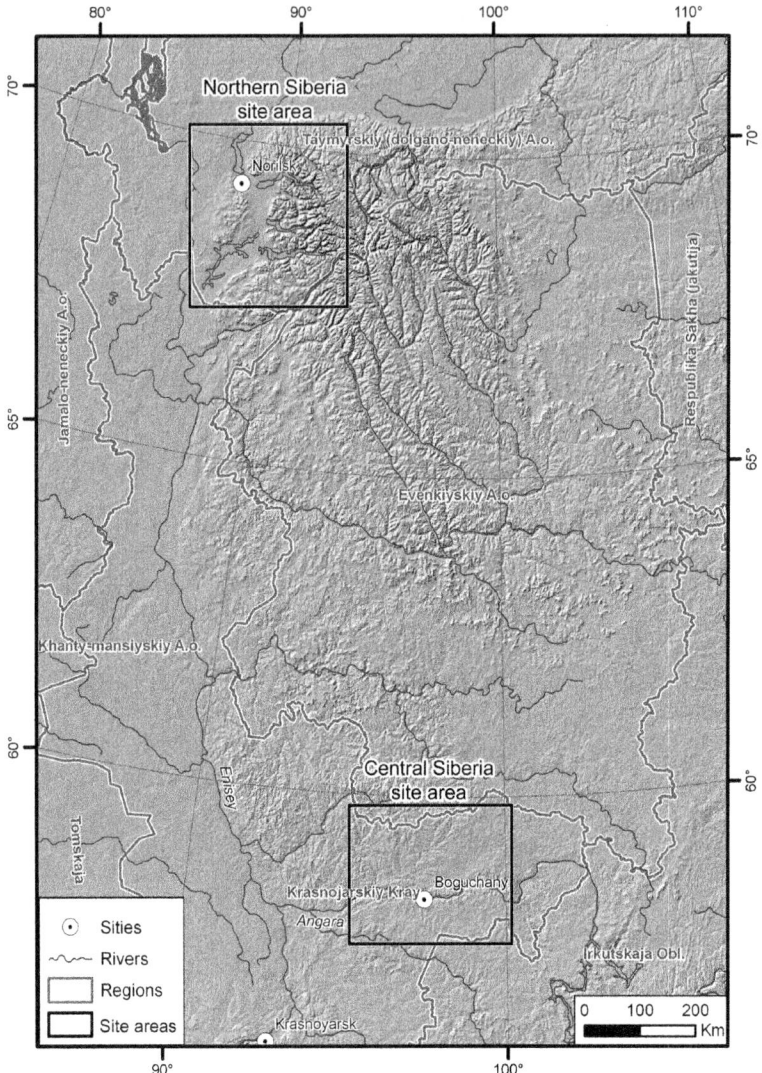

Fig. 1.1 The location of the test sites

2003 and 2004 contained locations of over 40 sample plots laid out on burned sites, fire dates, as well as stand species composition and tree mortality at the time of observation.

MERIS images passed trough the geometric and radiometric corrections in accordance with "Level 2 Products" specification (http://envisat.esa.int/instruments/meris/data-app/dataprod.html). For our space-scale analysis, we chose a minimum-cloud or minimum mist satellite image taken on 21 August 2003 (as close as possible to the 2003 ground observation), which was representative of the regional growing season peak (June–August). NDVI, REP, and MTCI were calculated for this image using the above equations. NDVI was based on data from MERIS channels 8 (681.25 nm) and 13 (865 nm). Sample plot locations were laid over this image and test sites 15–20 pixels each were then selected in the image. Test site size and shape followed a criterion of fire scar reflectance uniformity in both MERIS image and the superposed 35 m-resolution RGB composite images taken by Meteor MSU-E satellite in June–August 2004. Statistical signatures were calculated for test sites based on MERIS reflectances and NDVI, REP, and MTCI values. As a result, five post-fire tree mortality levels were identified, which differed significantly in the chlorophyll index averages (Table 1.1).

Figure 1.2 shows vegetation reflectance spectra obtained at different tree mortality levels. Increasing tree mortality induces a decrease in forest canopy chlorophyll concentration and is, hence, associated with decreasing energy absorption in the red

Table 1.1 NDVI, MTCI, and REP at different post-fire tree mortality

Tree mortality (%)	NDVI	REP (nm)	MTCI
80–100	0.25–0.45	712.00–718.07	1.20–1.69
60–80	0.42–0.56	718.08–719.77	1.70–2.00
40–60	0.55–0.61	719.78–721.24	2.01–2.30
20–40	0.61–0.65	721.25–722.45	2.31–2.65
0–20	0.61–0.67	722.46–723.81	2.66–3.00

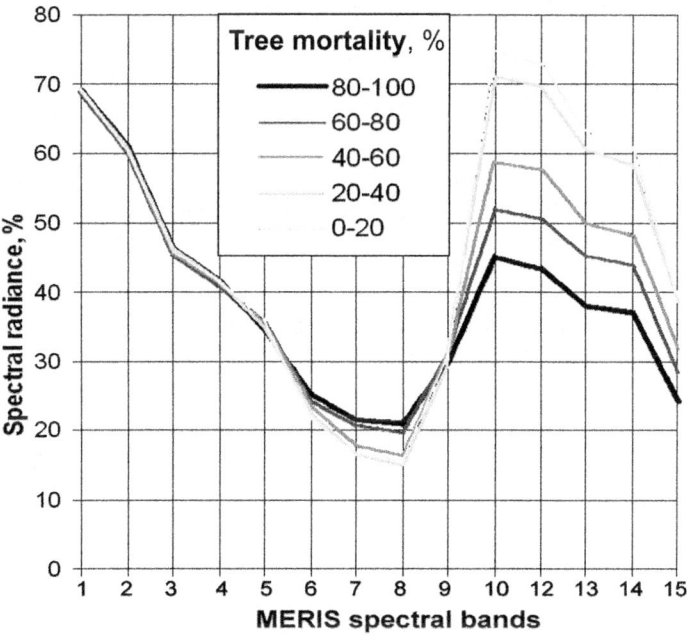

Fig. 1.2 Spectral radiance of vegetation cover at different tree mortality derived from MERIS standard band setting

spectral band, as well as decreasing reflectance in the near-IR band. As tree mortality increases and chlorophyll decreases, the reflectance difference between red band channels 7 and 8 and that between near IR channels 9–13 decreases to result in decreasing NDVI, REP, and MTCI.

As is clear from the behaviour of the indexes represented in Fig. 1.3, tree mortality is related almost linearly with MTCI, unlike with NDVI and REP, where a logarithmic dependence is observed. Consequently, NDVI and REP values would be less accurate in the saturation zone at high chlorophyll, i.e. at low tree mortality, in our case. The differences in the indexes behaviour become more apparent from their interaction shown in Fig. 1.4. The relationship between REP and MTCI is of logarithmic character, however, it is much steadier than that for NDVI-MTCI and REP-NDVI pairs. The range (spread) of values obtained in the two latter cases is most probably induced by NDVI, which index, unlike REP and MTCI, is more susceptible to "external" influences not related with chlorophyll concentration, such as woody species composition, stand structure, the presence of under-canopy or background objects including non-vegetation ones.

Among the three indexes of interest, MTCI thus appears to be the sensitive and simple chlorophyll-based indicator of forest stand condition or level of disturbance. Figure 1.5 presents a MERIS image fragment classified by tree mortality level using MTCI.

1.4 Vegetation Condition in the Industrial Emission Zone

In order to assess spatial and temporal forest disturbance patterns in the zone under long-term industrial pollution, we used satellite images of moderate (ENVISAT MERIS) and low (SPOT Vegetation) resolution. We chose five MERIS FR ("Level 2 Products" specification) images taken over the same area within the region of interest on July 24, 25, 28, and 30, 2004. In attempt to carry out visual analysis of these scenes using the reflectances in the basic visual spectral channels (R1–R7), we built RGB composites:

$$R = \log(0.05 + 0.35 * R2 + 0.6 * R5 + R6 + 0.13 * R7)$$
$$G = \log(0.05 + 0.21 * R3 + 0.5 * R4 + R5 + 0.38 * R6)$$
$$B = \log(0.05 + 0.21 * R1 + 1.75 * R2 + 0.47 * R3 + 0.16 * R4)$$

These RGB composites based on a markedly wide coverage of the short wavelength- (blue) spectral band enabled estimation of the visible smoke plume length and pattern (shape) (Fig. 1.6). The plume direction was found to vary within an angle close to 180° south of Norilsk. It appeared to be 300 km long (starting from Norilsk) and to cover about 2 million hectares.

We used MTCI (Dash and Curran 2004) to assess vegetation disturbance. Figure 1.7 presents reflectances of sites covering a range of industrial pollution-caused vegetation disturbance levels obtained from MERIS channels. MTCI was calculated per pixel in each of five initial (source) MERIS scenes.

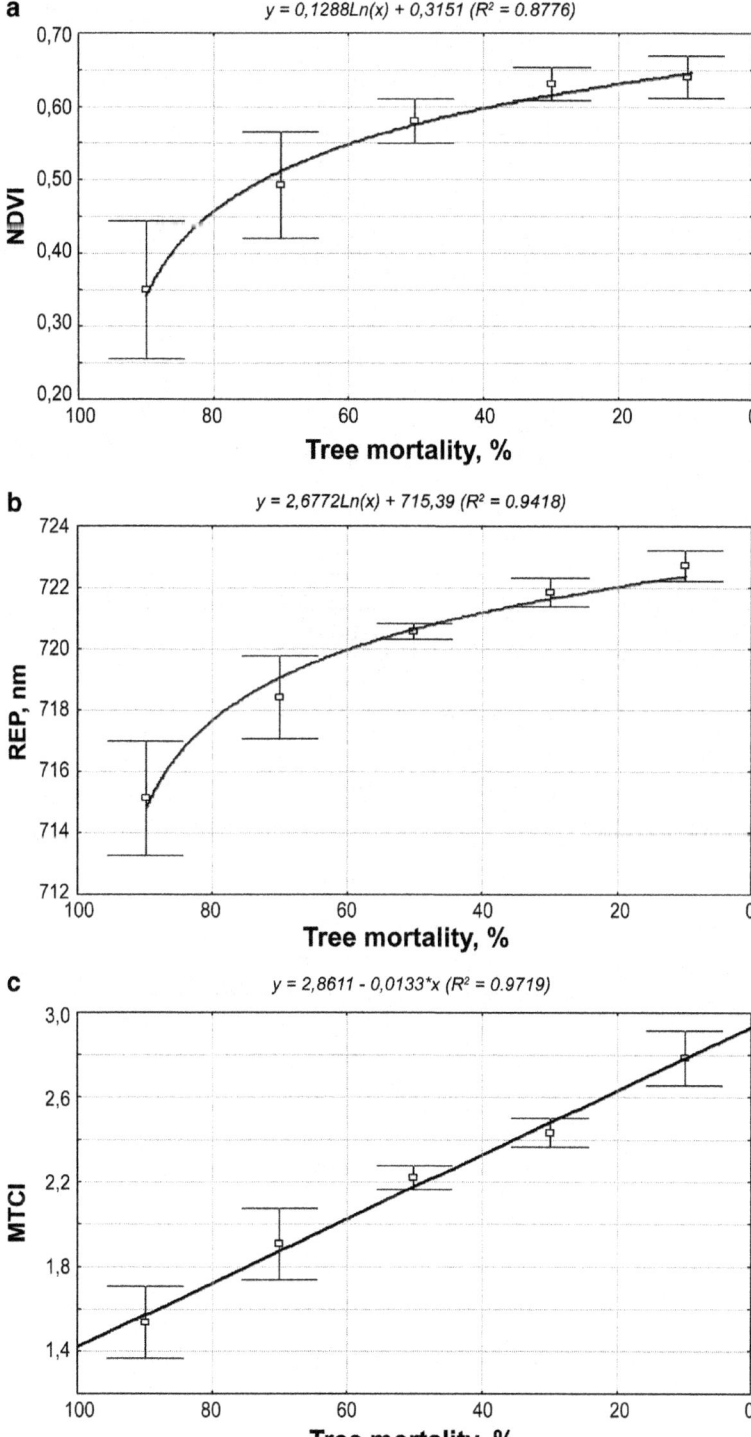

Fig. 1.3 Tree mortality relationship with (**a**) NDVI, (**b**) REP, and (**c**) MTCI

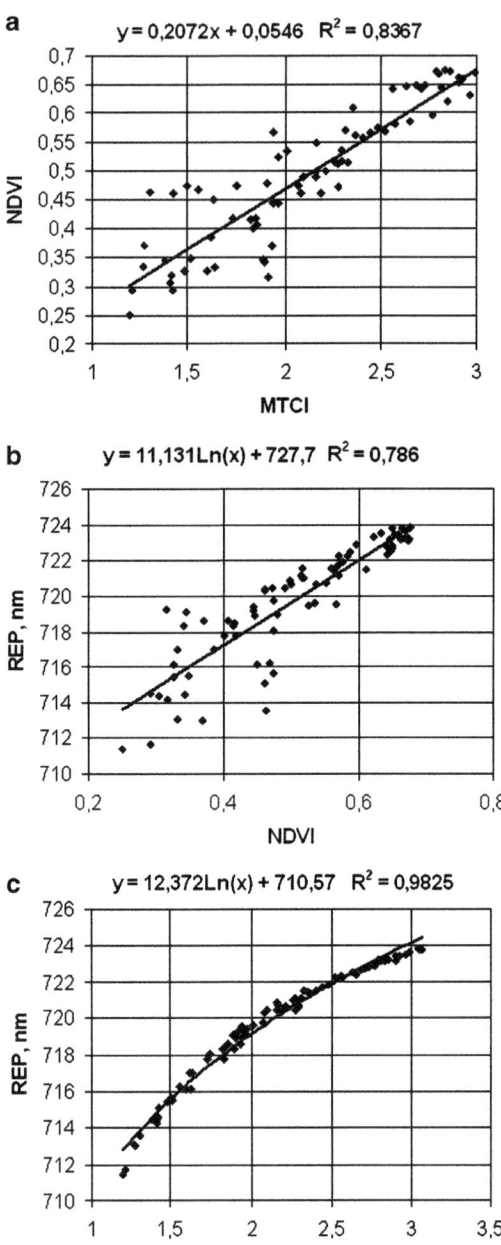

Fig. 1.4 The relationship between (**a**) NDVI and MTCI, (**b**) REP and NDVI, and (**c**) REP and MTCI

The scenes were superposed and the resultant MTCI value was obtained for each image pixel (i) for all the scenes, in effort to reduce atmospheric interference (mist, clouds, shadows):

1 Forest Disturbance Assessment Using Satellite Data of Moderate and Low Resolution 11

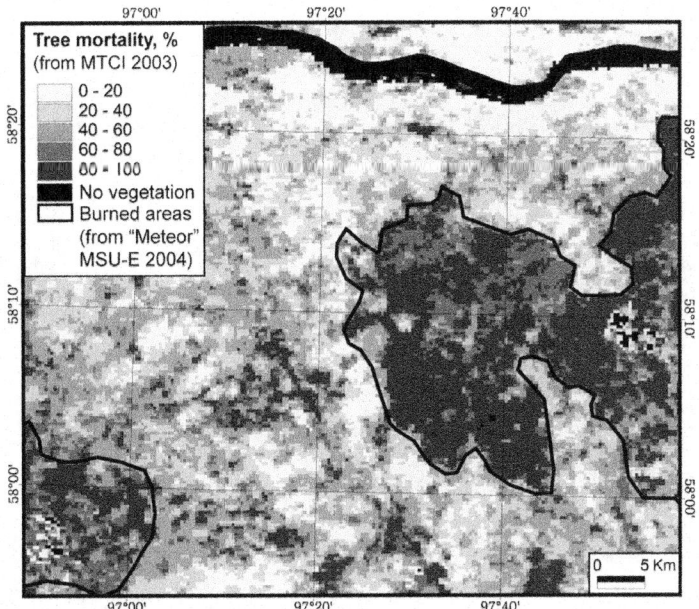

Fig. 1.5 A MERIS image fragment classified by tree mortality level using MTCI

$$MTCI^i = \max(MTCI_1^i, MTCI_2^i, ..., MTCI_5^i)$$

This MTCI values served as a basis for complex zoning of areas by vegetation condition (Fig. 1.12). The low MTCI values and, hence, low chlorophyll concentration, found for around Norilsk and in Rybnaya river valley indicate that these are the most heavily disturbed areas. However, decreasing chlorophyll can be accounted for by orographic factors, for example, in mountain landscapes northeast of Norilsk.

In order to assess temporal forest disturbance patterns, we used SPOT Vegetation 10-day composite images (s10 product) taken during the period between 1998 and 2005. The images passed thought the geometric and radiometric corrections in accordance with "Product P" specification (http://www.spot-vegetation.com).

For each of these images, NDVI was calculated from near-IR and red spectral bands (SPOT Vegetation channels 3 and 2, respectively). Ten-day composites covering the growing season (April 1 through October 1) were analyzed. Thirteen 10-day NDVI composites were thus used (analyzed) for each year between 1998 and 2005. These 8 years totalled 104 10-day periods were chosen for calculating NDVI trend (13 10-day periods × 8 years).

The spatial NDVI trend was determined for each pixel using a network of 91 images ordered by 10-day period times (dates). The percentage change of NDVI as compared to the initial NDVI found on the starting date in the 1998–2005 period was calculated as a linear trend for each image pixel.

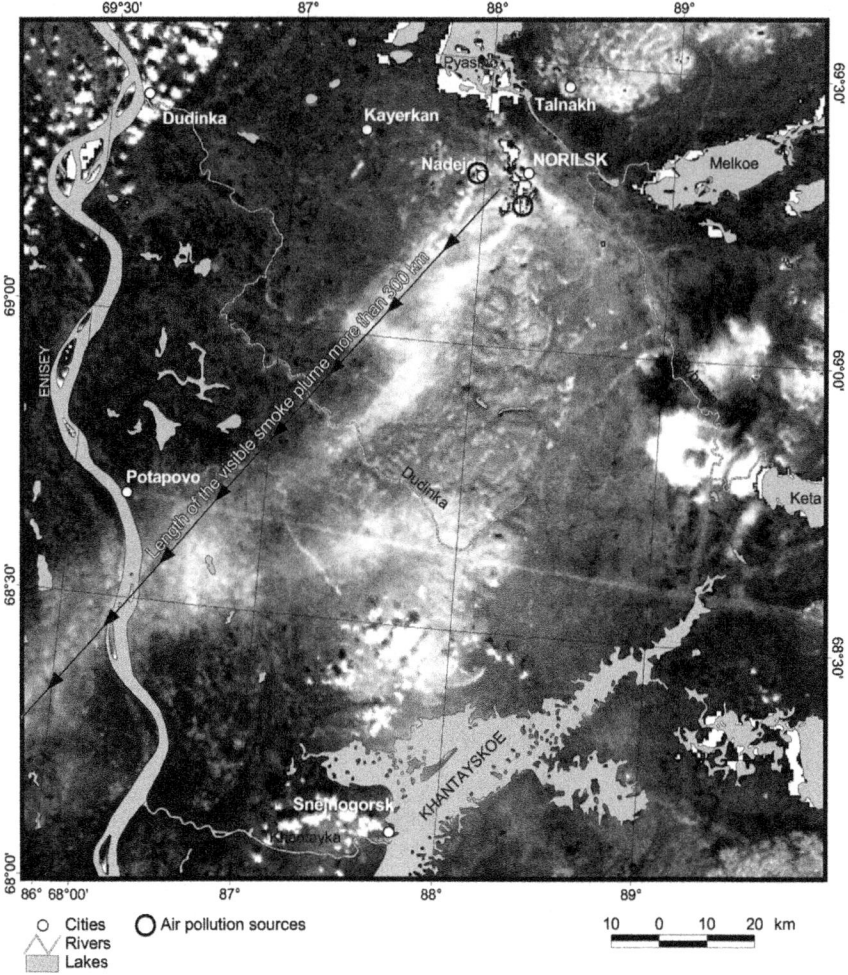

Fig. 1.6 The visible smoke plume from the Norilsk industrial complex observed by Envisat-MERIS satellite sensor

As a result, the 1998–2004 raster spatial NDVI trend map was built (Fig. 1.8). As is clear from this map, NDVI and, hence, chlorophyll concentration, decrease in the area stretching south-westward within 30 km from Norilsk. Average NDVI exhibits a steady decrease in this area (black box in Fig. 1.8) over the entire 8-year period of interest (Fig. 1.9).

The 2001 and 2003 ground observation data collected on 33 sample plots laid out within test sites at different distances from the chemical pollution source were used to quantify MTCI and NDVI links with forest stand disturbance levels. The proportion of the dead tree crown part in the total crow weight was taken as a stand disturbance criterion. This relative indicator was calculated for each sample plot as:

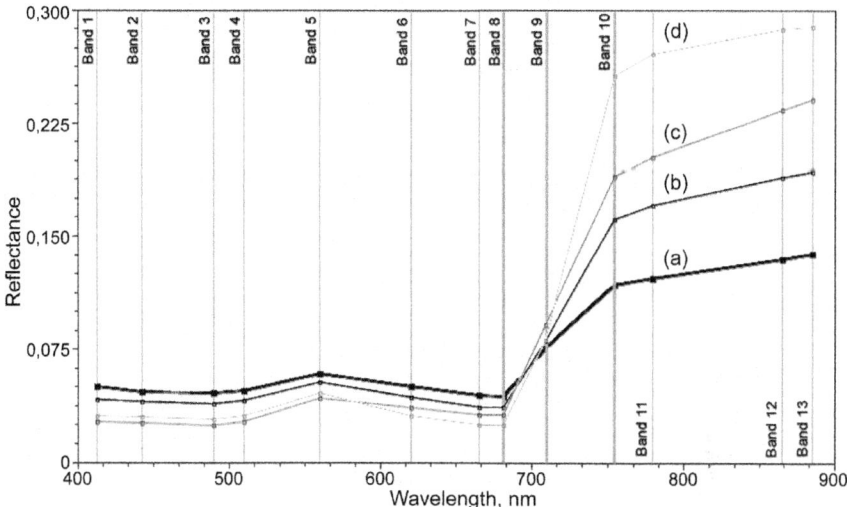

Fig. 1.7 MERIS channel-derived reflectances of sites differing in vegetation disturbance level situated (**a**) 8 km (Yergalah site), (**b**) 30 km (Rybnaya river), (**c**) 100 km (Tukulanda site) south of Norilsk, and (**d**) 98 km (Lower Agapa river) north of Norilsk

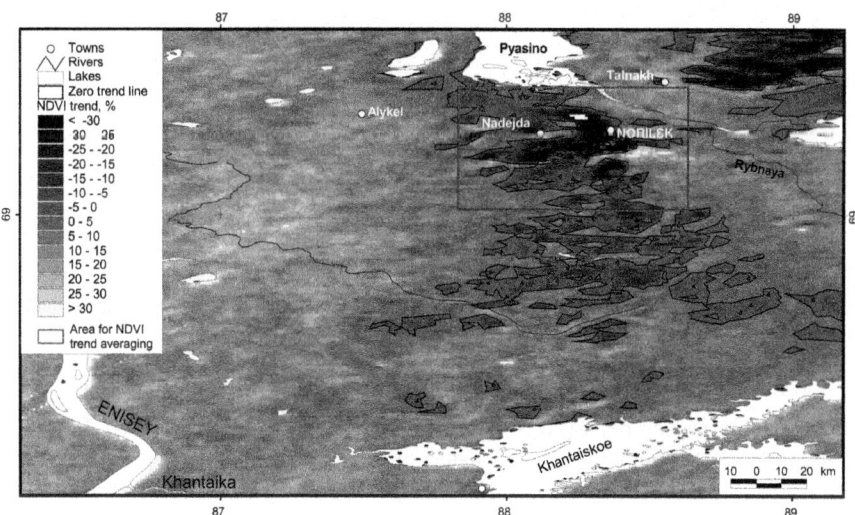

Fig. 1.8 The 1998–2005 spatial NDVI trend distribution based on SPOT Vegetation satellite data

$$D = \frac{W_{dead}}{W_{dead} + W_{green}}$$, where W_{dead} is dead tree crown branches (abs. dry wt.), t/ha; W_{green} is living foliage (abs. dry wt.), t/ha.

Ground sample plots were laid over the 1994 MTCI image (Fig. 1.12) and an NDVI image (SPOT Vegetation) averaged over three 10-day periods close to the

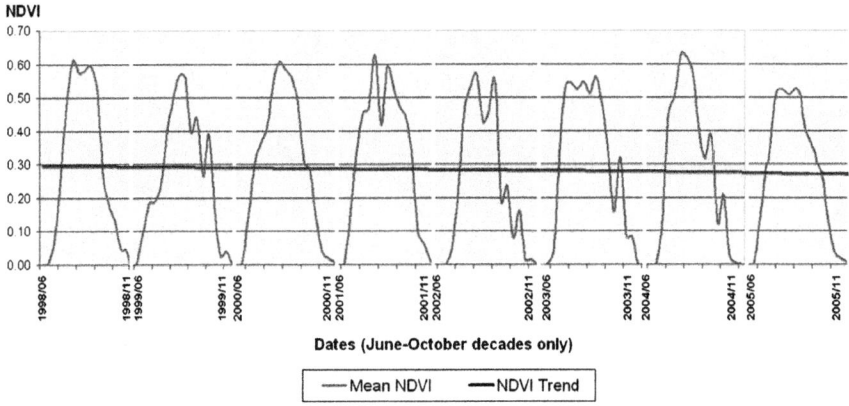

Fig. 1.9 Growing-season average NDVI and its 1998–2005 linear trend within 30 km from Norilsk derived from SPOT Vegetation satellite data

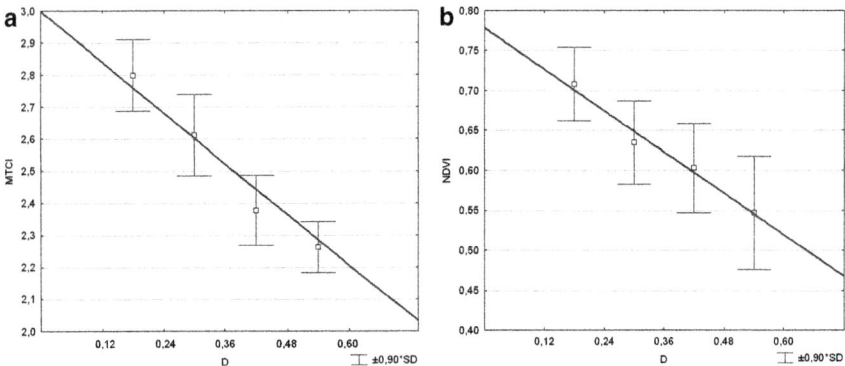

Fig. 1.10 The relationship of the relative vegetation disturbance factor (D) with (**a**) MTCI (ENVISAT MERIS) and (**b**) NDVI (SPOT Vegetation)

2004 growing season peak (dated July 11, July 21, and August 1, 2004). For each sample plot, MTCI and NDVI were determined for the pixels falling within 600 m around a sample plot. The resulting MTCI and NDVI relationships with four relative vegetation disturbance levels (D) are presented in Fig. 1.10.

The images of these two vegetation indexes classified (ordered) by relative stand disturbance level allowed to obtain the 1998–2005 distribution of vegetation zones differing in disturbance severity. It is clear from Fig. 1.10 that MTCI provides a more accurate severity class distribution. Furthermore, NDVI values averaged over three 10-day periods occurring within the peak of the growing season permitted to assess the spatial dynamics of these zones during the period of interest (Fig. 1.11).

The areas characterized by a decrease in NDVI values (a negative trend) over the past 8 years combined with the 2004 MERIS data on MTCI distribution were based

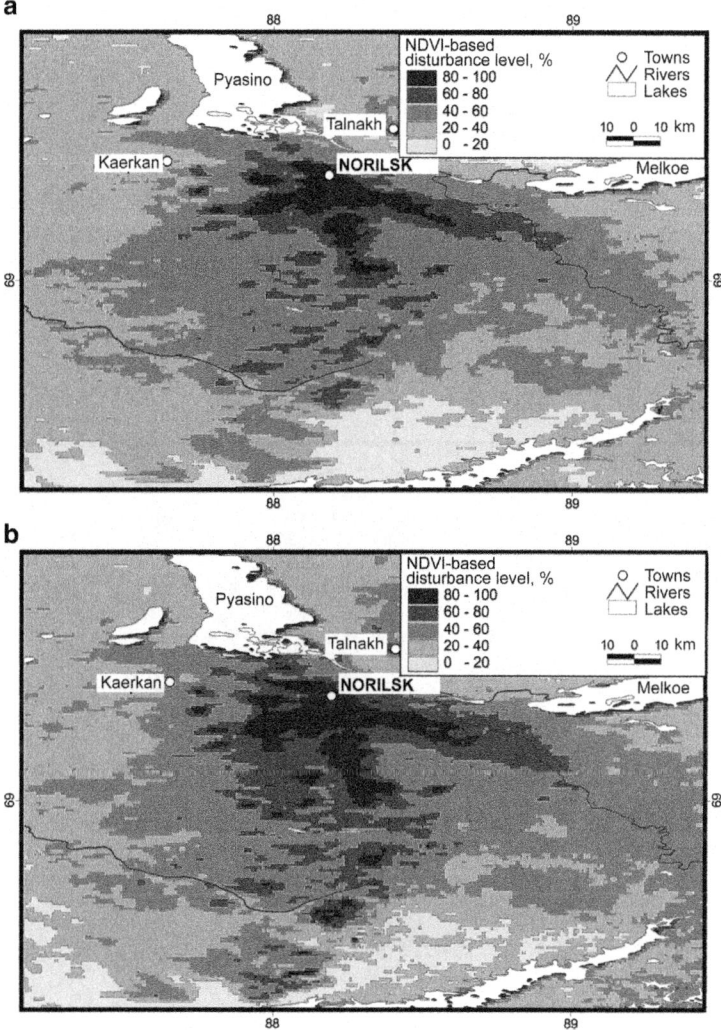

Fig. 1.11 Forest Stand disturbance levels based on NDVI from (**a**) the 1998 and (**b**) the 2005 SPOT Vegetation data

upon zoning the forest area by industrial pollution severity, as well as identifying ecologically accepted levels of industrial emissions.

The interpretation of the results of satellite imagery processing involved the use of thematic GIS databases, which contained field data, literature data, archive information, as well as original and GIS-based thematic maps. Qualitative estimation of vegetation cover, including a range of ecosystem types (forest-tundra, tundra, bogs) found in mountain and plain landscapes, was carried out by analyzing the ecosystem parameters indicative of the level vegetation decay due to industrial pollution. Fifteen such indicators covering all vegetation layers (i.e., overstory, tall shrub, small shrub-grass, and feather moss-lichen layers) were selected. Using the results of comparative analysis, a scale of 5 points (scores) describing vegetation disturbance rate was developed. Undisturbed (background) vegetation communities were assigned to 1, slightly disturbed communities to 2, moderately disturbed to 3, heavily disturbed to 4, and extremely (totally) disturbed communities were assigned to scale point 5.

All the above materials were used at the final step, i.e. in zoning the area of interest by vegetation cover rate of disturbance (Fig. 1.13). MTCI distribution map with SPOT Vegetation-based areas of a negative NDVI trend were used as the base for this zoning (Fig. 1.12).

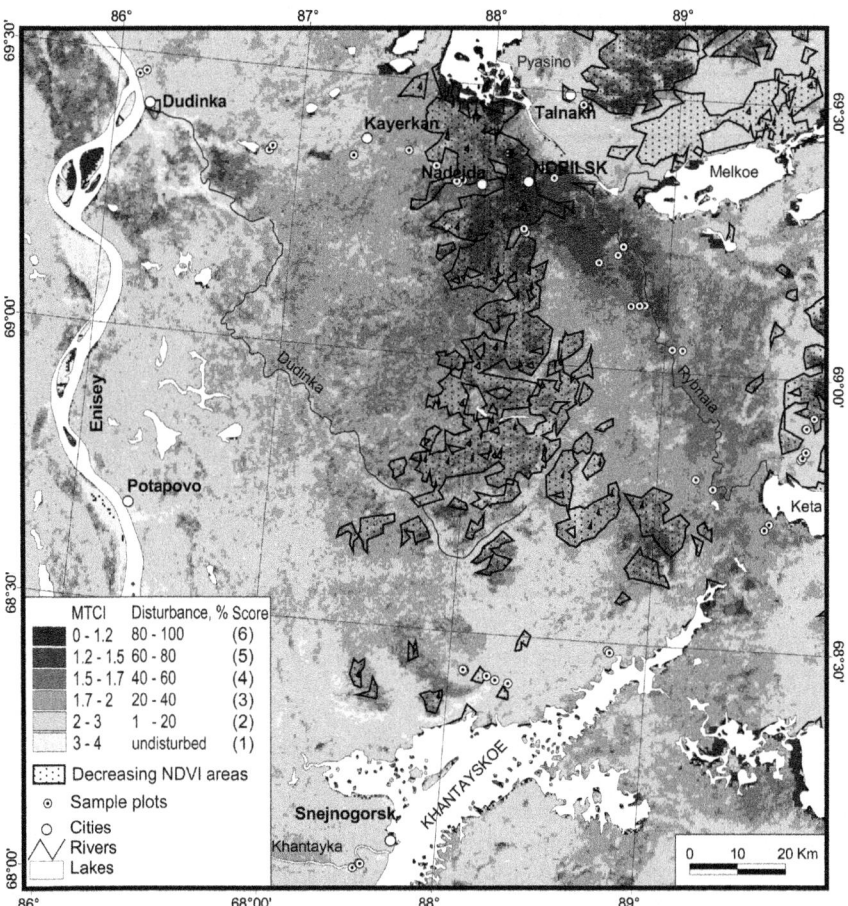

Fig. 1.12 The 2004 MTCI (ENVISAT MERIS)-based forest stand disturbance levels and the 1990–2004 SPOT Vegetation based areas of a negative NDVI trend

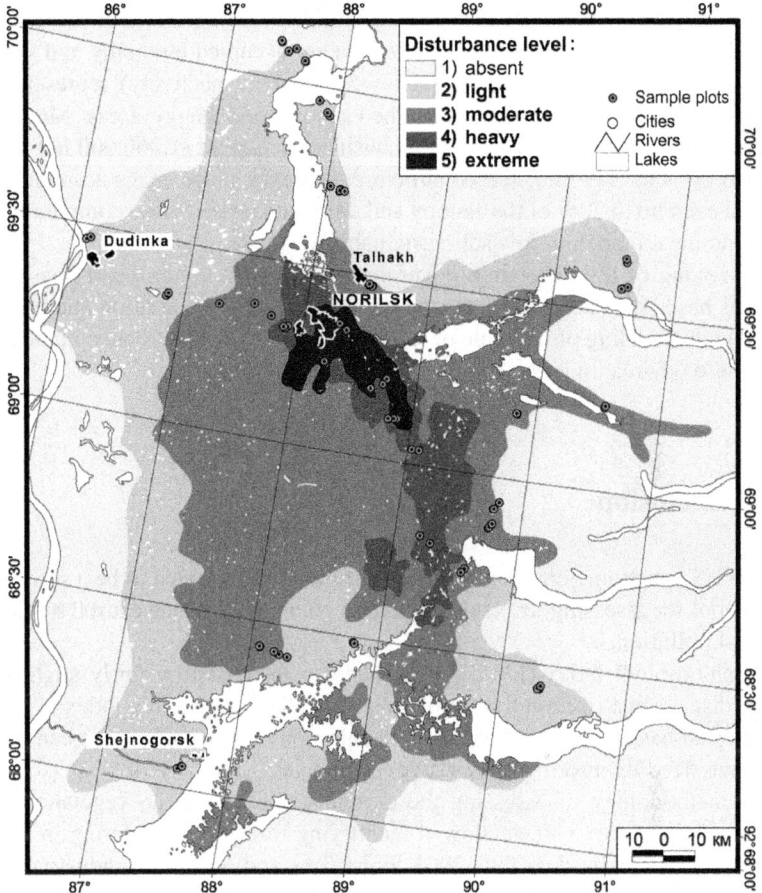

Fig. 1.13 Vegetation disturbance zones

As is clear from Fig. 1.12, low MTCI values occur for the most severely disturbed area in the vicinity of Norilsk and along Rybnaya river valley. Low chlorophyll and, particularly, its continuous decrease (a negative trend) might be indirect indicators of vegetation condition worsening. Complete overstory and tall shrub mortality or decreasing canopy closure of these layers, progressive mineral soil exposure, and decreasing extent of small shrubs and grasses – all these are common in zones 4 and 5 (Figs. 1.12 and 1.13). In mountain landscapes, however, decreasing chlorophyll concentration can be the result of the influence of natural factors, like, for example, in the north-eastern mountainous sites. Burned areas can cause a similar effect. For this reason, vegetation disturbance-based zoning considered all available material including the 1950–1960s topographic maps showing the landscapes of interest before they began to experience industrial pollution.

The total area of sites with different vegetation disturbance levels was estimated to be almost 2,400,000 ha, thereof 240,000 ha are occupied by totally and severely disturbed vegetation communities (zones 5 and 4, respectively) represented by completely dead forest, severely disturbed tundra, and boggy areas. Moderately disturbed vegetation (zone 3) having a considerable extent (1,060,000 ha) is characterized by generally undisturbed structure, however, since snags account for ca. 50% and even up to 70% of the canopy and tall shrub layers, these communities are deemed to have markedly low self-sustainability.

This zoning of the vegetation covering forest-tundra, tundra, and bog ecosystems was based on satellite and ground data on vegetation condition and it gives a general understanding of the scale of terrestrial ecosystem disturbance in the industrially polluted area under study.

1.5 Conclusion

The MERIS spectrometer, an ENVISAT instrument, has proved to be a sufficiently reliable tool for assessing levels of vegetation cover disturbance caused by fire and industrial pollution.

The chlorophyll index (MTCI) was found to respond to a fairly slight forest canopy disturbance (tree mortality of less than 20%). Since this index is easy to calculate and has a linear relationship with chlorophyll concentration, it can be used in computerized monitoring of forest cover changes.

This methodology of assessing the current condition of the vegetation cover from vegetation index values allowed identifying forest areas differing in level of disturbance caused by the 1996–2003 forest fires and long-term industrial pollution. Analyzing satellite images with the help of the available GIS data (in an overlaying manner) permitted to delineate zones differing in vegetation disturbance level. The maps built as a result of this study can be used in developing a system of monitoring of forest cover experiencing a variety of influences.

Acknowledgments This study was supported by FEMINE project (Forest Ecosystem Monitoring in Northern Eurasia) ESA-IAF and Russian Foundation for Basic Research project 07-04-00515-a.

References

Clevers JGPW et al (2002) Derivation of the Red Edge Index using MERIS standard band setting. Int J Remote Sens 23:3169–3184

Curran PJ, Steele CM (2005) MERIS: the re-branding of an ocean sensor. Int J Remote Sens 26:1781–1798

Curran PJ et al (1997) Remote sensing the biochemical composition of a slash pine canopy. IEEE Trans Geosci Remote Sens 35:415–420

Dash J, Curran PJ (2004) MTCI: the MERIS Terrestrial Chlorophyll Index. ENVISAT Symposium Proceedings, Austria, Salzburg

Dawson TP, Curran PJ (1998) A new technique for interpolating the reflectance red edge position, Int J Remote Sens 19:2133–2139

Jeffrey A (1985) Mathematics for engineers and scientists. Van Nostrand Reinhold, Workingham, pp 708–709

Vinogradov BV (1994) Aerospace ecosystem monitoring. Nauka, Moscow, 320 pp

Chapter 2
Fire/Climate Interactions in Siberia

H. Balzter, K. Tansey, J. Kaduk, C. George, F. Gerard, M. Cuevas Gonzalez, A. Sukhinin, and E. Ponomarev

Abstract This paper presents an intercomparison of two burned area datasets, the L3JRC daily global burned area dataset derived from SPOT-VEGETATION and the FFID burned area dataset from MODIS. Burned area dynamics are presented and the influence of climate on the fire regime is discussed. Feedbacks of the fire dynamics to the climate system are evaluated. The Russian fire danger index is presented and compared to satellite observations of fires.

Keywords Climate • Fire • Temperature • Arctic oscillation • Remote sensing

2.1 The Fire Regime in Siberia

The circumpolar boreal forest covers approximately 1.37 billion hectares, or 9.2% of the world's land surface. Siberia is a hotspot for climate change. As a temperature controlled region it is particularly sensitive to even small increases in temperatures. In addition to this heightened vulnerability, the observed warming trend is more than twice as high as the global average, and climate model predictions show that this faster regional warming is likely to continue. Annual temperature anomalies

---

H. Balzter (✉), K. Tansey, and J. Kaduk
Department of Geography, Centre for Environmental Research, University of Leicester, University Road, Leicester LE1 7RH, UK
e-mail: hb91@le.ac.uk; kjt7@le.ac.uk; j.kaduk@leicester.ac.uk

C. George, F. Gerard, and M.C. Gonzalez
Centre for Ecology and Hydrology, Maclean Building, Benson Lane, Crowmarsh Gifford, Wallingford, Oxfordshire OX10 8BB, UK
e-mail: ctg@ceh.ac.uk; ffg@ceh.ac.uk; cuevasgonzalez@gmail.com

A. Sukhinin and E. Ponomarev
Siberian branch of Russian Academy of Sciences, VN Sukachev Institute of Forest, Academgorogok, Krasnoyarsk 660036, Russia
e-mail: boss@ksc.krasn.ru; evg@ksc.krasn.ru

since 1850 over central Siberia show a trend towards warmer temperatures at a higher rate than the global average, and with a faster increase after 1990 (Balzter et al. 2007).

The boreal forest is governed by fires, which generate a patchy mosaic of regenerating forest types. Lightning frequency, litter layer fuel mass and fuel moisture content all impact on the fire regime and are linked to meteorological conditions. Under scenarios of climate change many predictions show an acceleration of the fire regime. Many fires are also human-induced. Both climate and human population effects have been documented by Jupp et al. (2006). Greenhouse gas emissions from fires are an important component in the global carbon cycle. Fire is arguably the most important ecological disturbance worldwide releasing approximately 3.5 Pg C per year to the atmosphere (van der Werf et al. 2004). For the 1997/1998 carbon dioxide anomalies it is thought that 66% of the growth rate anomaly can be attributed to global biomass burning, of which 10% originated from the global boreal biome (van der Werf et al. 2004). It has been hypothesised that increasing greenhouse gas emissions from an accelerating fire regime could lead to a positive feedback with global warming (Amiro et al. 2001). Anticipated future climate change in the Northern Hemisphere with an increasingly dry and hot summer climate and an extended growing season could potentially lead to increased insect infestations and increased susceptibility of boreal trees to fire (Ayres and Lombardero 2000; Kobak et al. 1996).

Some authors have suggested that the fire regime in the boreal biome is coupled to the climate system through large-scale atmospheric circulation patterns, e.g. (Balzter et al. 2005, 2007; Hallett et al. 2003). Atmospheric oscillation patterns have an impact on regional climatic variability and consequently vegetation activity. Los et al. (2001) and Buermann et al. (2003) found that two predominant hemispheric-scale modes of covariability are related to teleconnections associated with the El Niño Southern Oscillation (ENSO) and the Arctic Oscillation (AO): The warm event ENSO signal is associated with warmer and greener conditions in far East Asia, while the positive phase of the AO leads to enhanced warm and green conditions over large regions in Asian Russia.

In the recent past Siberia has experienced extreme fire years (Sukhinin et al. 2004), which coincided with years in which the AO was in a more positive phase (Balzter et al. 2005). Jupp et al. (2006) found that regional clusters of fire scars in Siberia occurred in places with dry precipitation anomalies at scales of tens of kilometers. An analysis of surface air temperature and precipitation at ten meteorological stations in West Siberia by Frey and Smith (2003) showed that West Siberia shows increases in temperature and precipitation, particularly springtime warming and more winter precipitation. Frey and Smith (2003) found an association of autumn and winter temperatures with the AO. On average, the AO was linearly correlated with 96% (winter), 19% (spring), 0% (summer), 67% (autumn), and 53% (annual) of the warming (Frey and Smith 2003).

The AO has shown a statistically significant trend towards the positive phase between 1950 and the present day (Balzter et al. 2007), which is likely to indicate global climate change trends. Overland et al. (2002) observed a shift in wind fields from anomalous north-easterly flows in the 1980s to anomalous south-westerly flows in the 1990s during March and April in Siberia, coinciding with a systematic shift in the AO near the end of the 1980s. These hemispheric-scale changes in the heat transport from the oceans to continental parts of Siberia could have major repercussions for the fire regime (Balzter et al. 2005, 2007). The AO is also influenced by intense volcanic eruptions, which inject aerosols into the stratosphere and via an enhanced temperature gradient between the pole and the tropics lead to an acceleration of the polar vortex (Stenchikov et al. 2006). This acceleration expresses itself as a positive phase of the AO.

The following sections describe two remotely sensed burned area datasets, followed by a discussion of the impacts of climate on fire, and the feedbacks of fire on the climate system.

2.2 The L3JRC Global Daily Burned Area Dataset

Due to the extent and remoteness of Siberia the only cost effective way of monitoring the fire regime is using remote sensing. A global daily burned area dataset at 1 km spatial resolution is available from the VEGETATION sensor aboard the SPOT satellite. A single algorithm was used to classify burnt areas from the spectral reflectance data. SPOT 4 was launched in 1998 into a polar sun synchronous orbit at 832 km. The algorithm is described in Tansey et al. (2008), and is based primarily on the 0.83 µm near-infrared (NIR) channel.

Burned forest area statistics were extracted by overlaying administrative regions as vectors, reprojecting the L3JRC datasets to the Albers equal area projection and calculating polygon statistics in the programming language R. Forest areas were defined using the Global Land Cover 2000 map (Bartalev et al. 2003) as any of the land cover classes "Evergreen Needle-leaf Forest" (class 1), "Deciduous Broadleaf Forest" (3), "Needle-leaf/Broadleaf Forest" (4), "Mixed Forest" (5), "Broadleaf/Needle-leaf Forest" (6), "Deciduous Needle-leaf Forest" (7), "Broadleaf deciduous shrubs" (8), "Needle-leaf evergreen shrubs" (9), "Forest-Natural Vegetation complexes" (21) or "Forest-Cropland complexes" (22). On the assumption that the fire season is constrained by the winter time to be between Julian dates 161 and 272, any burned areas that were detected outside this date range were masked out. This matches the date range used in generating the FFID burned area dataset (next section). Table 2.1 gives the L3JRC burned forest area for each administrative region (oblast) obtained in this way. It shows that some oblasts have a stable fire regime but in others a large interannual variability is observed. The standard deviation between years as a measure of interannual variability reveals that Yakutia Republic, Evenk a.okr., Irkutsk oblast, Chita oblast, Buryat Republic, Khabarovsk Kray, Amur oblast, Magadan oblast, Chukchi a.okr., Krasnoyarsk Kray

Table 2.1 Annual burned area statistics (km^2) per oblast (administrative region) based on the L3JRC global daily burned area dataset. Only forest areas (based on GLC2000) and Julian dates 161–272 were analysed

OBLAST	2000	2001	2002	2003	2004	2005	2006
Adigei Republic	27	54	6	27	8	25	51
Aga-Buryat a.okr.	64	19	3	327	121	15	54
Altai Kray	115	92	124	88	82	142	164
Amur oblast	2,493	869	2,632	3,708	1,841	1,333	5,048
Arkhangelsk oblast	4	4	9	2	5	9	3
Astrakhan oblast	0	0	0	1	3	0	9
Bashkortostan Republic	288	304	154	166	97	444	549
Belgorod oblast	112	58	65	47	47	57	181
Bryansk oblast	8	0	29	0	0	9	5
Buryat Republic	4404	1,656	1,235	7,695	2,771	2,964	4,918
Checheno-Ingush Republic	0	0	0	0	0	0	0
Chelyabinsk oblast	22	111	23	82	85	108	63
Chita oblast	5,625	2,128	1,176	9,505	4,590	4,212	6,493
Chukchi a.okr.	995	986	1,587	3,025	1,829	488	2,752
Chuvash Republic	21	74	31	2	3	12	12
Daghestn Republic	0	0	0	0	0	0	4
Evenk a.okr	1,026	713	804	10,895	2,960	8,002	10,582
Gorno-Altai Republic	202	78	649	548	490	539	409
Irkutsk oblast	2,916	1,464	1,715	4,868	1,461	7,127	9,744
Ivanovo oblast	0	1	20	0	0	0	0
Kabardino-Balkarian Republic	3	0	0	0	1	0	0
Kaliningrad oblast	0	0	13	2	0	0	1
Kalmyk-Khalm-Tangch Republic	2	2	1	4	2	1	1
Kaluga oblast	0	1	29	0	0	0	0
Kamchatka oblast	686	50	153	153	398	245	77
Karachai-Cherkess Republic	4	6	2	2	0	2	3
Karelia Republic	6	3	0	4	0	4	4
Kemerovo oblast	5	20	196	59	39	23	99
Khabarovsk Kray	6,469	2,344	4,232	6,130	4,482	6,171	4,740
Khakass Republic	12	15	38	49	27	73	60
Khanty-Mansi a.okr.	166	79	82	200	216	167	303
Kirov oblast	9	3	0	0	1	9	4
Komi Republic	216	214	211	33	96	73	60
Koryak a.okr.	940	761	311	1,085	343	331	529
Kostroma oblast	0	4	5	0	0	1	0
Krasnodar Kray	563	846	312	642	469	537	986
Krasnoyarsk Kray	999	660	539	2,495	1,988	949	1,528
Kurgan oblast	104	149	46	225	164	90	130
Kursk oblast	96	35	37	10	23	42	46
Leningrad oblast	0	0	4	0	2	0	24
Lipetsk oblast	95	159	93	54	146	235	135

(continued)

Table 2.1 (continued)

OBLAST	2000	2001	2002	2003	2004	2005	2006
Magadan oblast	5,186	3,329	3,265	6,878	3,574	3,097	4,499
Mari-El Republic	0	1	1	0	0	0	0
Mordovian SSR	30	50	49	2	12	24	8
Moscow oblast	1	9	47	0	0	6	2
Murmansk oblast	7	59	65	164	93	58	22
Nenets a.okr.	9	13	38	13	17	14	20
Nizhni Novgorod oblast	14	47	110	15	8	34	13
North-Ossetian SSR	0	0	0	0	0	0	0
Novgorod oblast	0	0	0	1	0	0	0
Novosibirsk oblast	59	74	31	109	91	105	229
Omsk oblast	22	174	66	21	16	18	23
Orenburg oblast	63	133	116	79	98	219	185
Oryel oblast	91	108	44	15	36	79	15
Penza oblast	168	173	108	32	75	93	44
Perm oblast	12	69	10	22	10	50	14
Primorski Kray	1	16	6	253	41	50	57
Pskov oblast	0	0	19	1	0	0	1
Rostov oblast	215	319	315	220	394	296	324
Ryazan oblast	137	96	238	19	92	112	56
Sakhalin oblast	66	14	8	208	23	39	12
Samara oblast	159	328	309	149	123	319	184
Saratov oblast	208	318	184	198	313	429	312
Smolensk oblast	0	0	22	0	0	0	0
Stavropol Kray	86	212	66	123	119	155	315
Sverdlovsk oblast	19	55	76	143	86	374	28
Tambov oblast	181	316	241	113	238	348	251
Tatarstan Republic	484	431	554	172	158	282	201
Taymyr a.okr.	45	37	1	287	164	193	187
Tomsk oblast	42	152	395	110	689	66	225
Tula oblast	59	188	206	14	20	97	30
Tuva Republic	1,055	812	2,464	1,557	757	827	1,667
Tver oblast	2	2	47	0	0	1	1
Tyumen oblast	71	260	128	298	146	150	129
Udmurt Republic	3	2	0	0	21	2	0
Ulyanovsk oblast	243	291	146	73	56	173	117
Ust-Orda Buryat a.okr.	67	38	29	254	42	131	87
Vladimir oblast	0	2	21	0	5	0	0
Volgograd oblast	38	79	72	64	72	60	78
Vologda oblast	1	10	7	2	0	2	0
Voronezh oblast	287	334	214	187	272	214	274
Yakutia Republic	18,684	19,623	38,307	44,691	29,326	73,500	56,497
Yamalo-Nenets a.okr.	474	263	95	497	713	386	500
Yaroslavl oblast	1	2	22	1	0	0	0
Yevrey a.oblast	14	9	4	62	6	15	198
Russia	57,001	42,410	64,712	109,180	62,696	116,457	116,576

and Tuva Republic (in descending order) show the highest variability between years, with standard deviations exceeding 500 km² year⁻¹. Yakutia, the largest oblast covering more than 3,100,000 km² of the ~17,000,000 km² of Russia, also shows the highest mean burned forest area over the observed years.

2.3 Forest Fire Intensity Dynamics (FFID) Daily Burn Scar Identification

Using moderate resolution sensors (approx. 1 km² pixels 2,000 km swath width) that have a repeat time of 1 day or less in boreal regions, it is possible to determine the date when a fire occurred during cloud-free conditions. This method was investigated in the FFID project (Forest Fire Intensity Dynamics). For the FFID Daily Burned Area product, instead of using thermal sensors for detecting active fires which can then be missed due to cloud or smoke for example, a vegetation index differencing approach is used which is able to discriminate disturbances long after the event has occurred. The parameter used was the Normalised Difference Short-Wave Infrared Index (NDSWIR), a combination of the near-infrared (NIR) and short-wave infra-red (SWIR) signals, which is sensitive to vegetation water content, and so can be used as a proxy for canopy density (George et al. 2006).

$$NDSWIR = \frac{(\rho 858\,\text{nm} - \rho 1640\,\text{nm})}{(\rho 858\,\text{nm} + \rho 1640\,\text{nm})} \quad (2.1)$$

The satellite data used was the Terra-MODIS Nadir BRDF-Adjusted Reflectance (NBAR) 16-Day composite (MOD43B4) (Friedl et al. 2002), which has reduced view angle effects that are present in wide view-angle sensors. The NBAR data provide a nadir adjusted value of reflectance in each of seven bands once in every 16-day period. The removal of view angle effects and the adjustment to the mean solar zenith angle (of the 16-day period) produce a stable, consistent product allowing the spatial and temporal progression of phenological characteristics to be easily detected (Schaaf et al. 2002). A MODIS data granule is 1,200 × 1,200 pixels, each pixel being 927.4 m on a side.

At the northern reach of the boreal zone (approx. 70°N) the growing season is very short so only the composites from mid July to mid September were included to reduce any phenological effects. To keep the methodology consistent the same period was used at the lower latitudes even though these areas had a much longer growing season. The four composites within this time period were used to produce the NDSWIR layers. For each of the four NDSWIR layers within a year, a NDSWIR difference layer was calculated by subtracting that layer from the corresponding layer from the previous year. This difference layer would then show a high value where there was a large decrease in biomass, and a low value for those areas of little change. The four difference images for each year were then combined to give one annual difference image (ADI). This annual difference greyscale image, ranged from low values of no change to higher values showing missing biomass compared with the previous year. To set the threshold to separate out burned areas, MODIS thermal anomalies (TA) (Justice et al. 2002), which give the location and Julian Day of active fires, were used. This assumed that if a TA were present, then that ADI pixel had burned. Then for each of the IGBP woody land covers (classes 1–8) within a granule, the mean ADI value under the TA's were calculated, and this value was used to set the threshold for that land cover class. The result is a binary mask, with 1's representing disturbance scars. However, this layer will also show other disturbances apart from burning, such as insect infestations, wind blow or logging. It also doesn't show the date of burning. To identify and date any burns, the TA's are used again. Any scars not overlain with TA's are discarded. For the remaining scars, the pixels corresponding to the TA's are assigned the Julian Day of that TA. This leaves many of the burned areas being a combination of dated pixels and undated pixels, the undated pixels being where perhaps there was too much cloud or smoke for an active fire to be detected, but where there was still a significant reduction in vegetation biomass. These undated pixels are then dated by extrapolating from the dated pixels. The result is a raster with each burnt pixel having a value of the Julian Day when it was burnt.

Table 2.2 shows the FFID burned area for each administrative region (oblast).

Table 2.2 Annual burned forest area statistics (km^2) per oblast (administrative region) based on the FFID dataset

OBLAST	2001	2002	2003	2004	2005	2006
Adigei Republic	0	0	0	0	0	0
Aga-Buryat a.okr.	473	58	3,452	298	243	205
Altai Kray	7,637	8,594	9,485	6,087	5,289	5,049
Amur oblast	13,278	20,096	33,445	5,972	9,817	20,172
Arkhangelsk oblast	530	274	173	292	189	317
Astrakhan oblast	0	0	0	0	0	0
Bashkortostan Republic	2,126	1,217	1,424	1,816	510	2,087
Belgorod oblast	1,189	1,124	96	120	373	408
Bryansk oblast	422	1,780	256	259	463	1,388
Buryat Republic	1,035	1,617	43,649	1,165	2,616	2,457
Checheno-Ingush Republic	0	0	0	0	0	0
Chelyabinsk oblast	4,628	1,806	2,080	3,062	845	2,197
Chita oblast	4,947	5,436	78,097	5,226	5,031	11,432
Chukchi a.okr.	2,177	3,295	10,944	500	587	106
Chuvash Republic	142	75	24	80	148	342
Daghestn Republic	0	0	0	0	0	0
Evenk a.okr	80	623	167	102	964	6,731
Gorno-Altai Republic	275	190	309	129	16	30
Irkutsk oblast	3,837	6,756	26,583	2,578	3,080	13,194
Ivanovo oblast	40	559	32	28	60	681

(continued)

Table 2.2 (continued)

OBLAST	2001	2002	2003	2004	2005	2006
Kabardino-Balkarian Republic	0	0	0	0	0	0
Kaliningrad oblast	88	299	329	281	192	561
Kalmyk-Khalm-Tangch Republic	0	0	0	0	0	0
Kaluga oblast	30	1,392	156	103	109	1,549
Kamchatka oblast	1,730	574	556	83	117	181
Karachai-Cherkess Republic	0	0	0	0	0	0
Karelia Republic	66	82	181	28	144	234
Kemerovo oblast	1,192	3,906	2,394	3,306	2,365	1,296
Khabarovsk Kray	6,423	7,375	16,696	3,020	11,260	4,086
Khakass Republic	588	1,671	594	992	1,225	390
Khanty-Mansi a.okr.	691	597	1,914	7,569	5,434	3,703
Kirov oblast	522	344	218	172	241	743
Komi Republic	941	68	57	242	127	97
Koryak a.okr.	1,294	1,276	3,759	200	287	390
Kostroma oblast	178	258	39	32	68	482
Krasnodar Kray	0	0	0	0	0	0
Krasnoyarsk Kray	3,925	6,859	10,013	7,868	7,336	11,214
Kurgan oblast	1,002	774	1,383	5,046	421	2,212
Kursk oblast	1,895	2,895	243	1,206	2,089	1,071
Leningrad oblast	68	1,397	183	277	303	2,143
Lipetsk oblast	1,866	2,002	378	1,361	2,106	1,018
Magadan oblast	6,248	1,993	9,871	762	365	564
Mari-El Republic	78	167	21	55	67	226
Mordovian SSR	681	729	187	464	528	1,283
Moscow oblast	83	2,339	237	208	101	1,755
Murmansk oblast	162	127	174	121	130	67
Nenets a.okr.	7	0	5	38	6	26
Nizhni Novgorod oblast	796	1,113	152	394	659	1,711
North-Ossetian SSR	0	0	0	0	0	0
Novgorod oblast	94	710	106	269	40	1,107
Novosibirsk oblast	9,184	8,082	6,641	9,180	7,415	16,584
Omsk oblast	5,436	3,237	2,568	7,551	1,777	6,784
Orenburg oblast	5,112	4,398	4,968	4,815	5,165	3,931
Oryel oblast	1,417	2,337	142	1,303	1,225	1,335
Penza oblast	1,701	1,434	532	1,023	1,052	2,812
Perm oblast	439	98	83	99	135	482
Primorski Kray	4,275	1,675	4,759	4,069	2,191	2,874
Pskov oblast	283	2,010	251	668	222	2,922
Rostov oblast	17	13	1	1	11	3
Ryazan oblast	775	1,929	261	876	1,188	2,142
Sakhalin oblast	208	540	1,169	102	68	100

(continued)

Table 2.2 (continued)

OBLAST	2001	2002	2003	2004	2005	2006
Samara oblast	2,105	3,432	1,187	1,735	1,549	2,161
Saratov oblast	3,402	4,459	1,976	3,439	5,775	3,696
Smolensk oblast	206	3,652	966	559	58	3,916
Stavropol Kray	0	0	0	0	0	0
Sverdlovsk oblast	558	796	673	2,938	716	3,275
Tambov oblast	3,147	3,082	1,005	1,687	2,402	2,156
Tatarstan Republic	1,694	1,733	962	1,480	706	1,435
Taymyr a.okr.	68	29	28	43	39	176
Tomsk oblast	1,144	1,177	4,413	5,117	4,307	4,192
Tula oblast	791	1,515	163	851	1,005	1,814
Tuva Republic	1,184	8,383	1,771	221	736	532
Tver oblast	74	2,515	667	187	117	1,736
Tyumen oblast	1,194	638	2,288	7,676	741	5,560
Udmurt Republic	124	108	90	38	65	265
Ulyanovsk oblast	838	1,192	590	996	930	1,818
Ust-Orda Buryat a.okr.	186	708	3,010	39	482	836
Vladimir oblast	144	1,232	49	106	58	529
Volgograd oblast	2,713	2,403	905	1,553	2,822	1,398
Vologda oblast	173	581	99	54	116	532
Voronezh oblast	2,972	3,131	780	1,526	2,275	1,248
Yakutia Republic	36,534	58,789	22,535	1,875	11,259	3,793
Yamalo-Nenets a.okr.	539	1,015	774	1,145	3,717	3,067
Yaroslavl oblast	68	735	201	35	60	1,102
Yevrey a.oblast	2,769	1,945	3,193	3,847	3,510	1,878
Russia	164,940	221,451	329,761	128,643	129,841	191,992

2.4 Burned Forest Area Intercomparison

An intercomparison of the L3JRC and FFID datasets with other published burned area data by Soja et al. (2004) and George et al. (2006) was carried out, the results of which are shown in Fig. 2.1. The study region "SIBERIA-2" is the same as in George et al. (2006) since this was the largest common area coverage. The SIBERIA-2 region covers over 3 million km^2 of Central Siberia, and includes Irkutsk Oblast, Krasnoyarsk Kray, Taimyr, Khakass Republic, Buryat Republic and Evensky Autonomous Oblast (approximately 79–119°E, 51–78°N). Figure 2.1 shows several catastrophic fire years in the Central Siberian region: 1992–1993, 2003 and 2006 showed large forest fires. When comparing the different datasets it becomes apparent that while in most cases the interannual variability is similar, but in particular years there are large uncertainties in the estimates.

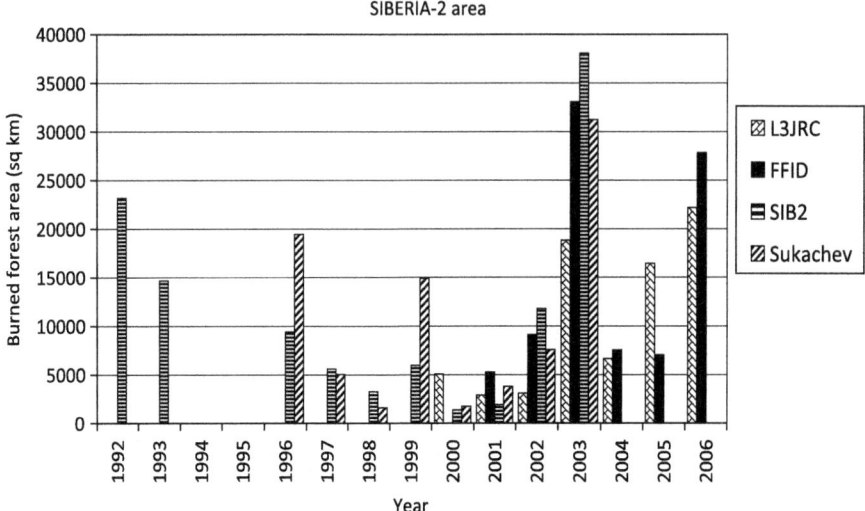

Fig. 2.1 Intercomparison of annual burned forest area estimates from the datasets L3JRC, FFID, L3JRC, SIBERIA-2, and SUKACHEV. The datasets cover different time ranges, only 2001–2003 is the common temporal coverage

2.5 Climate Impacts on Fire

Observations from remote sensing have shown that large-scale climate oscillations, in particular the Arctic Oscillation, are thought to have an impact on forest fire frequency in Central Siberia (Balzter et al. 2005, 2007). Climate data have shown and climate models predict that the Arctic Oscillation responds to large-scale volcanic eruptions such as the Mount Pinatubo eruption in 1991, which injected large amounts of aerosols into the lower stratosphere and changed global climate for several years (Stenchikov et al. 2002, 2006). Volcanic eruptions can lead to a positive phase of the Arctic Oscillation (Stenchikov et al. 2002, 2006), which in turn provides conditions that are conducive to extreme forest fires (Balzter et al. 2005).

Central Siberia contains several climatic and ecological zones. As a result many authors have noted specific fire regimes influencing different forest types in the region. The fire regime influences the duration of the fire season and the spatial patterns of forest fires locations (Ivanova et al. 2005, Kurbatski and Ivanova 1987, Valendick and Ivanova 2001). The degree of forest fine fuel to be ignited is determined by the variation of fuel moisture content, which is dependent on the length of the dry period. Forest fire initiation and fire spread across the ground cover is possible if the moisture content of fine fuels reaches a fixed low value after which this parameter changes only slightly. In particular, for the needles of conifers (except larch) the balanced moisture content is 11–26% depending on relative humidity, and for leaves of deciduous trees, needles of larch and grasses it is 9–31% (Kurbatski et al. 1987).

Mass forest fire ignitions are caused mostly under the influence of atmospheric anticyclones. The moisture content of fine fuels decreases to 9–30% and an extreme fire danger state evolves after 85–150 h under these conditions without precipitation. An uncontrollable situation develops if forest fires cannot be localized and extinguished at an early stage.

Experimental data of the last 10 years show the interconnection between local fire activity and local weather conditions forming at the same point in time. This interconnection is determined by a formation of stable anticyclones with lifetimes up 30–90 days over the region. Usually the process can be observed over regions where mass forest fires burned at the same time. The exact physical processes have not yet been described. However, it can be hypothesised that stable anticyclone weather formations are influenced by convective heat flow from the epicentre of active forest fires. This formed high-pressure zone ejects other cyclones and cumulonimbus clouds.

The forest fire danger condition is characterized by the Russian fire danger index (FD) that can be calculated using daily air temperature and dew point temperature measurements during the fire season. This index forecasts the degree of forest fine fuel dryness and fire ignition ability indirectly. At the same time the value of this index and the persistence of high values of the fire danger characterize not only the forest fire danger state but also weather condition features formed by fire convection flow.

According to experimental data, certain values of the FD index were identified by Russian researchers for different stages of forest fire danger. An extreme fire danger level in the forests of Central Siberia is present when FD reaches values of 3,000–4,200. However, during last 10 years this index has been observed to be much higher after long droughts. For example, the rain-free period in the Angara river forests in 2006 was over 50 days (Fig. 2.2). In Yakutia in the middle of the summer anticyclone periods are dominating over 60 days annually. During these times the fire danger index can be between 14,000 and 20,000. As Fig. 2.2 shows, the Russian fire danger index is correlated with the Duff Moisture Code (DMC) of the Canadian Forest Fire Weather System, although a slight temporal phase is noticeable.

Consequences of long droughts affect fire locating and extinguishing statistics. Wildfires should be detected at the early stage of burning to enable efficient and effective fire prevention measures. However, in a case of an extreme fire situation non-localized fires are uncontrollable when fire fighting cannot extinguish them efficiently anymore. Under these conditions forest fires can be active for about 30 days. In 2007 the percentage of fires that was located during the first day of activity was about 88% (see Fig. 2.3).

Figure 2.3 is illustrating the opportunity of forest fire prevention measures according to material and technical support level. The annual part of large fires (area more than 1,000 ha) that amount to not more than 5% of the total fire statistics but up to 90% of the total damaged forest area – provides an objective appraisal for the region.

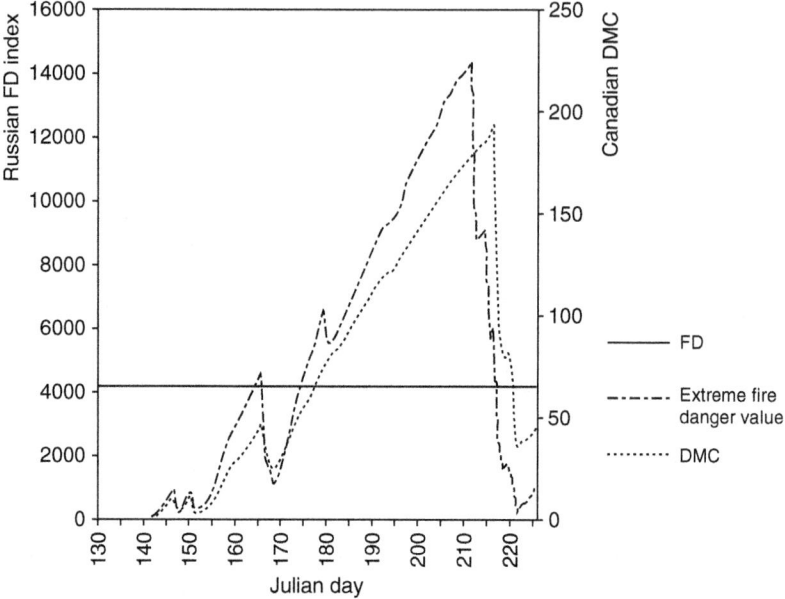

Fig. 2.2 Extreme fire danger index dynamics in the Angara River region, from data recorded at Kezhma meteostation for the fire danger season of 2006. The Canadian Duff Moisture Code (DMC) is shown for comparison

Fig. 2.3 Frequency distribution of the duration of active forest fires in the Krasnoyarsk region, 2007. About 97% of the fires burned only for 1–2 days, and only 1% of fires burned for longer than 5 days

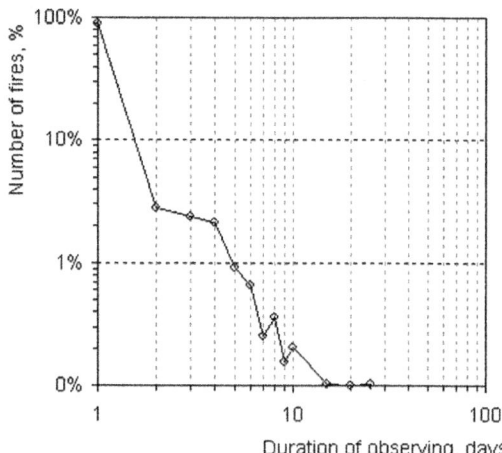

The FD index is effective at detecting conditions that enhance extreme fire activity. The number of days on which the FD index exceeds 4,200 explains about half the interannual variability in burned area in the Krasnoyarsk administrative region determined from the FFID remotely sensed dataset (Fig. 2.4).

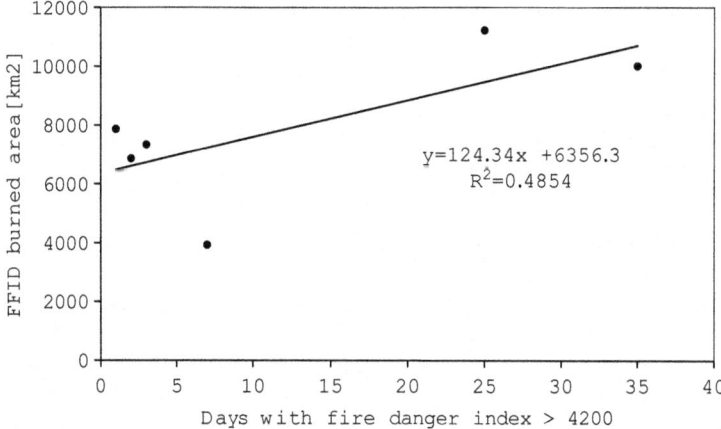

Fig. 2.4 Regression analysis of remotely sensed burned area from the FFID project (km^2) and the number of days with a fire danger index exceeding 4,200 for the Krasnoyarsk region. *Data points* represent the years 2001–2006

Thus, weather conditions are determining the characteristics of the fire season in Siberia. The frequency of prolonged droughts has been observed to increase. Mass forest fire activity is influenced by extreme weather conditions forming at a regional level.

2.6 Fire Feedbacks to the Climate System

Depending on the dominant processes, biosphere feedbacks to the climate system can accelerate or slow down climate change (Cox et al. 2000). Fluxes of heat, water, carbon, and other greenhouse gases between the land surface and the atmosphere interact in complex nonlinear ways (Delworth and Manabe 1993). Siberian forest fires feed back to the climate system by (i) emitting trace gases that contribute to the greenhouse effect, (ii) emitting aerosols that reflect incoming solar radiation back to space having a net cooling effect, (iii) disrupting carbon sequestration by destroying vegetation that would otherwise take up carbon dioxide through photosynthesis, (iv) changing the heterotrophic respiration in the soil, (v) depositing char and charcoal particles and dust on the ground that can be subject to infiltration into the soil or erosion after rainfall and sedimentation downstream, (vi) changing the water balance because of vegetation destruction leading to dryer conditions and increased repeat fire risk in the fire scar, (vii) changing the albedo (proportion of reflected incoming radiation).

Quantitative trace gas emission estimates from forest fires in Siberia are still subject to considerable uncertainty. Soja et al. (2004) estimate that from 1998 to 2002 direct carbon emissions during forest fires quantified by a mean standard emission scenario amount to 555–1031 Tg CO_2, 43–80 Tg CO, 2.4–4.5 Tg CH_4 and 4.6–8.6 Tg carbonaceous aerosols. These emissions represent between 10% and 26% of the global emissions from forest and grassland fires (Soja et al. 2004).

A study of post-fire photosynthetic activity using MODIS fraction of absorbed photosynthetically active radiation (fAPAR) data over Siberian burn scars found that in the years immediately following a fire, fAPAR was reduced between 3% and 27% compared to unburned control plots (Cuevas-González et al. 2008). The amount of photosynthetic reduction depended on forest type and an interaction term of forest type/latitude of the site.

Randerson et al. (2006) studied one particular boreal forest fire in Alaska and quantified the effects of greenhouse gas emissions, aerosols, black carbon deposition on snow and sea ice, and post-fire changes in surface albedo on climate. The net radiative forcing effect was a net warming of 34 Wm^{-2} of burned area during the first year, but a net cooling effect of -2.3 Wm^{-2} over an 80 year period. The reason for this is that long-term increases in surface albedo can have a larger radiative forcing impact than greenhouse gas emissions from the fire (Randerson et al. 2006). However, whether these results are applicable to the entire boreal biome is questionable.

2.7 Conclusions

Siberian forest fires are significant as a factor in the global carbon cycle because of their large interannual variability. Climate impacts on the frequency and extent of forest fires, and fires in turn feed back to the climate system via the atmosphere. Current scenarios of global change indicate that we are likely to see changes in the vegetation patterns and fire regime in Siberia. Satellite remote sensing has an important role to play in monitoring the evolving fire regime from space.

Acknowledgments The Global Land Cover 2000 database was generated by the European Commission, Joint Research Centre, 2003, http://www-gem.jrc.it/glc2000.

References

Amiro BD, Stocks BJ, Alexander ME, Flannigan MD, Wotton BM (2001) Fire, climate change, carbon and fuel management in the Canadian boreal forest. Int J Wildland Fire 10:405–413

Ayres MP, Lombardero MJ (2000) Assessing the consequences of global change for forest disturbance from herbivores and pathogens. Sci Total Environ 262:263–286

Balzter H, Gerard F, George C, Weedon G, Grey W, Combal B, Bartholome E, Bartalev S, Los S (2007) Coupling of vegetation growing season anomalies and fire activity with hemispheric and regional-scale climate patterns in central and east Siberia. J Climate 20:3713–3729

Balzter H, Gerard FF, George CT, Rowland CS, Jupp TE, McCallum I, Shvidenko A, Nilsson S, Sukhinin A, Onuchin A, Schmullius C (2005) Impact of the Arctic Oscillation pattern on interannual forest fire variability in Central Siberia. Geophys Res Lett 32:L14709

Bartalev SA, Belward AS, Erchov DV, Isaev AS (2003) A new SPOT4-VEGETATION derived land cover map of Northern Eurasia. Int J Remote Sens 24:1977–1982

Buermann W, Anderson B, Tucker CJ, Dickinson RE, Lucht W, Potter CS, Myneni RB (2003) Interannual covariability in northern hemisphere air temperatures and greenness associated with El Nino-Southern Oscillation and the Arctic Oscillation. J Geophys Res-Atmos 108:4396

Cox PM, Betts RA, Jones CD, Spall SA, Totterdell IJ (2000) Acceleration of global warming due to carbon-cycle feedbacks in a coupled climate model. Nature 408:184–187

Cuevas-González M, Gerard F, Balzter H, Riaño D (2008) Studying the change in fAPAR after forest fires in Siberia using MODIS, Int J Remote Sens, 29:23: 6873–6892. DOI: 10.1080/01431160802238427

Delworth T, Manabe S (1993) Climate variability and land-surface processes. Adv Water Resour 16:3–20

Frey KE, Smith LC (2003) Recent temperature and precipitation increases in West Siberia and their association with the Arctic Oscillation. Polar Res 22:287–300

Friedl MA, McIver DK, Hodges JCF, Zhang XY, Muchoney D, Strahler AH, Woodcock CE, Gopal S, Schneider A, Cooper A, Baccini A, Gao F, Schaaf C (2002) Global land cover mapping from MODIS: algorithms and early results. Remote Sens Environ 83:287–302

George C, Rowland C, Gerard F, Balzter H (2006) Retrospective mapping of burnt areas in Central Siberia using a modification of the normalised difference water index. Remote Sens Environ 104:346–359

Hallett DJ, Lepofsky DS, Mathewes RW, Lertzman KP (2003) 11000 years of fire history and climate in the mountain hemlock rain forests of southwestern British Columbia based on sedimentary charcoal. Can J Forest Res 33:292–312

Ivanova GA, Volosatova NA, Kukavskaya EA, McCrae DD, Conard SG (2005) Fire emission of carbon in pines of Central Siberia. Remote sensing in forestry. Devises and techniques. Institute for Forest, Krasnoyarsk, pp 51–54

Jupp TE, Taylor CM, Balzter H, George CT (2006) A statistical model linking Siberian forest fire scars with early summer rainfall anomalies. Geophys Res Lett 33:L14701

Justice CO, Giglio L, Korontzi S, Owens J, Morisette JT, Roy D, Descloitres J, Alleaume S, Petitcolin F, Kaufman Y (2002) The MODIS fire products. Remote Sens Environ 83:244–262

Kobak KI, Turchinovich IY, Kondrasheva NY, Schulze ED, Schulze W, Koch H, Vygodskaya NN (1996) Vulnerability and adaptation of the larch forest in eastern Siberia to climate change. Water Air Soil Pollut 92:119–127

Kurbatski NP, Ivanova GA (1987) Fire danger of pine forests of forest-steppe and its decreasing technique. Institute for Forest, Krasnoyarsk, 112 p

Los SO, Collatz GJ, Bounoua L, Sellers PJ, Tucker CJ (2001) Global interannual variations in sea surface temperature and land surface vegetation, air temperature, and precipitation. J Climate 14:1535–1549

Overland JE, Wang MY, Bond NA (2002) Recent temperature changes in the Western Arctic during spring. J Climate 15:1702–1716

Randerson JT, Liu H, Flanner MG, Chambers SD, Jin Y, Hess PG, Pfister G, Mack MC, Treseder KK, Welp LR, Chapin FS, Harden JW, Goulden ML, Lyons E, Neff JC, Schuur EAG, Zender CS (2006) The impact of boreal forest fire on climate warming. Science 314:1130–1132

Schaaf CB, Gao F, Strahler AH, Lucht W, Li X, Tsang T, Strugnell NC, Zhang X, Jin Y, Muller J, Lewis PE, Barnsley M, Hobson P, Disney M, Roberts G, Dunderdale M, Doll C, d'Entremont RP, Hug B, Liang S, Privette JL, Roy D (2002) First operational BRDF, albedo nadir reflectance products from MODIS. Remote Sens Environ 83:135–148

Soja AJ, Cofer WR, Shugart HH, Sukhinin AI, Stackhouse PW, McRae DJ, Conard SG (2004) Estimating fire emissions and disparities in boreal Siberia (1998–2002). J Geophys Res-Atmos 109:D14S06

Stenchikov G, Robock A, Ramaswamy V, Schwarzkopf MD, Hamilton K, Ramachandran S (2002) Arctic Oscillation response to the 1991 Mount Pinatubo eruption: Effects of volcanic aerosols and ozone depletion. J Geophys Res-Atmos 107:4803

Stenchikov G, Hamilton K, Stouffer RJ, Robock A, Ramaswamy V, Santer B, Graf HF (2006) Arctic Oscillation response to volcanic eruptions in the IPCC AR4 climate models. J Geophys Res-Atmos 111:D18101

Sukhinin AI, French NHF, Kasischke ES, Hewson JH, Soja AJ, Csiszar IA, Hyer EJ, Loboda T, Conrad SG, Romasko VI, Pavlichenko EA, Miskiv SI, Slinkina OA (2004) AVHRR-based mapping of fires in Russia: new products for fire management and carbon cycle studies. Remote Sens Environ 93:546–564

Tansey K, Grégoire J-M, Defourny P, Leigh R, Pekel J-F, van Bogaert E, Bartholomé E (2008) A new, global, multi-annual (2000–2007) burnt area product at 1 km resolution. Geophys Res Lett 35:L01401. doi:10.1029/2007GL031567

Valendick EN, Ivanova GA (2001) Fire regimes in forests of Siberia and Far East. Lesovedenie 4:69–76

van der Werf GR, Randerson JT, Collatz GJ, Giglio L, Kasibhatla PS, Arellano AF, Olsen SC, Kasischke ES (2004) Continental-scale partitioning of fire emissions during the 1997 to 2001 El Nino/La Nina period. Science 303:73–76

Chapter 3
Long-Term Dynamics of Mixed Fir-Aspen Forests in West Sayan (Altai-Sayan Ecoregion)

D.M. Ismailova and D.I. Nazimova

Abstract Space-temporal dynamics of secondary mixed fir-aspen forest types is considered by the example of the communities typical of chern forest zone (*Abies sibirica – Populus tremula + Athyrium filix-femina – Matteuccia struthiopteris – Anemone baicalensis*) in perhumid and moderate-continental climate of the Altai-Sayan Ecoregion (South Siberia). Different variants of 40-year-long succession typical of one native forest type are described on permanent plots: (1) Aspen forest with large ferns and herbs. (2) Fir-aspen forest with less developed floor of the same species, which is transformed rapidly into mixed forest with dominance of fir and a poorly developed layer of boreal and nemoral herbs. (3) Siberian pine stand with well developed herbaceous layer. The last variant was formed with the help of cleaning cutting; species diversity and composition remain similar to Aspen forest. The comprehensive data on dynamics of the communities' structure are presented. To estimate the rate of successions in the secondary fir-aspen forests qualitative and quantitative methods were applied. Index of succession rate, Sorenson's dissimilarity coefficient and the Shannon index of species diversity were used. The study highlights the importance of spatial heterogeneity, especially within herb layer, for the regeneration and long-term coexistence of Fir, Aspen and Siberian pine. Dense herb layer of tall forbs and ferns form the main barrier to Siberian pine regeneration and further immature stages even more than shade-tolerant Fir bio-groups. The permanent plots allow us to research all the transformations in space pattern and to learn intra- and inter-specific relationships and trends of succession in chern tall-forbs mixed forests.

Keywords Mixed fir-aspen (chern) forest zone (belt) • Succession • Space-temporal dynamics • Permanent plots

---

D.M. Ismailova (✉) and D.I. Nazimova
V.N. Sukachev Institute of Forest, SB RAS, 50 Akademgorodok, Krasnoyarsk 660036, Russia
e-mail: dismailova@mail.ru; inpol@mail.ru

3.1 Introduction

The study of community structure and its dynamics is important relating to direction and mechanisms of forest succession (Vasilevich 1993; Bakker et al. 1996). The changes in vertical and horizontal structural patterns reflect the main stages of succession induced by local and global factors (Svensson and Jeglum 2001). Possible changes in the structure, composition and biomass of mountain forests have been a focus over the past years (Bugmann 1997; Bugmann and Solomon 2000).

Diversity of succession dynamics is a way to sustain a stable equilibrium in forest ecosystems with a high level of biodiversity. In spite of the large amount of work, the problems of space-temporal ordering of forest structure are not clearly understood.

The mixed dark coniferous-small leaved formations of the Altai-Sayan Mountain with nemoral relics in herbaceous layer are known as "chern forest" which represent the most humid and the warmest variant of Siberian boreal forest (Hytteborn et al. 2005). They attract the attention of ecologists and botanists in the context of global change because they perform a native model of sustainable forest ecosystem structure in conditions of an active cyclonic regime on windward slopes of the South Siberian Mountains (Smagin 1980). Their long-term monitoring has been conducted since 1966 in West Sayan in the Ermakovsky forest station of Institute of Forest SB RUS. Some results of 40-year long research on experimental permanent plots are represented in this chapter showing the dynamics of phytocenotic structure and composition in the low mountain chern forest.

The term "chern" is applied to differentiate the herbaceous wet forest with nemoral species and some endemic forms among them from the typical boreal forest with feather moss and a dwarf-shrub layer. The characteristic features of chern forests are: dominating of Aspen (*Populus tremula*), Siberian fir (*Abies sibirica*) and Siberian pine (*Pinus sibirica*), tall herbaceous layer (1.2–1.8 m) formed by large herbs, ferns and grasses. Well developed synusia of nemoral relic species (*Brunnera sibirica, Anemonoides baicalensis, Galium odoratum, G. Krylovii, Festuca sylvatica, F. gigantea, Polystichum braunii* and some other species) add the specific composition of the plant communities. Besides, there is no thick moss layer but some of hygrophilous species (*Mnium* spp., *Bryum* spp., *Cyrriphyllum* spp., *Drepanocladus* spp., *Rhitidiadelphus* spp., etc.) covering less than 10% of ground surface.

Due to the relative openness of the crown space, the bushes (*Padus asiaticus, Salix caprea, Viburnum opulus, Ribes hispidulum, Rubus ideaus*) form a mosaic layer which is able to occupy the space after disturbances and dominate on the first stages of succession.

Aspen is the main tree species, successfully competing with Fir, therefore their combination forms a typical composition of chern forest.

Fir competes with Siberian pine, especially in the course of natural succession. After cutting, windfall, and wildfires fir is faster to occupy the space and to dominate for a long period (up to 200–300 years), while Siberian pine accumulates the stock slower and becomes a dominant on the latest stages of succession.

The term "chern" is applied to mixed dark-coniferous forests (*Abies sibirica* and *Pinus sibirica*) with *Populus tremula*, characterized by a high field layer (1–2 m) of herbs and ferns, such as *Athyrium filix-femina, Matteuccia struthiopteris, Dryopteris expansa, D. filix-mas, Aconitum septentrionale, Cirsium heterophyllum, Cacalia hastata, Geranium albiflorum* and others, more than 40 species of herbs. Their composition and structure are not typical of boreal Siberian taiga, and resemble the wet subnemoral dark coniferous forest which occurs in the moderate continental climate of East Europe.

An important characteristic is the presence of spring ephemeroides. They are supposed to be nemoral relicts, such as *Anemone baicalensis, Brunnera sibirica, Cruciata krylovii, Galium odoratum, Festuca sylvatica, F. gigantea* and *Polystichum braunii* (Nazimova 1975; Stepanov 1999). The common soil types are grey forest soils (greyzems) in the low mountain belt and mountain brown forest soils (cambisols) in the middle mountain belt.

Various complex investigations in chern forests of West Sayan have been carried out according to the program of *Pinus sibirica* forests restoration since the 1960th on Ermakovsky station of IF SB RAS (Polikarpov 1970; Nazimova and Ermolenko 1980).

The research aim is the study of space-temporal dynamics of secondary mixed fir-aspen forest, their restoration after clear cutting on two control plots (natural succession) and one experimental plot with the experience of effective silviculture. The aim was to create Siberian Pine forest-garden from mixed stands using cleaning cutting.

For the first time on the basis of the data collected on permanent plots from 1966 to 2006 the space-temporal dynamics and rate of succession are analyzed.

3.2 Materials and Methods

The study was conducted in the low mountain relief of West Sayan (Altai-Sayan Ecoregion). The climate is moderate-continental and per-humid. The mean temperature in the coldest month (January) is $-18°C$ and the mean temperature in the warmest month (July) is $+18°C$. Sums of active temperature ($>10°C$) are rather high, comparing with typical boreal taiga and vary from $1,700°C$ to $1,900°C$, precipitation ranges from 800 to 1,300 mm year^{-1}, that is much more than in the Siberian taiga zone. The potential evapotranspiration rate is 0.6–0.4, and the Budyko aridity index is close to 0.5–0.4 (Polikarpov et al. 1986). Locations of fir-aspen forest are presented in the diagram of climatic ordination in Fig. 3.1.

The main plant community of fir-aspen forests is the *Querco-Fagetea* class (Ermakov 2003) in the order *Fagetatia sylvaticae* (sub-order *Abietenalia sibiricae*), typical of low mountains of West Sayan.

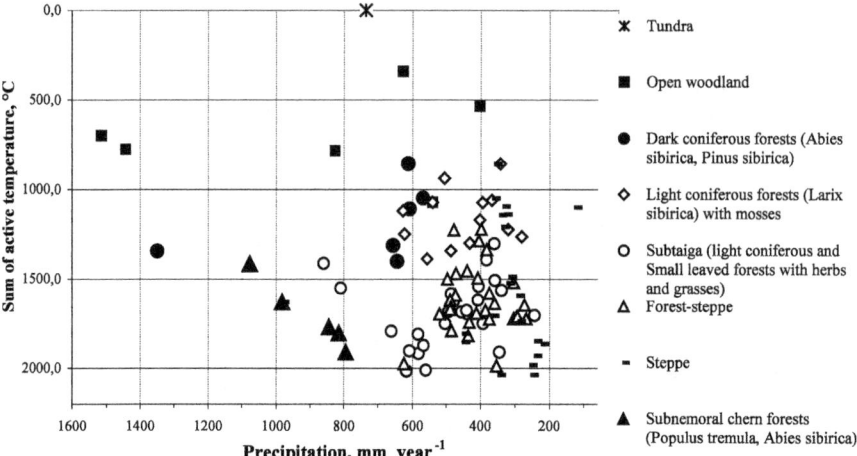

Fig. 3.1 Location of fir-aspen chern forest in the climatic ordination (data of climatic stations of Altai–Sayan Ecoregion)

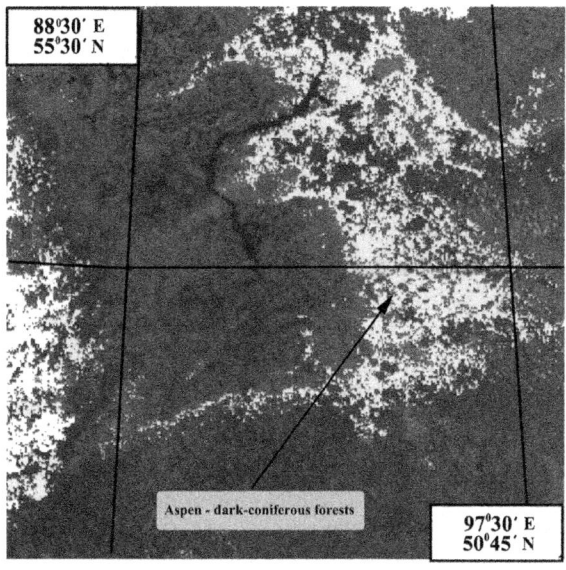

Fig. 3.2 Location of dark-coniferous chern forests in the Altai-Sayan mountains according to remote sensing data (northern part of the ecoregion with the Enissey River in the center) (Nazimova et al. 2005)

The chern forests form a well recognized altitudinal belt only on west and north-west macroslopes of mountains while subtaiga is identified as an altitudinal belt everywhere in the low mountains surrounding the forest-steppe Minusinskaya depression (Fig. 3.2).

Four communities include the same tree species (*Populus tremula, Betula pendula, Abies sibirica, Pinus sibirica, Pinus sylvestris*) and shrubs (*Salix caprea, Padus avium, Sorbus sibirica, Viburnum opulus, Ribes acidum, Ribes nigrum*). Ground-level flora is dominated by *Matteuccia struthiopteris, Dryopteris assimilis, Dryopteris carthusiana, Anemone baicalensis, Brunnera sibirica, Cirsium helenioides, Heracleum dissectum, Angelica sylvestris*.

3.2.1 Permanent Plots

The system of permanent plots for documenting vegetation development after cleaning cutting has been established since 1966 by N.P. Polikarpov, D.I. Nazimova and P.M. Ermolenko. Four plots 0.25 ha (A-III, A-I, C-I– 40 × 62.5 m; C-II – 50 × 50 m) were selected within the young mixed fir-aspen forests. Plot coordinates were recoded by means of a global positioning system (GPS) (53°08′ N, 92 54′ E). The plots are located between 400 and 420 m.

Permanent plots of 2,500 m^2 in secondary stands have been established to study space-temporal dynamics of the stand composition, undergrowth, herbaceous layer and the space-temporal heterogeneity (vertical and horizontal patterns) in mixed fir-aspen forests.

In each sample year, the plots were monitored for the presence and cover of vascular plant species. Tree height, stem diameter and parameters of crowns were periodically measured. To define direction of successions the data of regular mapping of tree, shrub and herbaceous layers were analyzed. Plant cover was estimated according to the visual method. This method demonstrated that visual estimates provided quite accurate, sensitive, and precise values of vegetation cover (Sukachev and Dilis 1966). The names of species are given according to Synopsis of Siberian flora (2005).

The index of succession rate (RS) (Lewis 1978; Nazimova and Ismailova 2007), Sörenson's dissimilarity coefficient (S) and the Shannon index (H) of species diversity were used (Magurran 1988; Roberts and Zhu 2002; Foster and Tilman 2000) to evaluate quantitatively the changes in composition and structure of communities.

3.3 Results and Discussion

3.3.1 Stand Characteristics

The natural course of succession dynamics in mixed fir-aspen stands goes in two directions: through fir-aspen through aspen and birch-aspen stages.

3.3.2 Tree Layer

At the similar stand compositions (A-III – 7Ос2Б1П+С,К and A-I – 5Ос2Б2П1К+С, the communities develop in two directions after cutting: through aspen (A-III) and fir-aspen phases (Table 3.1) (Ismailova and Nazimova 2007). Changes of vertical and horizontal structure in aspen stand from 1966 to 2006 are shown in Figs. 3.3 and 3.5. The proportion of aspen on A-III decreased as a result of its felling. Crown projections of overstory increased from 43.5% to 57%. The number of woody stem trees decreased by 72% as a result of falling and drying of trees lagging in the growth in the aspen stands and by 57% in fir-aspen stand (A-I) as compared to 1966. With the increasing height and diameter of growing trees the forest yield progressively accumulates from 100 to 226 m^3 h^{-1} (A-III), and from 106 to 337 m^3 ha^{-1} (A-I) (Fig. 3.4).

The forming second tree layer from fir on plot A-I (Fig. 3.3) continues its influence in community which changes microclimate: the shading extends, warmth and humidity regimes change that leads to structural transformation of community layers (Fig. 3.4).

3.3.3 Shrub Layer

The share and distribution of shrub species change simultaneously with falling of trees (Fig. 3.5). On plot A-III shrub layer coincides with gaps in overstory.

Table 3.1 Characteristics of the stands in the secondary fir-aspen chern forests in low mountain of West Sayan[a]

Characteristics of the stands	Year	Plots		
		A-III (control)	A-I (control)	C-I (experiment)
Density (stems ha^{-1})	1966	1,144	3,472	3,436
	1984	624	3,392	336
	2005	320	1,508	372
Mean tree height (m)	1966	12	11.5	6.7
	1984	17.2	17.5	–
	2005	18.6	23.5	20.2
Mean DBH (cm)	1966	14.2	15.6	6.2
	1984	20.5	19	–
	2005	37.4	35	32.2
Canopy closure (%)	1966	60	100	100
	1984	60	100	95
	2005	50	100	70
Growing stock (m^3 ha^{-1})	1966	100	106	70.8
	1984	216	320	110
	2005	226	337	173

[a]Only trees > 11.4 cm in diameter at breast height (1.37 m) were included in the estimates of density, mean tree height

3 Long-Term Dynamics of Mixed Fir-Aspen Forests in West Sayan

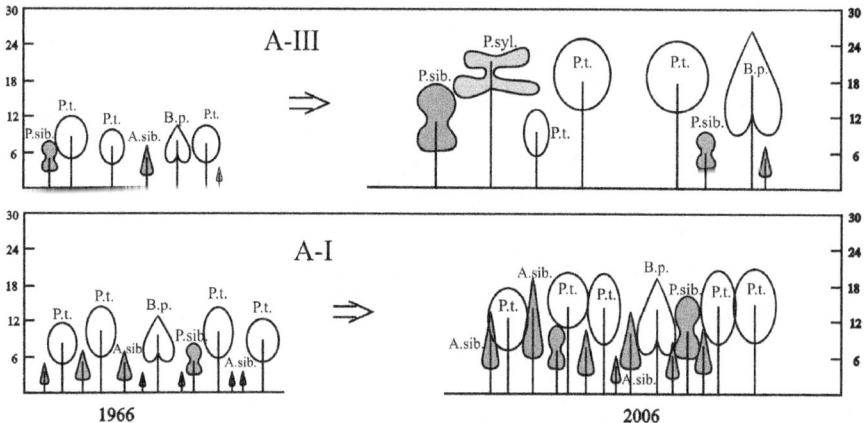

Fig. 3.3 Dynamics of tree layer vertical structure on plots A-III and A-I. P.t. – Populus tremula L.; B.p. – Betula pendula Roth.; A. sib. – Abies sibirica Ledeb.; P.sib. – Pinus sibirica Du Tour.; P.syl. – Pinus sylvestris L

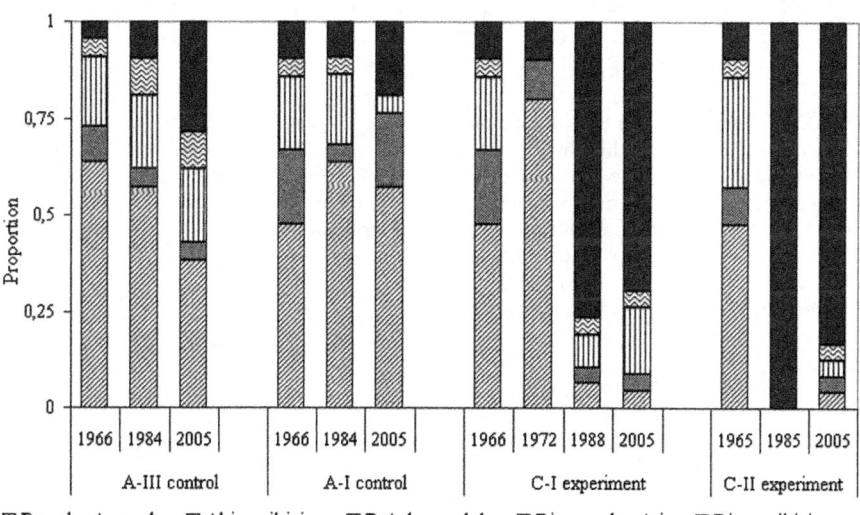

Fig. 3.4 Dynamics of tree species proportion (m^3 ha^{-1}) in the secondary stands. Control plots: A-III (aspen stand), A-I (fir-aspen stand); experimental plots: C-I and C-II (Siberian pine stands)

During the succession process the level of crown closure and vitality of shrub species shifted. With developing of tree layer the vitality of *Padus avium*, *Rubus idaeus*, *Viburnum opulus* has been decreasing. *Salix caprea* fell out. On plot A-I shrub layer was absent from 1966 to 2006.

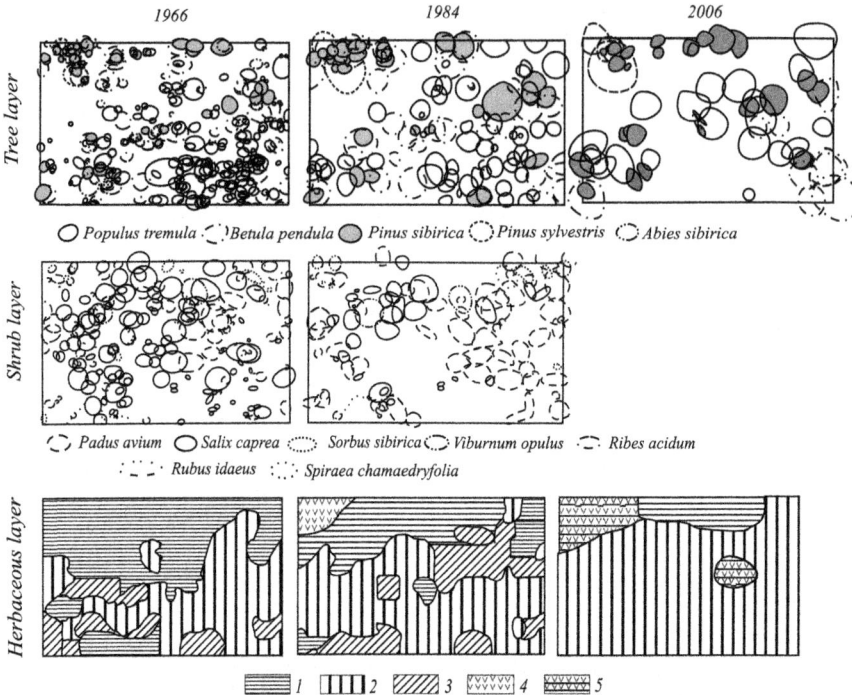

Fig. 3.5 Crown projection map of tree, shrub layer and structure of herbaceous layer on plot A-III (aspen stand). The distribution patterns of dominant species combinations (synusia): (1) Filipendula ulmaria – 15%, Brunnera sibirica – 15%, Pulmonaria mollis 3%; (2) Matteuccia struthiopteris – 60%, Athyrium filix-femina – 8%, Aconitum septentrionale – 10%, Brunnera sibirica – 20%, Pteridium pinetorum – 10%; (3) Pteridium pinetorum – 5%, Brachypodium pinnatum – 10%, Brunnera sibirica – 5%, Calamagrostis obtusata – 5%, Milium effusum – 0,5%; (4) Carex macroura – 55%; (5) Carex macroura – 35%, Brachypodium pinnatum – 25%, Calamagrostis obtusata – 10%

3.3.4 Herbaceous Layer

In case of forming tree layer from aspen and birch trees (A-III) during 40 years *Matteuccia struthiopteris*, *Brunnera sibirica* and tall herbs dominate (coverage – 90–100%, height – 80–100 cm). The distribution patterns of dominant species transformed in response to changing overstory (Fig. 3.5). Under *Pinus sylvestris* crown the pattern of *Carex macroura* (55%) was distinguished, it had transformed into pattern of *Carex macroura* (35%), *Brachypodium pinnatum* (25%), *Calamagrostis obtusata* (10%) by 2006.

The small presence of *Pinus sibirica* doesn't influence the pattern structure of the herbaceous layer though under *Pinus sibirica* crown the development of the herb layer decreases to some extent. In the more closed stand (A-I) with a less developed herbaceous layer (A-I), the share of fir in the stand structure increased from 1966 to 2005 (Fig. 3.4). It had been accompanied by the further herb coverage

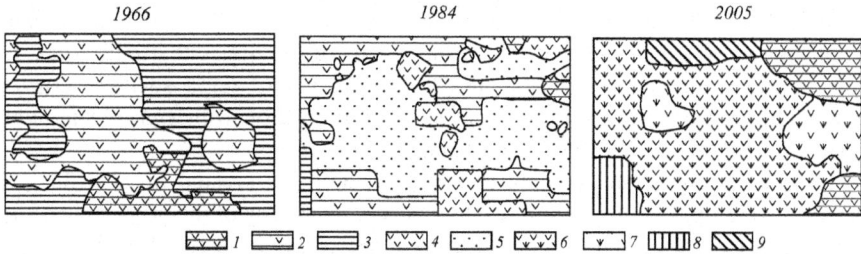

Fig. 3.6 Dynamics of herbaceous layer structure on plot A-I (fir-aspen stand). The distribution patterns of dominant species combinations (synusia): (1) Anemone baicalensis – 25%, Brunnera sibirica – 10%, Carex macroura – 10% (coverage 40-60%); (2) Anemone baicalensis – 25%, Brunnera sibirica – 10%, Carex macroura – 10% (coverage 5–20%); (3) Anemone baicalensis – 30%, Athyrium filix-femina – 5%, Matteuccia struthiopteris – 15%, Carex macroura – 3%, Brunnera sibirica – 5%; 4) Carex macroura – 15% (coverage – 10–20%); 5) Anemone baicalensis – 0,5%, Maianthemum bifolium – 0,5%, Carex macroura– 0,5% (coverage 1–5%); 6) Oxalis acetosella – 15%, Carex macroura – 30% (coverage 60%); 7) Oxalis acetosella – 2%, Carex macroura – 1% (coverage 3-5%); 8) Gymnocarpium dryopteris – 10%, Athyrium filix-femina – 5%, Aconitum septentrionale – 3%, Matteuccia struthiopteris – 3%, Crepis sibirica – 1%; 9) Carex macroura – 40%, Calamagrostis obtusata – 10%, Stellaria bungeana – 10%

digression (distribution pattern of dominant species: *Anemone baicalensis* – 0.5%, *Maianthemum bifolium* – 0.5%, *Carex macroura* – 0.5% (coverage 1–5%) occupied 64% of the plot in 1984). The pattern of dominant species distribution transforms (Fig. 3.6). This process takes a short period of time (from 1985 to 2005) and is expressed in the elimination of large forest ferns (*Athyrium filix-femina*, *Matteuccia struthiopteris*) and in the expansion of *Carex macroura*. Under the fir crowns, the *Carex macroura* actively displaces *Anemone baicalensis*, *Anemone altaica*, *Anemone reflexa*, *Cruciata krylovii*, *Paris quadrifolia*, *Stellaria bungeana*.

3.3.5 Regeneration

In the aspen stand (A-III) the regeneration of *Pinus sibirica*, *Pinus sylvestris*, *Abies sibirica* has been poor for 40 years (100–500 number ha^{-1}) and young undergrowth of coniferous species is almost absent. Its direct cause is a strong competition of large forest ferns and tall forest herbs. On the contrary, the regeneration of small-leaved trees with energetically growing sprouts is successful in the aspen stand (A-III). Most favourable conditions for successful fir regeneration are formed in fir-aspen stand (A-I). Here the regeneration of *Pinus sibirica* is very poor. Thus, the changes of the space-temporal structure have dynamic character on the boundary of chern and sub-taiga forests (light coniferous and small leaved forests with herbs and grasses). The mosaics combining together sub-taiga, nemoral and taiga species have appeared. Great significance in transformation of space-temporal structure belongs to *Pinus sylvestris*, *Carex macroura*, *Brachypodium pinnatum*. Their activation can be a sign of displacement of chern species by sub-taiga species. It can be connected with anthropogenic influence more than with climate changes. On the other hand, fir is a strong competitive species that intensifies its role in some micro-sites. Fir favours the expansion of taiga species (*Oxalis acetosella* and others). The taiga species in combination with *Carex macroura* displace nemoral species that were dominants 40 years ago.

Favourable conditions of light and mineral nutrition created in small-leaved stands lead to light-requiring species dominance. Increasing of the proportion of fir gives rise to an expansion of taiga species, to decreasing of nemoral species, large forest ferns and tall herbs. Two alternative ways of dynamics illustrate the stochastic pattern of anthropogenic succession in mixed fire-aspen forests of the low-mountain relief in the region.

As it is known after cleaning cutting the quantity of heat and soil moisture increase, biological processes activate, metabolic processes between soil and plants accelerate, which results in increase of general stand productivity. In the first years after felling intensive nitrification takes place. In the first 2 years after felling the air temperature increases by 3–4.5°C.

The problems of space-temporal dynamics after cleaning cutting for *Pinus sibirica* undergrowth were studied on experimental plots C-I and C-II. The aspen and fir trees were cut with the help of cleaning cutting twice (in two hops) that influenced the direction of the communities' restoration.

The aim of the experiment was to grow the stand-garden with early bearing trees. From 1966 up to 1988 the stock of *S. pine* increased 11 times (from 8.9 to 99.6 m^3 ha^{-1} and in 2005 was 173 m^3 ha^{-1} (Fig. 3.4).

Long-term silvicultural experiments from 1966 to 2006 have shown the effects of alternative forest management strategy on the forest restoration processes after cleaning cutting of mixed forests (Fig. 3.7). Results have shown that a cleaning

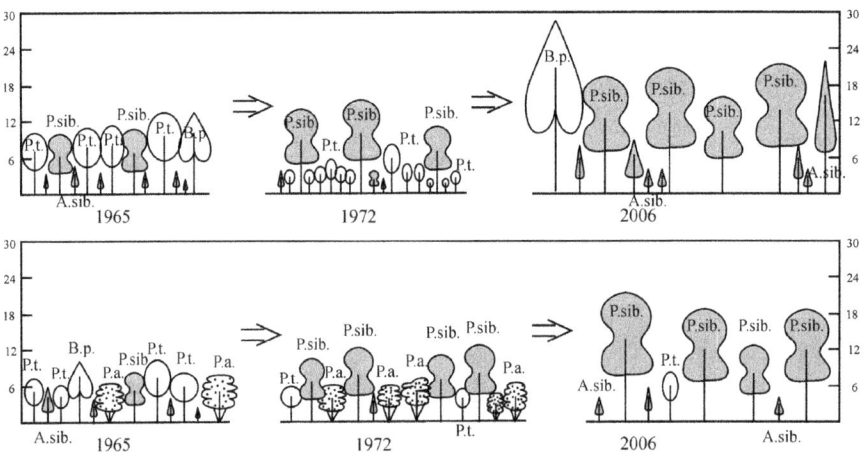

Fig. 3.7 Dynamics of tree layer vertical structure after cleaning cutting on plots C-I and C-II. P.t. – Populus tremula L.; B.p. – Betula pendula Roth.; A.sib. – Abies sibirica Ledeb.; P.sib. – Pinus sibirica Du Tour.; P.a. – Padus avium Mill

cutting for *Pinus sibirica* in young stands of mixed fir-aspen forests had a positive impact on the successful restoration of *Pinus sibirica* forests in the low mountains of West Sayan.

The long-term studies allow us to preserve a unique relic population of chern *Pinus sibirica*, which occurs only on windward slopes of the Altai-Sayan Mountains with perhumid continental climate.

The indices of the succession rate for estimation the actual changes in overall floristic composition, ecological groups of species (nemoral species, tall herbage, large forest fern, taiga species etc.) taking into account relative abundance species of herbs between 1966 and 2005 years were calculated. The rates of species turnover were computed by the Sorenson's dissimilarity coefficient. Each index emphasizes a different aspect of change in plant community structure.

As measured by the index of succession rate in communities, without changes in stand structure and composition (dominance of small-leaved trees – aspen and birch) the succession rate remained slow: the index of succession rate varies from 0.017 to 0.007, the Sorenson's dissimilarity coefficient from 0.2 to 0.3 (Fig. 3.8). If the role of fir in stand structure becomes stronger, the succession rate accelerates: the index of succession rate varies from 0.067 to 0.012 (A-I). Such great values are explained by replacement of dominant species in herbaceous layer. The species of tall herbage (*Aconitum septentrionale*), nemoral species (*Anemone baicalensis*,

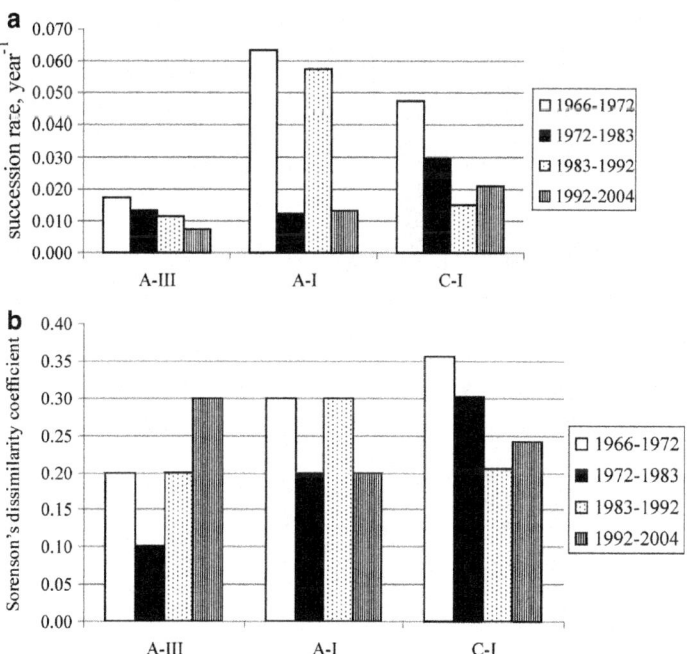

Fig. 3.8 Index of succession rate (**a**) and Sörenson's dissimilarity coefficient (**b**) vs. time for control and experimental plots. Control plots: A-III (aspen stand), A-I (fir-aspen stand); experimental plots: C-I and C-II (Siberian pine stands)

Brunnera sibirica, Cruciata krylovii), large forest ferns (*Matteuccia struthiopteris, Pteridium aquilinum*, et al.) decrease. The taiga species (*Oxalis acetosella, Cerastium pauciflorum*) increase their relative cover under fir crowns. The abundance of *Carex macroura* raised on the plot A-I can be explained by location of the network near the boundary with belt of pine-small-leaved herbaceous forests which are classified as subtaiga.

On experimental plot the index of succession rate such as Sörenson's dissimilarity coefficient decreased during 40 years (Fig. 3.9).

The index of succession rate as an element of an empirical statistical descriptive model allows us to estimate the composition and structure change of fir-aspen forests quantitatively. The succession rate of species herb abundance and species turnover is connected with stand structure. The succession rates are accelerated with increasing proportions of fir in the stands.

The Shannon index and the number of species similarly changed on control and experimental plots (Fig. 3.9). The Shannon index and the number of species strongly decreased as a result of herbs competition with fir groups (A-I). Under aspen overstory the Shannon index didn't change during 40 years. After cleaning cutting for Siberian pine the Shannon index and the number of species increased from 3.2 to 4.2 and 29 to 53 species correspondingly.

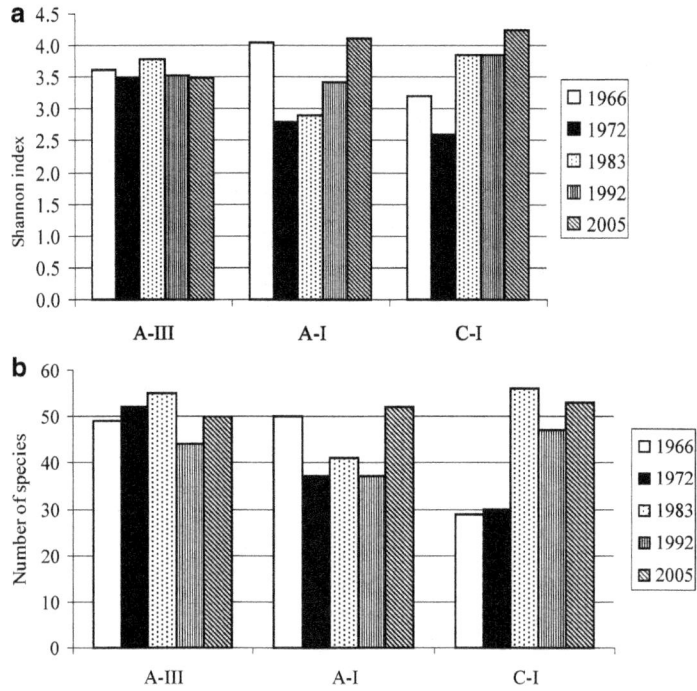

Fig. 3.9 Shannon index (**a**) and the number of species (**b**) vs. time for control and experimental plots. Control plots: A-III (aspen stand), A-I (fir-aspen stand); experimental plots: C-I and C-II (Siberian pine stands)

3.4 Discussion

For the first time 40-year long dynamics of fir-aspen forests after cutting has been monitored on permanent plots in chern low-mountain zone of West Sayan. The observed succession embraced several age stages, from young to mature stands, and two directions of regeneration, through fir-aspen and aspen stands: from 30–40 to 70–80 years old. Diversity of lower layers composed by boreal and nemoral species forms mosaics of horizontal and vertical structure within the given type of forest ecosystem.

Well developed tall forbs layer plays a role of subdominant in functioning of chern forest which influences soil processes and regeneration of trees. It makes impossible successful regeneration of *Siberian pine* which remains poor (500 trees ha^{-1}) during all period of monitoring and restricts fir regeneration in the gaps with tall forbs. Thus it favours regeneration of the aspen offshoots growing faster than other tree species.

The rate and direction of the space-temporal structure changes depend on the dominant trees' influencing force. Fir is the greatest; aspen is the less strong edificator. In the aspen stand the space-temporal structure was relatively constant during 40 observing years. On the contrary, under fir crowns the transformation of lower layer occurs faster, which is reflected in the changes of composition and structure of communities. The data of mapping fix stage-by-stage changes of mosaics. The rate of succession varies from 0.012 up to 0.067 per year under fir crowns (A-I) and it gradually decreases from 0.017 to 0.007 per year under aspen crowns (A-III).

Multivariance of synusia structure of fir-aspen mixed forests is the result of species richness and the sign of sustainability of the given forest ecosystem. The permanent synusia of herb layer are found out: climax, derivative, or temporary and sporadical. Their set is quite special for chern forests with tall forbs and ferns but their presence varies on different stages of succession.

The investigations carried out are a part of complex study. They can be used to make *Siberian pine* stands in their potential sites with high natural productivity taking into account the specificity of this process in chern forests.

The developed database on the dynamics of floristic composition and phytocenotic structure of permanent plots can be applied to long-term monitoring.

The results are of interest for prognosis of composition, structure and sustainability of coniferous and small-leaved formations at different anthropogenic disturbances and natural dynamics in the current environment.

3.5 Conclusions and Future Research Directions

Long-term monitoring of forest ecosystems highlights the tendencies of formations' composition change. Besides, it may be useful to estimate resistance and flexibility of tree populations on the border of mixed fir-aspen and light-coniferous-birch forests in the past, present and future.

They are interesting as a basis of empirical modeling forest community dynamics in conditions of climate moistening if the warming is compensated by additional rainfall. From the other side they may be vulnerable if the balance between warmth and water supply reduces.

The lessons of long term study convince us in great adventures of sustainable growing aspen populations comparing to the other species (fir, Siberian pine, birch) in the zone (altitudinal belt) of chern forest ecosystems. Nevertheless, taking into account large value of Siberian pine forests in this conditions it is important to restore dark coniferous forests with *P. sibirica* possessing high productivity and protective functions, improving mountain landscapes.

The long-term studies on the permanent plots provide an extensive basis for future monitoring. Moreover, the complexity and sensitivity of mountain areas require concerted research efforts and combining experimental field measurements.

Acknowledgments This research was financially supported by RFBR (grant 09-04-98040, 08-04-00600a).

References

Bakker FA, Olff HJ, Willems JH, Zoebel M (1996) Why do we need permanent plots in the study of long-term vegetation dynamics? J Veg Sci 7:145–156

Bugmann H (1997) Sensitivity of forests in the European Alps to future climatic change. Climate Res 8:35–44

Bugmann H, Solomon AM (2000) Explaining forests biomass and species composition across multiple biogeographical regions. Ecol Appl 10:95–114

Ermakov NB (2003) Diversity of boreal vegetation of North Eurasia. Gemiboreal forests. Classification and ordination. SBR RAS, Novosibirsk

Foster BL, Tilman D (2000) Dynamic and static views of succession: testing the descriptive power of the chronosequence approach. Plant Ecol 146:1–10

Hytteborn HH, Rysin LP, Nazimova, DI, Maslov, AA (2005) Boreal forest of Eurasia. In: Anderssen F (ed.) Ecosystems of the World. Coniferous forests. Elsevier, Amsterdam/Boston/London/New York/Oxford/Paris/San Diego/Singapore/Sydney/Tokyo, pp 23–99

Ismailova DM, Nazimova DI (2007) Long-term dynamics of phytocenotic structure of chern fir-aspen forests. Lesovedenie 3:3–10

Lewis WMJr (1978) Analysis of succession in a tropical phytoplankton community and a new measure of succession rate. Am Nat 112(984):401–414

Magurran A (1988) Ecological diversity and its measurement. Princeton University Press, Princeton, NJ

Nazimova DI (1975) Mountain dark-coniferous forests of West Sayan (experience of ecological-phytocenotic classification). Nauka, Leningrad

Nazimova DI, Ermolenko PM (1980) Dynamics of synusial structure during progressive succession of Siberian forest biogeocenosis. In: Smagin V (ed) Dynamics of forest biogeocenosis of Siberia. Science. Nauka, Novosibirsk, pp 54–87

Nazimova DI, Ponomarev EI, Stepanov NV, Fedotova EV (2005) Chern dark coniferous forests in Southern Krasnoyarsk krai and problems of their general mapping Lesovedenie 1: 12–18

Nazimova DI, Ismailova DM (2007) Direction and rate of secondary successions in contact zone of chern and subtaiga forests (humid low Sayan). Bot J 8:1203–1214

Polikarpov NP (1970) Complex investigations in mountain forests of West Sayan. In: Zukov A (ed.) The problems of forestry. Institute of Forest and Wood, Krasnoyarsk, pp 29–79

Polikarpov NP, Tchebakova NM, Nazimova DI (1986) Climate and mountain forests of the Southern Siberia. Nauka, Novosibirsk

Smagin VN (ed) (1980) Forest types of South Siberia Mountains. Nauka, Novosibirsk

Roberts MR, Zhu L (2002) Early response of the herbaceous layer to harvesting in a mixed coniferous–deciduous forest in New Brunswick, Canada. For Ecol Manag 155:17–31

Stepanov NV (1999) Floristic peculiarities of vascular plants in the North-East part of the West Sayan. Bot J 84(5):95–101

Sukachev VN, Dilis NV (1966) Program and methods of biogeocenotic researches. Nauka, Moscow

Svensson JS, Jeglum JK (2001) Structure and dynamics of an undisturbed old-growth Norway spruce forest on the rising Bothnian coastline. For Ecol Manag 151:67–79

Malyshev LI, Peshkova GA, Bajkov KS et al (2005) Synopsis of Siberian flora: vascular plants. Nauka, Novosibirsk

Vasilevich VI (1993) Some new directions in studying of vegetation dynamics. Bot J 78(10):1–15

Chapter 4
Evidence of Evergreen Conifers Invasion into Larch Dominated Forests During Recent Decades

V.I. Kharuk, K.J. Ranson, and M.L. Dvinskaya

Abstract Dark needle coniferous (DNC: Siberian pine, spruce, fir) expansion into larch dominated area was investigated along transects, oriented from the west and south borders of the larch dominated communities to its centre. The expected invasion of DNC into larch habitat was quantified as an increase of the proportion of those species both in the overstory and regeneration. Abundance and invasion potential was expressed using the following variables: (1) N_i and n_i – the proportion of a given species in the overstory and regeneration, respectively, and (2) K_i – "the normalized propagation coefficient" defined as $K_i = (n_i - N_i)/(n_i + N_i)$. The results show that Siberian pine and spruce have high K_i values both along the margin and in the centre of zones of absolute larch dominance, where their presence in the overstory is <1%. There is a tendency of K_i to increase for DNC and birch from south to north and from west to east. The age structure of the regeneration showed that it was formed mainly during the last 2–3 decades. Regeneration number correlates with winter temperature increase, showing winter temperatures importance for regeneration survive. The DNC invasion into larch habitat is wildfire dependant. Fires promote an invasion of DNC due to better ecological conditions on the burns. On the other hand observed climate-induced fire retune interval reduction may complicate DNC invasion into larch habitat, because larch regenerates better after fire than DNC since larger seed-trees amount. The results obtained indicate DNC and birch invasion into the larch habitat and its relation to the climatic changes for the last 3 decades. At the same time larch stand crown closure and larch invasion into tundra observed in the northern forest-tundra ecotone.

Keywords Larch communities • Climate-induced species migration • Burns • Permafrost

---

V.I. Kharuk (✉) and M.L. Dvinskaya
V.N. Sukachev Institute of Forest, SB RAS, Krasnoyarsk 660036, Academgorodok 50, Russia
e-mail: kharuk@ksc.krasn.ru; mary_dvi@ksc.krasn.ru

K.J. Ranson
NASA Goddard Space Flight Center, Greenbelt MD 20771, USA
e-mail: jon.ranson@nasa.gov

4.1 Introduction

The zone of larch dominance (LDZ) spreads from the Yenisey ridge in the west to the Pacific, and from Lake Baikal in the south to 73° north, where it forms the world's northward Ary-Mas stand. In Central Siberia larch on its southern and western margins is contacting with dark needle conifers (DNC: Siberian pine, *Pinus sibirica*, pine, *Pinus silvestris*, spruce, *Picea obovata*, fir, *Abies sibirica*) and hardwoods (birch, *Betula pendula*, *B. pubescens*, and aspen, *Populus tremula*). Larch competes effectively with other tree species due to its higher resistance to harsh climatic conditions. On its northern border, larch survives where mean annual temperature is −14°C (with absolute minimum −68°C). Surpassing the other tree species in its water use efficiency, it could survive at the semi-desert level of precipitation (<250 mm/year) (Kloeppel et al. 1998). Moreover, larch can survive in the permafrost zone because its deciduous leaf habit and dense bark protect stems from winter desiccation and snow abrasion (Berg and Chapin 1994; Kharuk and Fedotova 2003). The larch area, including the vast forest-tundra ecotone zone, is considered a "carbon sink". However a temperature and precipitation increase in high latitudes (IPCC 2001) can turn this territory in to a greenhouse gas source. Better climatic conditions in general results (1) in larch invasion into tundra on its northern and altitudinal borders, stand crown closure and tree-ring width increasing (Hughes et al. 1999; Sturm et al. 2001; Kharuk and Fedotova 2003); (2) decreasing the current competitive advantage of larch in comparison with other species in its traditional dominance zone. This may result in invasion of "southern" species into larch habitat (Kharuk et al. 2007).

The purpose of this study is to answer the question: is there evidence for an invasion of DNC into the larch habitat? As an indicators of DNC invasion the regeneration are considered: its abundance, species composition, age structure, and the proportion of those species both in the overstory and regeneration.

4.2 Study Area

The main part of the investigated area is situated in the Central-Siberian plateau with heights from 200 to 700 m. LDZ is the zone of permafrost mainly. On the west it is bounded by the Yenisey ridge, with heights up to 1,000 m, and by the Putorana plateau, with heights up to 1,700 m in the north. The climate is severe continental with average annual temperatures between −8°C to −14°C. Mean annual precipitation level is 300–400 mm in the central part, 600–800 in the east and 400–500 mm in the south. Wildfires are typical for this territory with the majority occurring as ground fires due to low crown closure. Western and southern transect margins are the areas of "larch-DNC ecotone". The larch-dominated communities are composed mainly of larch with an admixture of birch. The investigations were made 2001–2003 year along two transects: the "West–East" (WE, 91°E – 106° 30′ E, ~800 km), and "South–North"

4 Evidence of Evergreen Conifers Invasion into Larch Dominated Forests 55

Fig. 4.1 Location of the transects in Central Siberia

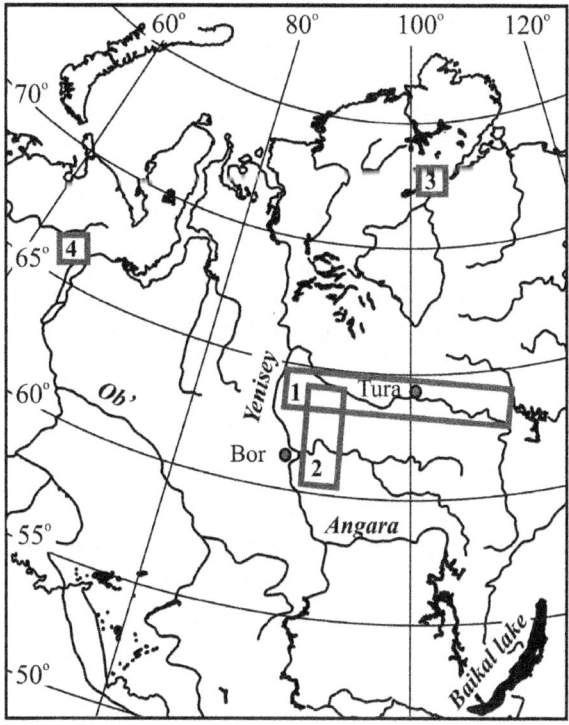

(SN, 57° 30′ N – 64° 30′ N, ~ 420 km) (Fig. 4.1). The WE transect is pointed along a gradient of climate severity – from Yenisey ridge mixed forests (which under the Atlantic cyclones influence) into mid of larch dominance. The SN transect south is the area of mixed woods of Angara river region (Fig. 4.1).

4.3 Material and Methods

Test sites (TS) were established according to Russian Forest Service (1995) inventory rules. For each TS (19.6 m in diameter) the following parameters were described: relief (height above mean sea level, slope aspect, slope steepness), vegetation cover type, forest stand structure (e.g., open, closed), tree diameters and heights, disturbance history (wildfire damage or logging), regeneration structure, description of shrub and ground cover, and soil type. Each TS centre point coordinates were also geo-referenced with ±15 m accuracy. The total amount of TS was 80 along WE and 58 along SN transects. Regeneration studies were made on the test plots 10 × 10 m size (three plots for each TS). Regeneration number ($h \leq 2.5$ m), its age and vigour were determined. Temperature and precipitation data were taken from the "Bor" and "Tura" weather stations (Fig. 4.1). The period from June to

August was conventionally considered taken as "summer", and the period from September to May was considered as "winter". Available forest inventory data were also used in this study (Anonymous 1990).

The expected invasion of DNC into larch habitat was quantified as an increase of the proportion of those species both in the overstory and regeneration. Abundance and invasion potential was expressed using the following variables: (1) N_i and n_i – the proportion of a given species in the overstory and regeneration, respectively, and (2) K_i – "the normalized propagation coefficient" defined as $K_i = (n_i - N_i)/(n_i + N_i)$. The K_i values are +1 in the case of absence of ith species in overstory, −1 in the opposite case and 0 when $n_i = N_i$. The non-parametric Kolmogorov-Smirnov statistics were used in the analysis (StatSoft Inc. 2003).

4.4 Results

4.4.1 Regeneration Distribution Along the Transects

Fig. 4.2 represents a "snap-shot" of the mature tree species regeneration, and propagation coefficient, K_i, values along WE transect: from the margin to the mid of LDZ. The data for the SN transect are similar; since the SN transect is shorter, and there are no principal differences between data for the both transects, in the following analysis the WE transect data will be discussed mainly. The data on the regeneration abundance were presented on Fig. 4.2b and Table 4.1. Figure 4.2a data were generated by a "sliding window" (10 × 150 km) of the forest inventory map (Anonymous 1990), which allows to obtain mean values of the species distribution in the overstory. These data correlates with the on-ground data along the transect (R = 0.58). The Siberian pine regeneration data are of special interest, since it's most wide-spread along the transect; data for spruce and fir on the chart are summarized for better visualization (Fig. 4.2b). Fir is eliminated from the overstory at the ~63°N since this species has a minimal resistance to the harsh climatic conditions. In general Scotch pine, fir and aspen do not significantly penetrate into the LDZ, growing fragmentarily on warmed southern slopes.

The "West–east" transect is directed from the mixed taiga on the east macro slope of Yenisey ridge to the larch dominated area. It corresponds to the climatic severity increase: in the transect west annual temperatures and precipitations were 3.7°C and 560 mm accordingly, in the area of larch domination those parameters were 9.0°C and 353 mm (weather station "Tura"). Climate variables were considered for the whole period of instrumental observations (1934–2003). With increase of a level of continentality larch increase its proportion in overstory and regeneration (Fig. 4.2a).

The regeneration distribution along the transect does not replicate species distribution in the overstory: the DNC regeneration predominates on a significant

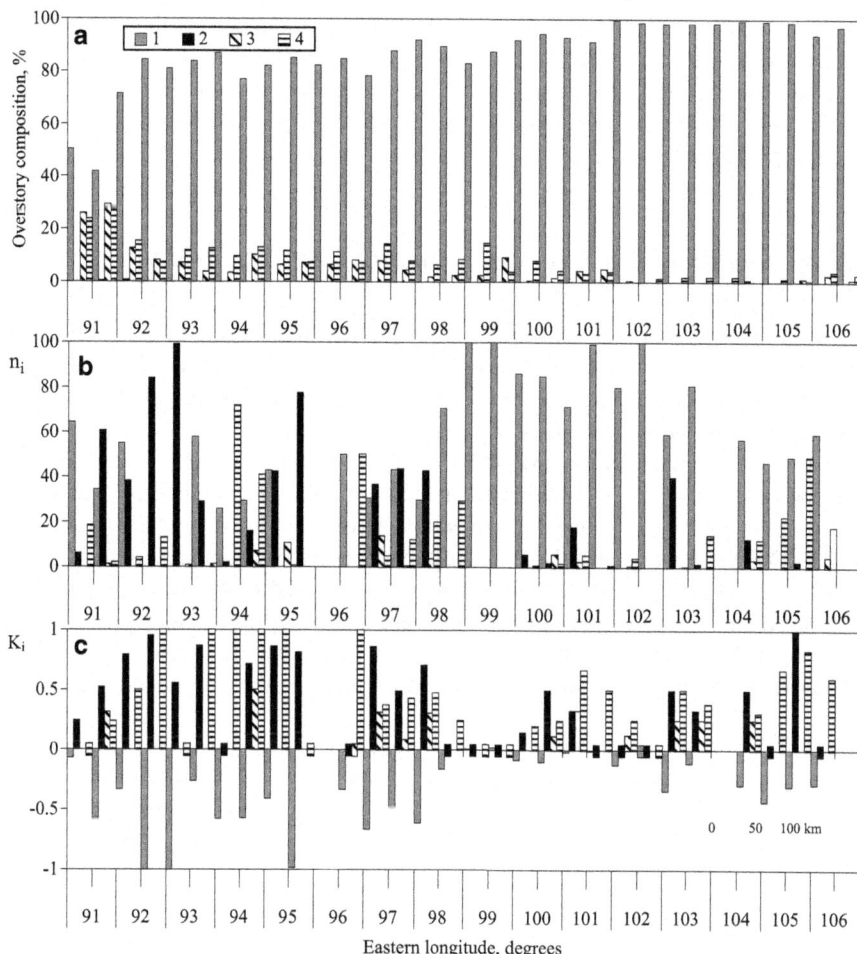

Fig. 4.2 WE transect. (**a**) Overstory. (**b**) Regeneration. (**c**) Propagation coefficient. 1 – larch, 2 – Siberian pine, 3 – spruce + fir, 4 – birch

Table 4.1 Regeneration number on the test sites along WE transect

Regeneration number, n/ha								
	All test sites (TS)				Test sites where n > 0			
	Siberian pine	Larch	Fir	Birch	Siberian pine	Larch	Fir	Birch
TS number	58	58	58	58	36	50	29	33
n/ha, max	13,200	530,000	2,000	62,130	13,200	530,000	2,000	62,130
n/ha, min	0	0	0	0	167	17	17	67
n/ha, mean	1,080	37,150	210	3,390	1,740	43,100	430	5,960

part of the transect (Fig. 4.2b). This effect is more pronounced for the propagation coefficient values: Ki for DNC regeneration (and a birch) is higher than for larch even in a zone of absolute larch domination where DNC presence in the overstory is <1% (Fig. 4.2b). This data supports the idea of DNC propagation into the typical larch habitat zone.

4.4.2 Regeneration Age Structure and Climate Variables

Siberian pine regeneration appeared mainly during last 3 decades (Fig. 4.3), and >90 % of it is of good vigour. The regeneration dying off over the last 30 years was about 10%. It was checked by dead saplings counting: under climatic conditions of high latitudes they could be identified for decades.

The driving forces of these events are the temperature and precipitation increase (Fig. 4.4a,b). Met data were counted in two variants: (1) for the last ~30 years (1970–2003 year) anomalies for the whole period of instrumental observation (1934–2003 year) was counted; (2) the differences between anomalies for the periods of 1970–2003 and of 1934–1969 year were considered (Table 4.2). For the last 30 years temperature and precipitation increase was observed for both Tura and Bor weather stations (the precipitations increase is significant at $p<0.05$, Table 4.2). Though temperature differences itself are not significant, the temperature trends are

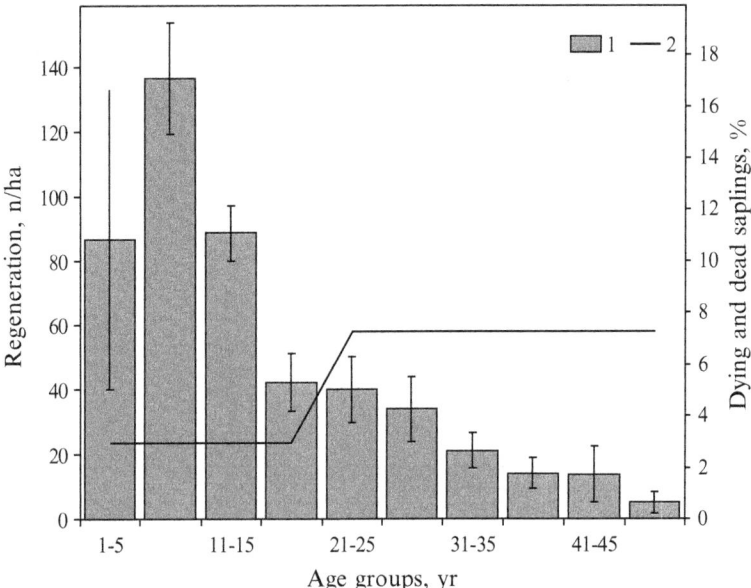

Fig. 4.3 Siberian pine regeneration age structure. 1. Alive Siberian pine saplings. 2. Dying and dead Siberian pine saplings

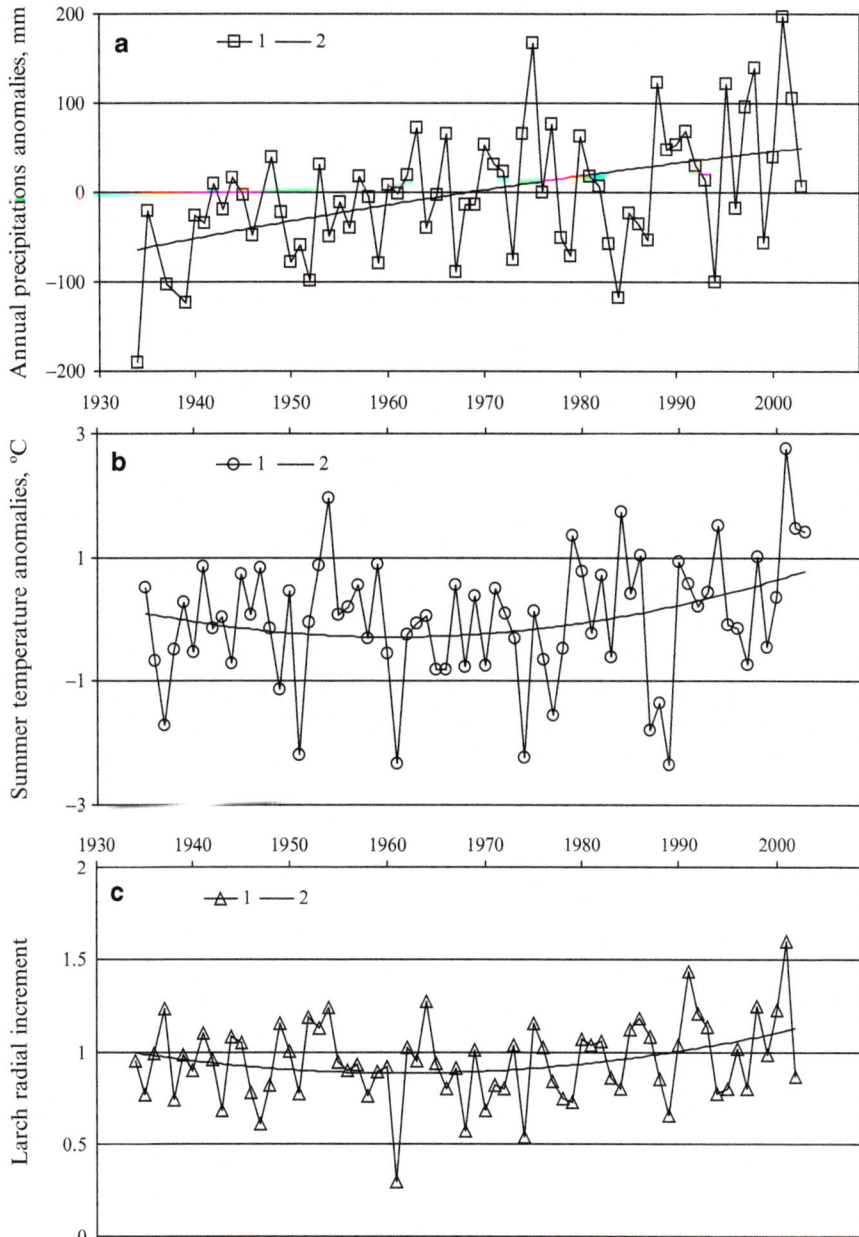

Fig. 4.4 The annual precipitation (**a**) and summer temperature (**b**) anomalies recorded at Tura station; (**c**) larch radial increment

Table 4.2 Temperature and precipitation dynamics (weather stations "Tura" and "Bor")

Weather station	Temperature changes Δt, °C				Precipitations changes Δp, mm			
	$\Delta t_{(70\text{-}03,\ 34\text{-}03)}$, °C		$\Delta t_{(70\text{-}03,\ 36\text{-}69)}$, °C		$\Delta p_{(70\text{-}03,\ 34\text{-}03)}$, mm		$\Delta p_{(70\text{-}03,\ 36\text{-}69)}$, mm	
	t_1	t_2	t_1	t_2	p_1	p_2	p_1	p_2
Tura	0.2	0.3	0.5	0.6	26	**28**	**48**	**51**
Bor	0.2	0.3	0.5	0.6	**41**	**51**	**85**	**99**

$\Delta t_{(70\text{-}03,\ 34\text{-}03)}$ – mean values differences for the periods 1970–2003 years and 1934–2003 years.
$\Delta t_{(70\text{-}03,\ 36\text{-}69)}$ – mean values differences for the periods 1970–2003 years and 1936–1969 years.
t_1 – mean annual temperatures (1–12 months of the current year), °C.
t_2 – mean winter-time temperatures (9–12 months of the previous year and 1–5 months of the current year), °C.
p_1 – mean annual summarized precipitations (1–12 months of the current year), mm.
p_2 – mean winter-time summarized precipitations (9–12 months of the previous year and 1–5 months of the current year), mm.
Data significant at $p < 0.05$ are marked bold.

significant ($p < 0.05$). The correlation between regeneration age structure for the last 3 decades and winter temperature anomalies was found (for "Tura": R = 0.83, and for "Bor": R = 0.78; $p < 0.05$). Correlations with summer temperatures are not significant. This fact shows the importance of the winter temperature regime for the Siberian pine regeneration survival. The maximum of the regeneration number corresponds to the last 10–15 years; it is remarkable, that last decade of the 20th century was the "warmest" for last millennium (IPCC 2001).

4.4.3 DNC Propagation into LDZ and Wildfires

The DNC invasion into larch habitat is wildfire dependant. Wildfires (its absolute majority are ground fires) may promote an invasion of DNC: fire scars represents a "starting place" for "southern species" invasion, since better thermal and soil conditions, enriched biogenic elements content, increased soil thawing depth and drainage. On the other hand, larch (and birch) regenerates better after fire than DNC because fires tend to eliminate DNC saplings, and higher post-fire presence of larch trees in the overstory.

Meanwhile the decrease of fire return interval in the twentieth century in comparison with 19th century (from about 100 to 65 years) (Kharuk et al. 2008) may interfere with the "southern species" invasion into LDZ. Data in Fig. 4.5 shows that Siberian pine regeneration is more abundant in old burns in comparison with fresh ones, whereas larch preferably occupies fresh fire scars, since there are more larch seed trees in the adjacent to fire scar territories than DNC trees. In the centre of LDZ larch dominates on the fire scars, also birch participates in this process too. At favourable conditions (the wildfire coincides with the year of intensive cone production) the amount of larch regeneration on the fire scars reaches up to

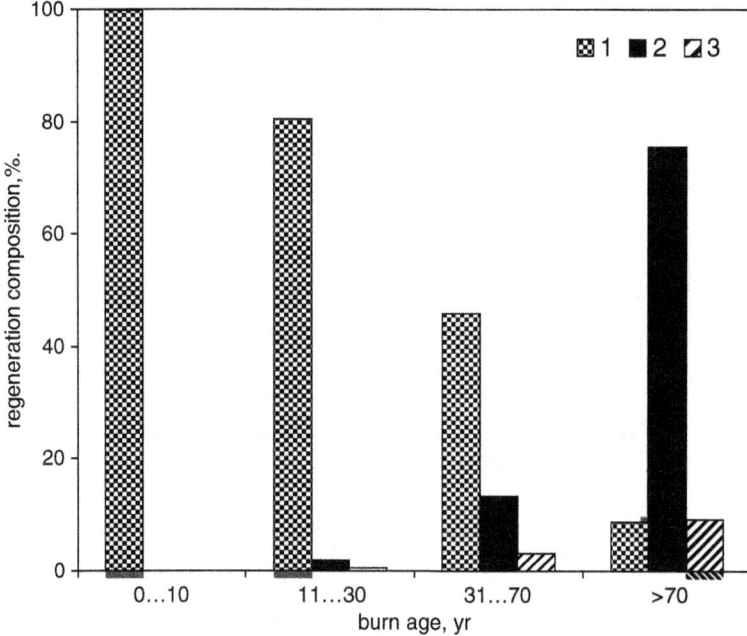

Fig. 4.5 Proportion of tree species vs burn age. Data averaged for the WE transect 1 – larch, 2 – Siberian pine, 3 – spruce

700,000 saplings/ha. It should be noted that an alder (its bush-like form, *Duschekia fruticosa*) is aggressively invading into fresh burns. On southern and western margin of larch domination the fire scars are intensively occupied by birch (up to ~1,000,000 saplings/ha).

4.5 Discussion

The detected differences in the regeneration number and overstory for DNC on the one hand and larch on the other hand as well as differences in Ki values indicate a DNC invasion into a larch domination zone. The phenomenon of Siberian pine invasion into larch habitat could be attributed to precipitation and temperature increases. There are also natural fluctuations in of Siberian pine regeneration number (and, in less extent, others coniferous). The Fourier-analysis of Siberian pine regeneration age structure shows that saplings number is cycling; the main peak of the regeneration number corresponds to 3.5 years (Fig. 4.6), which is similar to the fructification cycle. Since this cycle is much less than an analyzed time interval (last ~30 years), it could not impact on the reliability of results obtained, but it could decrease the correlation between Siberian pine regeneration number and meteorology data.

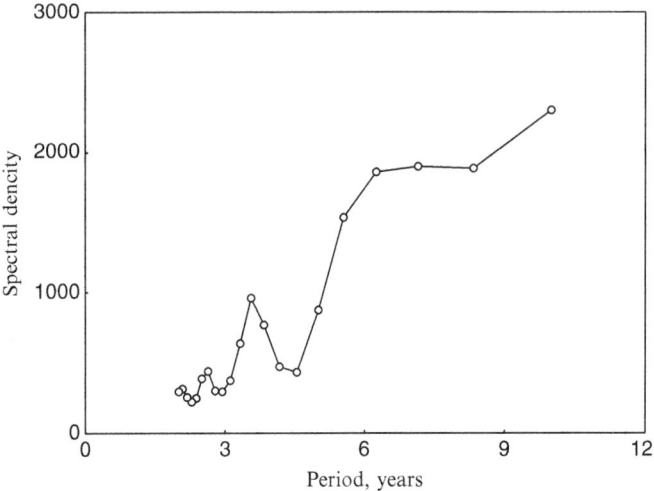

Fig. 4.6 Fourier-analysis of the Siberian pine regeneration age structure

Fig. 4.7 Siberian pine seedlings can grow while roots are within a thick (>40 cm, insert) moss cover. Mid of WE transect (*Color version available in Appendix*)

The finding of higher pine propagation into LDZ in comparison with spruce and fir could be attributed to the two main causes (1) Siberian pine saplings surpass spruce and fir in its resistance to permafrost. In the case of low permafrost thawing depth its root system could developing in the moss layer (Fig. 4.7). Some Siberian pine specimen could be found at latitudes above Polar circle. The dissemination of this species is facilitated by a specialized bird ("a cedar-bird") and some mammals (e.g., ground monks). Siberian pine regeneration was found up to a distance of several hundred meters from the seed sources. At the same time spruce could move to the higher latitudes along the rivers and creeks. In general, the hydro-net is one of the principal ways of DNC invasion into larch habitat. Along creeks and river valleys the microclimatic conditions (higher humidity, wind-protection) and higher soil drainage in the narrow (10–20 m) shore strips are favourable for DNC growth and development. Actually the strip of "alley forest" along the hydro-net represents an example what will happens with species composition in the LDZ with improvement of climatic conditions. The precipitation increase over the last years (Table 4.2) is one of the factors that promotes Siberian pine invasion into larch dominated area. It justifies the slang name of Siberian pine: a "tree of fogs". Wintertime precipitation increase (Table 4.2) may also play a positive role in Siberian pine propagation: deeper snow cover promotes seedlings surviving, since the critical periods of their development is the time when seedlings height exceeds snow cover. Once appeared over snow cover, the seedlings bark becomes actively affected by snowflakes abrasion during snowstorms, resulting in dying off the seedlings tops, and, consequently, sapling dying back, or transforming them into prostrated forms. Trees that effectively exceed this barrier have a typical crown shape: "tree-in-skirt". In general, wind impact (in combination with low temperatures) is an important factor of tree survival. Larch, due to its dense bark (>20% of stem volume) surpasses the other tree species in winter desiccation resistance.

If existing climatic trends (temperature and precipitation increasing) continue DNC regeneration in a larch dominated area may form the second layer that is observed on the west and south borders of the LDZ (Fig. 4.8). Formation of a DNC

Fig. 4.8 Siberian pine regeneration forming a second layer under the larch canopy (*Color version available in Appendix*)

overstory will cause albedo reduction and increase of solar radiation absorption, i.e. the positive feedback occurrence that strengthen greenhouse effect.

The DNC invasion into the larch habitat is a part of general phenomena of larch area "shrinking". In the former epoch the larch was dominating over southern Siberia too, where presently it occupied Sayany and Altai highlands only ("precipitation-shadowed" slopes). In the mixed forests where larch regeneration is negligible, larch survives due to its longevity. In the south taiga larch can reach an age of 600 years, and ~1,000 years in the northern taiga. For other conifers these values are lower: maximum fir age is ~300 years, spruce ~250, pine ~500 and Siberian pine ~600 years.

The results obtained indicate a DNC and birch invasion into a traditional larch domination zone, and the connection of this phenomenon with climatic trends during last decades. On the western and southern margins DNC regeneration formed a second layer in the forest canopies, which could eventually replace the larch in the overstory (Fig. 4.8). Larch as a species also responds to climate trends: its radial increment significantly increased during last 3 decades, and this increase correlates with summer temperatures ($r = 0.50$) and precipitation ($r = 0.43$; $p < 0.05$) (Fig. 4.4c). The other larch response is the migration into the tundra zone and a crown closure increase, as was found on the key-sites "Polar Ural" and "Ary-Mas" (Fig. 4.1) (Kharuk et al. 2006). The resulting effect of this process may be that larch could reach the Arctic shore, a phenomenon that has happened in a former epoch (Kind and Leonova 1982), whereas the traditional area of larch dominance will turn to mixed taiga forest.

Acknowledgments This research was supported by the NASA Science Mission Directorate, Terrestrial Ecology Program, and Russian Fund for Fundamental Investigations # 06-05-64939.

References

Anonymous (1990) Forest inventory map. State Forest Committee, Moscow
Berg EE, Chapin FS III (1994) Needle loss as a mechanism of winter drought avoidance in boreal conifers. Can J For Res 24:1144–1148
Hughes MK, Vaganov EA, Shiyatov S, Touchan R, Funkhouser G (1999) Twentieth-century summer-warmth in northern Yakutia in a 600-year context. Holocene 9(5):603–608
IPCC Third Assessment Report. V.1: Climate change (2001) The scientific basis. Cambridge University Press
Kharuk VI, Fedotova EV (2003) Forest-tundra ecotone dynamics. In: Bobylev LP, Kondratyev KY, Johannessen OM (eds) Arctic environment variability in the context of global change. Springer-Practice, Heidelberg, pp 281–299
Kharuk VI, Ranson KJ, Im ST, Naurzbaev MM (2006) Forest-tundra larch forests and climatic trends. Russ J Ecol 37:323–331
Kharuk V, Ranson K, Dvinskaya M (2007) Evidence of evergreen conifer invasion into larch dominated forests during recent decades in Central Siberia. Euras J Forest Res 10–2: 163–171
Kharuk VI, Ranson KJ, Dvinskaya ML (2008) Wildfires dynamic in the larch dominance zone. Geophys Res Lett 35:L01402. doi:10.1029/2007GL032291

Kind NV, Leonova BN (eds) (1982) The antropogenic impact on Taimyr Peninsula. "Nauka", Moscow, 182 pp

Kloeppel BD, Gower S, Trechel IW, Kharouk V (1998) Foliar carbon isotope discrimination in Larix species and sympatric evergreen conifers: a global comparison. Oecologia 114:153–159

Russian Forest Service (1995) The instruction on the forest inventory for the Russian forest fund. Moscow VNIIlesresurs Press

StatSoft Inc (2003) Nonparametric statistics. http://www.statsoft.com/textbook/stnonpar.html. Accessed 31 Jan, April 29 2008

Sturm M, Racine C, Tape K (2001) Climate change-increasing shrub abundance in the Arctic. Nature 411:445–459

Chapter 5
Potential Climate-Induced Vegetation Change in Siberia in the Twenty-First Century

N.M. Tchebakova, E.I. Parfenova, and A.J. Soja

Abstract Siberian climate change investigations had already registered climate warming by the end of the twentieth century, especially over the decade of 1991–2000. Our goal is to model hot spots of potential climate-induced vegetation change across central Siberia for three time periods: from 1960 to 1990, from 1990 to 2020 and from 1990 to 2080.

January and July temperature and annual precipitation anomalies between climatic means before 1960 and for the 1960–1990 period are calculated from the observed data across central Siberia. Anomalies for 2020 and 2080 are derived from two climate change scenarios HADCM3 A1FI and B1 of the Hadley Centre. Our Siberian bioclimatic model operates using three climate indices (degree-days above 5°C, degree-days below 0°C, annual moisture index) and permafrost active layer depth. These are mapped for 1990, 2020 and 2080 and then coupled with our bioclimatic models to predict vegetation distributions and "hot spots" of vegetation change for indicated time slices.

Our analyses demonstrate the far-reaching effects of a changing climate on vegetation cover. Hot spots of potential Siberian vegetation change are predicted for 1990. Observations of vegetation change in Siberia have already been documented in the literature. Vegetation habitats should be significantly perturbed by 2020, and markedly perturbed by 2080. Because of a dryer climate, forest-steppe and steppe ecosystems, rather than forests, are predicted to dominate central Siberian landscapes. Despite the predicted increase in warming, permafrost is not predicted to thaw deep enough to support dark taiga over the Siberian plain, where the larch taiga will continue to be the dominant zonobiome. On the contrary, in the southern mountains in the absence of permafrost, dark taiga is predicted to remain the dominant orobiome.

N.M. Tchebakova (✉) and E.I. Parfenova
VN Sukachev Institute of Forest, SB RAS, 50 Akademgorodok, Krasnoyarsk, 660036, Russia
e-mail: ncheby@ksc.krasn.ru; 02611@rambler.ru

A.J. Soja
Resident at NASA Langley Research Center, National Institute of Aerospace, 21 Langley Boulevard, Mail Stop 420, Hampton VA 23681-2199, USA
e-mail: Amber.J.Soja@nasa.gov

Keywords Climate warming • Twenty-first century • Siberia • Vegetation change

5.1 Introduction

From scientific assessments of the International Panel on Climate Change, global temperature increased by 0.6±0.2°C in the twentieth century, the warmest century of the last millennium(IPCC 2001). Regional studies in Siberia already registered a change in climate at the end of the twentieth century (see a review of Tchebakova and Parfenova 2006): in West Siberia the mean annual temperature increased 1°C; in the southern Urals winter temperatures increased 0.6–1.1°C over the last 30 years; winter temperatures increased 2–4°C in central Siberia and 3–10°C in central Yakutia; and in southern Siberia, annual temperature anomalies varied between 0.4°C and 1.5°C. Across the southern mountains from the Urals to Transbaikalia, annual precipitation anomalies from 1960 to 1990 were strongly positive in the west (20–25%) and negative in the east (−10% to −20%), but the pattern was decidedly complicated (Soja et al. 2007).

Evidence of landscape and biota change associated with the changing climate also accumulated by the end of twentieth century (IPCC 2001). Boreal ecosystems and mountain ecosystems in particular are predicted to be especially vulnerable. Our model predictions for Siberia demonstrate that climate warming should promote desertification in the south and lowlands, reduce tundra in the north and high mountains, and profoundly impact forest ecosystems at all hierarchical levels: from biome to species and to populations within species (Rehfeldt et al. 2004; Tchebakova et al. 2003, 2006). These predicted locations of hot spots where climate change has affected vegetation have been verified by evidence of change reported in publications (Soja et al. 2007). Additionally, Soja et al. (2007) explored the possibility of evidence of climate-induced change across the circumboreal region, and they found increases in fire regimes, infestations and vegetation change, all of which had been previously predicted by models. Some of these changes occurred more rapidly than models predicted, which suggests a potential non-linear response in the terrestrial environment to climate change.

The objective of this study is to examine the potential effect of two climate change scenarios on spatial vegetation redistribution in central Siberia (from 1990 to 2080) and to identify locations ("hot spots") where current and future change in climate might create new habitats to be followed by vegetation change.

5.2 Methods

5.2.1 Study Area

Two vast areas within the window studied are located in central Siberia. The first is east of the Yenisei River on the elevated Central Siberian Plateau, north of the 56th latitude (56–75° N and 85–105° E). The second is the mountains and foothills of southern Siberia, south of the 56th latitude (48–56° N and 89–96° E).

5.2.2 Climate Change Scenarios

Climate change is evaluated for three climatic variables characterizing the thermal conditions of winter and summer (January and July temperatures) and for annual precipitation for three successive time slices: from 1960 to 1990, from 1990 to 2020, and from 1990 to 2080.

Climatic anomalies (1990) are calculated from registered data as the differences of climatic means between two periods: before 1960 and 1960–1990. Normalized climatic means for the period before 1960 were derived from reference books on climate (Reference books on climate, 1969–1974). Climatic means for 1960–1990 were collated and calculated from monthly reference bulletins (Monthly reference bulletins on climate of the USSR 1961–1990).

Climatic anomalies from 1990 to 2020 and from 1990 to 2080 are derived from two climate change scenarios, the HadCM3 A1FI and B1 of the Hadley Centre in the U.K. based on SRES (the Special Report on Emission Scenarios). The SRES include various additional effects of sulphur emissions and revised economic and technological assumptions. We selected two scenarios, which differ by story lines and reflect opposite ends of the SRES range, the A1FI scenario represents the largest temperature increases and the B1 scenario represents the smallest temperature increase. As illustrated below, Fig. 5.1 shows temperature increases across the studied area do not markedly differ in the A1FI and B1 2020 scenario but doubles for 2080, with the A1FI yielding greater warming, 8–9°C versus 4–5°C in the B1 scenario.

5.2.3 Vegetation Models

We use two bioclimatic models for predicting vegetation zones (zonobiomes, Walter 1985) across the tablelands and plateaus of northern Siberia and the elevation belts (orobiomes, Walter 1985) over the southern mountains. Both of our vegetation models are "envelope-type" models (Box 1981) that determine a unique vegetation class (unique climatic limits for a vegetation class) from three bioclimatic parameters: Growing Degree Days above 5°C (GDD_5) represent plant requirements for warmth; GDD_0 characterize plant cold tolerance; and Annual Moisture Index (AMI) characterize plant drought tolerance. Vegetation classes are analogous in both models, although some (highland sub-alpine taiga, lowland "chern" taiga) are found only in the mountains, not across the plains, because of their unique mountain habitats: wet and cold in sub-alpine highlands or moist and warm in "chern" lowlands.

Our Siberian vegetation model (Tchebakova et al. 2003) considers a total of 11 current vegetation types (Shumilova 1962; Ogureeva 1999) and three types anticipated with ongoing warming. Each class is defined by unique climatic limits from the zonal vegetation ordination in the climate space of the three climatic variables (Tchebakova et al. 2003). Boreal vegetation classes are: Tundra (1); forest-tundra and sparse forest (2); dark-needled (*Pinus sibirica, Picea obovata, and Abies*

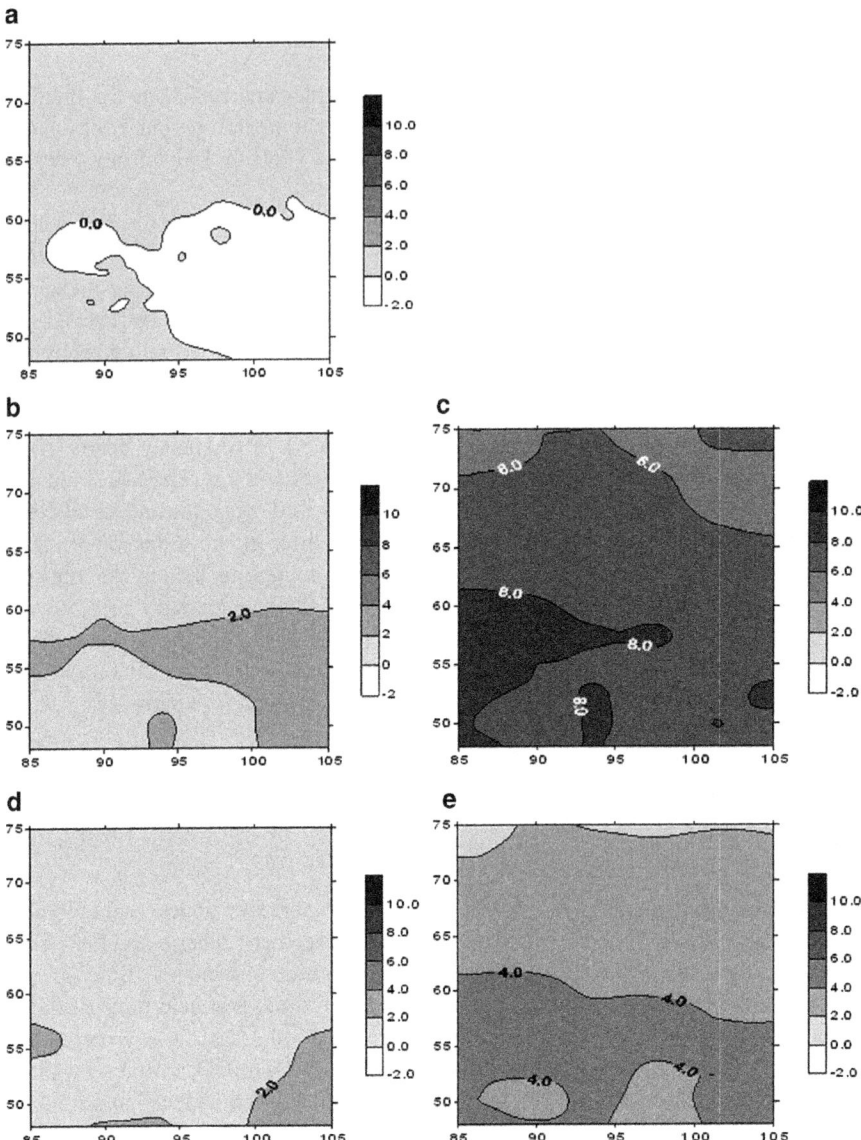

Fig. 5.1 July temperature anomalies over central Siberia for different periods: (**a**) from 1960 to 1990 based on registered data; (**b**) from 1990 to 2020 and (**c**) from 1990 to 2080 from the climate change scenario HadCM3 A1FI; (**d**) from 1990 to 2020; and (**e**) from 1990 to 2080 from the climate change scenario HadCM3 B1

sibirica) taiga: northern (3), middle (4), and southern with birch (*Betula pendula, B. pubescens*) and aspen (*Populus tremula*) subtaiga (5); light-needled taiga (*Larix sibirica* L_a *gmelinii, L. cajanderi and Pinus sylvestris*): northern (6), middle (7) and southern including birch, larch and pine subtaiga (8); birch and light-needled

Table 5.1 Climatic limits for the Siberian vegetation model of Tchebakova et al. (2003)

Vegetation type	GDD$_5$ Lower limit	GDD$_5$ Upper limit	AMI Lower limit	AMI Upper limit	NDDo Lower limit	NDDo Upper limit
Tundra	None	<350	None	None	None	None
Forest-tundra and sparse taiga	350	550	None	None	None	None
Northern dark-needled taiga	550	800	None	<1.5	>−4,500	None
Northern light-needled taiga	550	800	>1.5	None	None	<−4,500
Middle dark-needled taiga	800	1,050	None	<1.8	>−3,500	None
Middle light-needled taiga	800	1,050	>1.8	None	None	<−3,500
Southern dark-needled taiga	1,050	1,250	None	<2.2	None	None
Southern light-needled taiga and subtaiga	1,050	1,250	>2.2	None	None	None
Forest-steppe	1,250	1,600	None	<3.25	None	None
Steppe, Dry steppe	>1,250	1,600	>3.3	None	None	None

forest-steppe (9); steppe (10) and semidesert (11). Temperate vegetation classes are: Broadleaved forest (12), forest-steppe (13), and steppe (14) (Table 5.1). Broadleaved forests, found currently in Europe, existed in West Siberia in the warmer and moister climate of the mid-Holocene period (Khotinsky 1977).

Our mountain vegetation model considers ten current vegetation classes based on a classification of Nazimova (1975): mountain tundra (1); subalpine (2) and "subgolts" (3) sparse forest; dark-needled (*Pinus sibirica, Picea obovata*, and *Abies sibirica*) mountain taiga (4); light-needled (*Larix sibirica* and *Pinus sylvestris*) mountain taiga (5); dark-needled (*Pinus sibirica, Abies sibirica* with *Populus tremula*) "chern" forest (6); light-needled forest-steppe and subtaiga (7); steppe (8); dry steppe (9); and semidesert/desert (10). Additionally, with the prospect of climate warming, three classes of temperate broadleaved forest, forest-steppe, and steppe are included (Table 5.2).

In both models, vegetation distribution predicted only from climatic variables is then corrected for permafrost, which is the primary factor controlling vegetation distribution over interior Siberia. First, permafrost augments the forest's development across the cryozone, providing additional water from melting permafrost in the summer in the dry interior Siberian climate (Shumilova 1962). Secondly, permafrost controls the forest composition limiting the north- and eastward spread of dark-needled tree species (*Pinus sibirica, Abies sibirica, Picea obovata*) and some light-needled tree species (*Larix sibirica* and *Pinus sylvestris*). Only one tree species

Table 5.2 Climatic limits for the mountain vegetation model

Vegetation type	GDD$_5$ Lower limit	GDD$_5$ Upper limit	AMI Lower limit	AMI Upper limit	NDDo Lower limit	NDDo Upper limit
Tundra	None	<300	None	None	None	None
Subalpine and "subgolets" sparce dark-needled taiga	300	550	None	<1.0	None	None
"Subgolets" sparce light-needled taiga	300	550	>1.0	None	None	None
Mountain dark-needled taiga	550	900	None	<2.0	<−3,500	None
Mountain light-needled taiga	550	1,050	>2.0	None	None	>−3,500
"Chern" dark-needled taiga	>900	1,600	None	<2.0	None	None
Forest-steppe	1,050	None	2.0	3.3	None	None
Mountain Steppe	>300	None	3.3	6.0	None	None
Mountain Dry steppe	>300	None	6.0	8.0	None	None
Semidesert/Desert	>300	None	>8.0	None	None	None

Larix dahurica (recently split into *L. gmelini* and *L. cajanderii*) can survive continuous permafrost and dominates the forests in interior Siberia (Pozdnyakov 1993).

5.2.4 Mapping

The climate anomalies (differences of the means) of January and July temperatures and annual precipitation at 1990, 2020, and 2080 are mapped at roughly on 1 km^2 grid cell using the Surfer software (Fig. 5.1).

Contemporary climatic layers of GDD$_5$ and GDD$_0$ for 1990 are mapped on the 1 km^2 grid using Hutchinson's (2000) thin plate splines. The AMI layer is calculated by dividing the GDD$_5$ layer by the annual precipitation layer.

Future climatic layers of January and July temperatures and annual precipitation for each pixel were calculated by adding corresponding climate anomalies from the HadCM3A1FI and HadCM3B1 climate change scenarios to the baseline climate of 1960–1990. Future climatic layers of GDD$_5$ and GDD$_0$ for 2020 and 2080 are calculated using linear regressions determined from registered data: between the January temperature and GDD$_0$ ($R^2 = 0.96$, $n = 150$), between the July temperature and GDD$_5$ ($R^2 = 0.90$, $n = 150$). Future layers of AMI are calculated by dividing the future GDD$_5$ layers by future annual precipitation layers for corresponding time periods.

The continuous permafrost border is finely marked by an active layer depth (ALD) of 2 m on the Malevsky-Malevich's map (Malevsky-Malevich et al. 2001).

We mapped the current position of the permafrost border using the regression that predicted the ALD of 2 m from our three climatic indices ($R^2 = 0.70$, $n = 150$). For the future climates, we used Stefan's formula (Dostavalov and Kudriavtsev 1967) to calculate ALD for each pixel as a function of the ratio between GDD_5 in current and future climates.

Potential vegetation for contemporary 1990 and future 2020 and 2080 climates is mapped by coupling our zono- and orobiome bioclimatic models with climatic maps of GDD_5, GDD_0 AMI and the permafrost border map calculated for each time slice.

5.2.5 Climate Change

Climate change evaluated from registered January and July temperature anomalies across central Siberia showed that between 1960 and 1990 summer temperatures warmed on average 0.5°C in both the north and south (Fig. 5.1). Winter temperatures for this period appeared to warm even more: up to 1–2°C at some locations (Fig. 5.2). Temperature anomalies calculated with respect to the last decade of the twentieth century, the warmest decade of the century (IPCC 2001), are on average 1°C warmer in the north with even larger anomalies south of 56° N latitude, up to 2–4°C particularly in the mountains in winter (Soja et al. 2007). The pattern of precipitation change is more complicated, but in general, annual precipitation 5–10% decreased across central Siberia (not shown).

Climate change in the twenty-first century across the studied area is evaluated from climate change scenarios HadCM3 A1FI and B1. July temperature anomalies do not differ much for 2020 within the range of 0.7–2.0°C in the north and 1.2–2.2°C in the south (Fig. 5.1). January temperature anomalies for 2020 are in the range of 1.4–2.8°C in the HadCM3 B1 scenario and in the range of 1–1.6°C in the HadCM3 A1FI scenario for the area north of 56o N. Less warming (0.2–0.7) and even some cooling is predicted for the southern mountains.

From this analysis, we conclude that in the north, summer anomalies as observed for the 1960–1990 period are 20–100% smaller than those predicted for the 30-year period from 1990 to 2020 (Fig. 5.1). However, winter anomalies by 1990 already exceeded those predicted from the scenario of HadCM3 A1FI (Fig. 5.2). In the south, observed anomalies are 2–4°C, which is one order of magnitude greater than 0.2–0.7°C predicted from either scenario. The greatest difference between observed and predicted anomalies is found in the south-east with the anomaly of 4°C registered versus about 0°C or even negative anomalies predicted.

Comparison between precipitation anomalies by 1990 based on the record and by 2020 based on GCM's predictions showed that the trends are similar, showing a decrease in precipitation (Fig. 5.3). Negative precipitation anomalies by 1990 in the northern tablelands almost double predicted anomalies by 2020: 5% versus 10%. Anomalies both observed and predicted for the southern mountains are about the same, 10%, although in some dry intermountain hollows they are 30–40%.

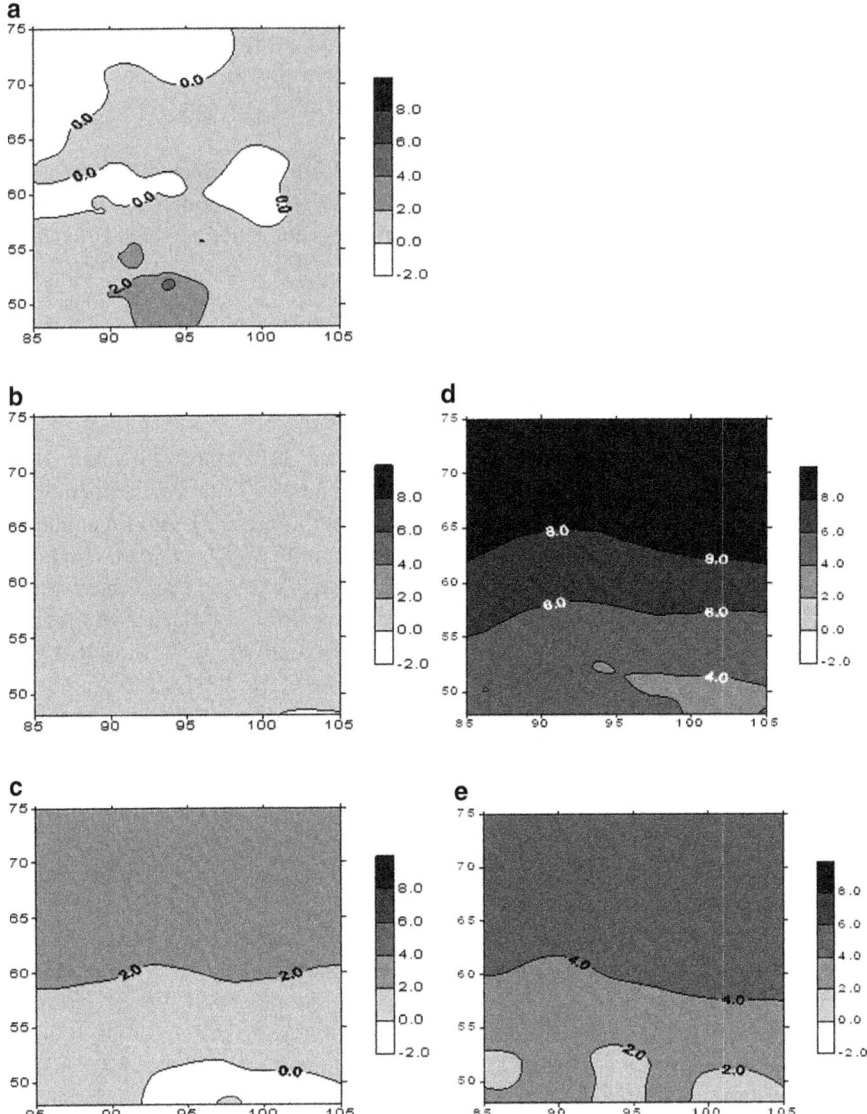

Fig. 5.2 January temperature anomalies over central Siberia for different periods: (**a**) from 1960 to 1990 evaluated registered data; (**b**) from 1990 to 2020; (**c**) from 1990 to 2080 from the climate change scenario HadCM3 A1FI; (**d**) from 1990 to 2020; and (**e**) from 1990 to 2080 from the climate change scenario HadCM3 B1

Precipitation anomalies by 2080 become positive over the central Siberian tablelands but stay slightly negative over the southern mountains according to both scenarios. Precipitation may increase over north-central Siberia by as much as 30% according to the A1FI scenario but only 5–7% according to the B1 scenario.

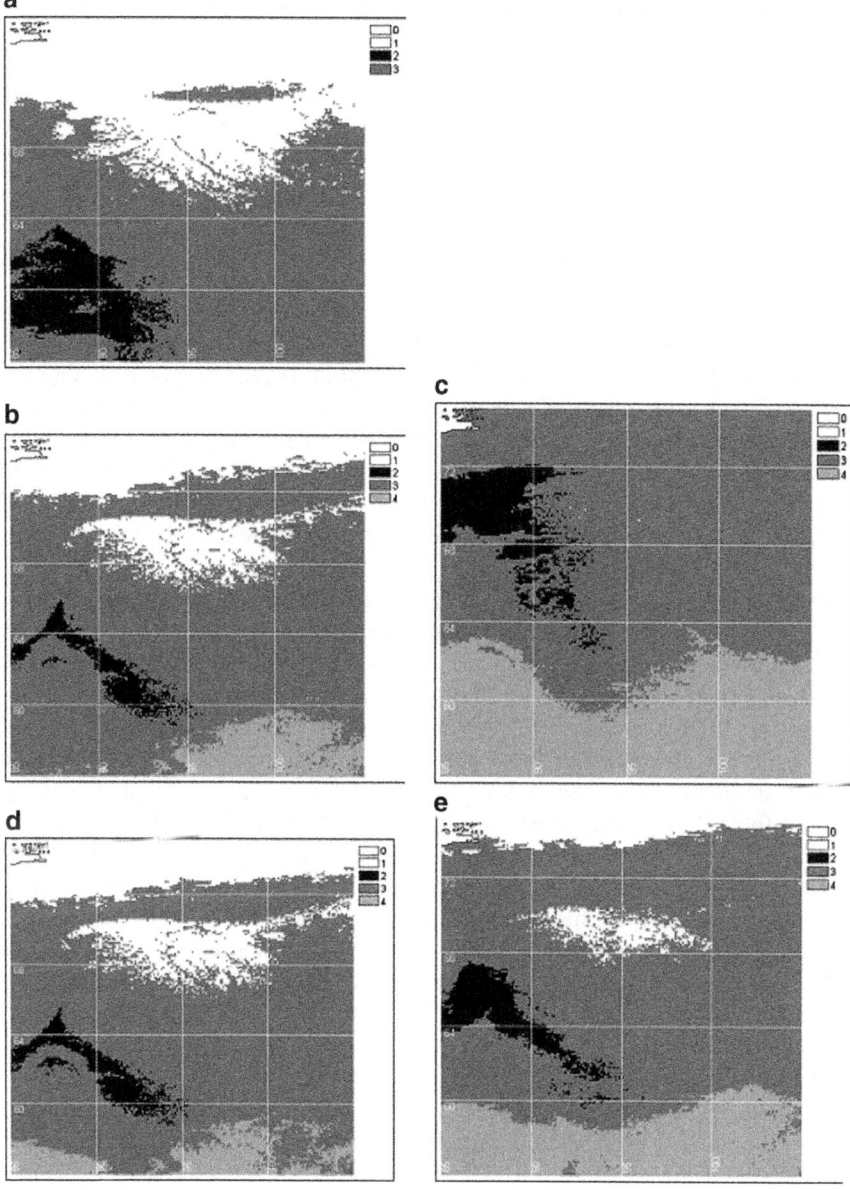

Fig. 5.3 Potential vegetation distributions over central Siberia, north of 56° N, by different time slices: (**a**) at 1990 predicted from registered data; (**b**) by 2020; (**c**) by 2080 predicted from the climate change scenario HadCM3 A1FI; (**d**) by 2020; and (**e**) by 2080 predicted from the climate change scenario HadCM3 B1. 0 – water: 1 – tundra; 2 – dark-needled forests; 3 – light-needled forests; 4 – grasslands, semi-desert

5.2.6 Climate-Induced Change in Vegetation Cover Predicted for the Twenty-First Century

Contemporary and future climate change in the vegetation structure across both the Siberian plains and the southern mountains are predicted using both our bioclimatic models and permafrost distribution.

Across the north-central tablelands of Siberia, north of 56° N, taiga prevailed on 60% of the area in 1990. Dark-needled taiga (about 10% of the area) appears only on elevated terraces with moist and warm climates, like the Yenisei Ridge at the mid-latitudes. Permafrost rather than climate restricts the advancement of dark-needled species into interior Siberia. Light-needled taiga with *Larix sibirica* in the south beyond the permafrost zone and *L. gmelini* in the north and east within the permafrost zone dominate the central Siberian taiga. *Pinus sylvestrisis* can be a component of taiga in the warmer climates of the south or in sandy soils in the middle and even northern taiga. *Picea obovata* and *Pinus sibirica* may be mixed with *Larix* in the large river valleys which tend to be warmer than the surrounding landscape. Tundra and forest-tundra occupy 40% of the area. No grasslands occur north of 56°N (Fig. 5.3; Table 5.3).

In a warmer 2020 climate, the taiga is predicted not to change in area (the HadCM3 B1 scenario) or to shrink slightly (HadCM3 A1FI) (Table 5.3, Fig. 5.3), although previously unobserved steppe and forests-steppe are predicted to appear and occupy more than one quarter of the area at the expense of taiga. Annual precipitation in 2020 is predicted to decrease by 50 mm causing the forests to retreat northwards and changing the forest structure. The light-needled component of the taiga is predicted to increase at the expense of the dark-needled taiga and forest-tundra (Table 5.3). In turn, forest-tundra is predicted to slightly increase at the expense of tundra. Both tundra and forest-tundra is predicted to decrease in area by 8–12%. The continuous permafrost border is expected to shift north- and eastwards as the climate warms. Warming predicted for 2020 by both scenarios is predicted to shift the permafrost border slightly from its current position and thus should not significantly change the boreal forest structure with the dominant larch (*Larix gmelinii*).

By 2080, the model predicts the tundra would fully disappear, displaced by northern and even middle taiga, as a result of increased warming (HadCM3A1FI). The forests is predicted to be replaced by forest-steppe and would decrease in area by as much as half. In fact, large areas of forest-steppe and steppe should cover about 40% of central Siberia and should reach the central Yakutian Plain and the Tungus Plateau, located more than 1,000 km north of the steppe's current location. More moderate changes should occur according to the HadCM3 B1 scenario, however, with the same trends in vegetation change: expanding forest-steppe and steppe at the expense of taiga, taiga decrease, and diminishing tundra (Table 5.3, Fig. 5.3).

Our Siberian vegetation model also estimates that new habitats for some temperate vegetation types such as temperate broadleaved forest, forest-steppe, and steppe

Table 5.3 Proportion of Siberia [% of the land within the window (56–75° N; 85–105° E)] expected for the trivariate climatic envelope of zonobiomes in the current climate 1960–1990 and the climates projected by the HADCM3A1FI and HADCM3 B1 climate change scenarios for 2020 and 2080

Zonobiome	Climate change scenarios				
	1960–1990	A1 2020	A1 2080	B1 2020	B1 2080
BOREAL:					
Tundra	27.1	14.3	0.0	17.4	5.1
Forest-tundra	12.6	13.5	0.2	14.4	10.2
Northern dark-needled taiga	0.0	0.0	0.0	0.0	0.0
Northern light-needled taiga	19.7	12.9	2.6	15.0	14.7
Middle dark-needled taiga	2.2	0.1	0.0	0.1	0.0
Middle light-needled taiga	20.9	17.3	9.5	19.8	11.6
Southern dark-needled taiga and birch subtaiga	8.6	3.6	2.2	4.5	1.0
Southern light-needled taiga and subtaiga	8.9	10.7	14.4	1.2	11.4
Forest-steppe	0.0	17.5	28.5	17.7	18.6
Steppe	0.0	9.8	7.8	9.8	23.5
Semidesert	0.0	0.0	0.0	0.0	0.0
TEMPERATE:					
Mixed and broadleaved forest	0.0	0.2	0.4	0.1	3.8
Forest-steppe	0.0	0.0	3.9	0.0	0.0
Steppe	0.0	0.0	30.5	0.0	0.0
Total	100	100	100	100	100

should occur in the warmed climate of 2080 (Table 5.3). Khotinsky (1977) reconstructed mid-Holocene vegetation for Siberia from pollen depositions and concluded that linden and other broad-leaved forests once were distributed east of the Ural Mountains as far as 70° E and 57° N into the West Siberian Plain.

Across the southern mountains, the model predicts large changes in montane vegetation under a warmer climate, which is similar to the change over the Siberian plain (Table 5.4), however there are some principal differences. Montane tundra is predicted to decrease by half in 2020 and disappear by 2080 according to both scenarios. The mountain forest is predicted to decrease, but its dark-needled portion of both montane and chern forests would remain the same for 2020. By 2080, light-needled forests are predicted to be replaced by forest-steppe in the lower elevations. It is predicted middle elevation mountain landscapes would be dominated by chern dark-needled forests in habitats with sufficiently warm and moist environments. Boreal forest-steppe is not expected to greatly change by 2020 but would decrease in area by two thirds by 2080, in contrast to the temperate forest-steppe, which is predicted to increase by an area three times greater than the boreal forest-steppe. Steppe of both boreal and temperate types, rather than forest-steppe, would prevail in the lowland mountains. Both forms of steppe are predicted to cover about 45–55% of the entire area by 2080 with a portion of dry steppe and semi-desert increasing from 2020 to 2080 (Fig. 5.4) because a combination of precipitation

Table 5.4 Proportion of southern montane Siberia (% of the land within the window [50–56° N; 89–96° E]) expected for the trivariate climatic envelope of orobiomes in the current climate 1960–1990 and the climates projected by the HadCM3A1FI and HADCM3B1climate change scenarios for 2020 and 2080

Orobiome	Climate change scenarios				
	1960–1990	A1FI2020	A1FI 2080	B1 2020	B1 2080
BOREAL:					
Mountain tundra and golets	10.9	4.6	0.0	4.6	1.0
Subalpine dark-needled taiga	10.0	6.3	0.2	6.5	2.6
Subsolets light-needled taiga	1.7	0.9	0.0	1.0	0.0
Mountain dark-needled taiga	18.5	14.3	1.8	14.6	9.3
Mountain light-needled taiga	8.9	4.3	0.0	5.0	1.8
"Chern" dark-needled taiga	12.5	16.3	14.2	14.8	21.5
Forest-steppe and subtaiga	14.9	16.3	4.9	14.6	12.3
Mountain steppe	13.1	19.4	2.2	18.9	8.5
Dry steppes	3.5	4.4	7.7	5.3	5.8
Semidesert, Desert	6.0	13.2	18.9	14.2	15.2
TEMPERATE:					
Mixed and broadleaved forest			6.1		0.2
Forest-steppe			27.0		15.0
Steppe			17.0	0.5	6.7
Total	100	100	100	100	100

decreased and summer temperature substantially increased would produce moisture conditions not suitable for forests at low and middle elevations of the mountains.

5.2.7 Evidence of Contemporary Changes in Vegetation in Central Siberia

A mounting body of evidence of the changes in Siberian vegetation and in the forests in particular related to climate warming is available in the literature and summarized by Soja et al. (2007) and Tchebakova and Parfenova (2006). Kharuk et al. (2004) found that during the last 40 years the most northern Siberian forest, Ary-Mas, shifted into tundra. This tundra is filled with trees and becomes a sparse forest which becomes densely stocked. In Evenkia, interior Siberia, within the permafrost zone, undergrowth of *Pinus sibirica, Picea obovata*, and *Abies sibirica*, which are not typically found on cold permafrost soils are emerging in *Larix gmelinii* taiga (Kharuk et al. 2005). At the northern mountains of the Putorana Plateau, at the Polar Circle, Abaimov et al. (2002) found 50-year-old trees at the upper treeline.

In the southern mountains, strong evidence for the upslope treeline shifts was found in West Sayan (Istomov 2005), in Kuznetsky Alatau (Moiseev 2002), and Altai (Timoshok et al. 2003). Treeline shifts varied from 50 to 120 m during a 50-year span in the mid-twentieth century. At the lower tree line in the West Sayan mountains, poor seed production in a *Pinus sibirica* forest was documented for the

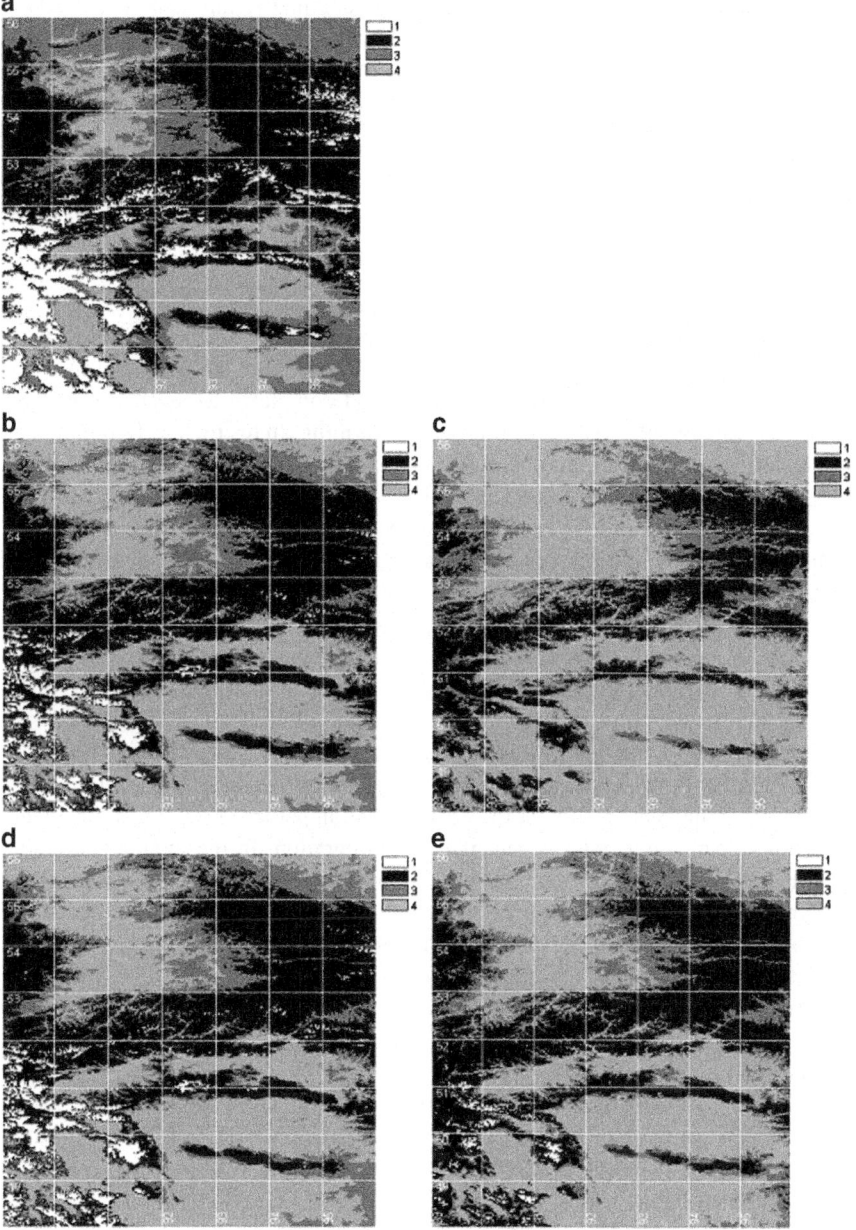

Fig. 5.4 Potential vegetation distributions over southern mountains in central Siberia, south of 56° N, at different time slices: (**a**) by 1990 predicted from registered data; (**b**) by 2020 and (**c**) by 2080 predicted from the climate change scenario HadCM3 A1FI; (**d**) by 2020 and (**e**) by 2080 predicted from the climate change scenario HadCM3 B1: 0 – water. 1 – tundra; 2 – dark-needled forests; 3 – light-needled forests; 4 – grasslands, semi-desert

warmest decade of the century, 1990–2000 (Ovchinnikova and Ermolenko 2004). This event Ermolenko (personal communication) related to increased moth (*Dioryctria abietella*) (Schft.) populations, which damages Siberian pine cones. A longer growing season allows two generations of moths thus increasing probabilities of cone damage.

5.3 Discussion

Significant vegetation shifts are predicted in central Siberia in both the northern tablelands and the southern montane regions. The impact of global warming on natural associations is predicted to be large and complex. However, natural processes are capable of accommodating global warming. In his review, Rehfeldt et al. (2004) discussed that migration and selection are the processes that will control the evolutionary adjustments. While extinction and immigration are expected at the margins of distributions, intra-specific adjustments should produce a wholesale redistribution of genotypes across the landscape according to the distribution of new climates. Calculations for *P. sylvestris* in Siberia (Rehfeldt et al. 2004) suggest that genetic responses to global warming may require as many as 10 generations. Analyses of migration rates, which tend to be slow, coupled with these estimates of genetic response suggest that in some regions, natural systems may require as many as ten centuries to adjust to global warming.

Fire and the melting of permafrost are considered to be the principal mechanisms that facilitate vegetation changes across Siberian landscapes (Polikarpov et al. 1998; Soja et al. 2007). At the northern and upper tree line, forest movement into tundra can occur only by means of tree migration. In the mountains, tundra may be replaced by forest more rapidly because migration rates of dozens meters per year (Kirilenko and Solomon 1998) are comparable with the tundra belt width of 500–1,000 m. In the plains, the tundra zone is commonly 500 km in width. Consequently, it may take a millennium for a tundra zone to be completely replaced by forest with the warming climate, although trees with broad climatic niches and high migration rates conceivably could adjust to a rapidly warming climate in the plains (Solomon et al. 1993).

Over the very vulnerable permafrost zone, many structural changes in vegetation and in forests in particular may happen due to permafrost melting. Forests might decline in extent and be replaced by steppe in well-drained habitats or by bogs in poorly drained habitats with the permafrost retreat (Velichko and Nechaev 1992; Lawrence and Slater 2005). Dark-needled species and *Pinus sylvestris* would be more competitive with *Larix daurica*, the dominant tree species of today's permafrost (Zavelskaya et al. 1993; Polikarpov et al. 1998). Excessive moisture caused by both melting permafrost and catastrophic fires as the climate warms could result in both solifluction and thermokarst formations across large areas, thereby disturbing forest landscapes (Abaimov et al. 2002).

The southern and lowland tree line is being shaped by forest fire which rapidly promotes equilibrium between the vegetation and the climate. Extreme and severe fire seasons have already occurred in Siberia. Tree decline in a dryer climate would facilitate the accumulation of woody debris which along with increased fire weather, would result in an increased potential for severe and large fires (Soja et al. 2007).

Acknowledgments The study was supported by grant # 06-05-65127 of the Russian Foundation for Basic Research. The authors thank Jerry Rehfeldt and Jane Bradford for helpful comments.

References

Abaimov AP, Zyryanova OA, Prokushkin SG (2002) Long-term investigations of larch forests in cryolithic zone of Siberia: brief history, recent results and possible changes under Global Warming. Euras J Forest Res 5-2:95–106

Box EO (1981) Macroclimate and plant Forms: An introduction to predictive modeling in phytogeography, W. Junk Publishers, The Hague, Boston et London, pp 258

Dostavalov BN, Kudriavtsev VA (1967) Basic permafrost science. Moscow University Press, Moscow

Hutchinson MF (2000) ANUSPLIN Version 4.1 User's Guide. Australian National University, Centre for Resource and Environmental Studies

IPCC (2001) Climate change 2001: the scientific basis. Contribution of Working Group I to the Third Assessment Report of the Intergovernmental Panel on Climate Change. Cambridge University Press

IPCC (2001) Emissions Scenarios. In: Nakicenovic N, Swart R (eds) Special report of the intergovernmental panel on climate change. Cambridge University Press, Cambridge, p 570

Istomov SV (2005) The current dynamics of the upper treeline in the West Sayan mountains. In: Actual questions of research and protection of the plant world. Transactions of the Reserve "Tigeretzky", pp 211–214

Kharuk VI, Im ST, Ranson KG, Naurzbaev MM (2004) A longterm dynamics of *Larix sibirica* in the forest-tundra ecotone. Trans Acad Sci Biol Ser 398(3):1–5

Kharuk VI, Dvinskaya ML, Ranson KG, Im ST (2005) Invasion of evergreen conifers into the larch dominance zone and climate trends. Russ J Ecol 3:186–92

Khotinsky NA (1977) Holocene of Northern Eurasia. Nauka, Moscow

Kirilenko AP, Solomon AM (1998) Modeling dynamic vegetation response to rapid climate change using bioclimatic classification. Climate Change 38:15–49

Lawrence DM, Slater AG (2005) A projection of severe near-surface permafrost degradation during the 21st century. Geophys Res Lett 32:L 24401

Malevsky-Malevich SP, Molkentin EK, Nadyozhina ED, Sklyarevich OB (2001) Numerical simulation of permafrost parameters distribution. Cold Regions Sci Tech 32:1–11

Moiseev PA (2002) Climate change impacts on radial growth and formation of the age structure of highland larch forests in Kuznetzky Alatau. Russ J Ecol 1:10–17

Nazimova DI (1975) Montane dark-needled forests of the West Sayan. Nauka, Moscow

Ogureeva GN (ed) (1999) Zones and altitudinal zonality types of vegetation of Russian and adjacent territories. Scale 1:8,000,000, Center "Integration". 2 plates

Ovchinnikova NF, Ermolenko PM (2004) Long-term forest vegetation inventories in the West Sayan mountains. World Resource Review

Polikarpov NP, Andreeva NM, Nazimova DI, Sirotinina AV, Sofronov MA (1998) Formation composition of the forest zones in Siberia as a reflection of forest-forming tree species interrelations. Russ J Forest Sci 5:3–11

Pozdnyakov LK (1993) Forest science on permafrost. Nauka, Moscow

Rehfeldt GE, Tchebakova NM, Milyutin LI, Parfenova EI, Wykoff WR, Kouzmina NA (2003) Assessing population responses to climate in *Pinus sylvestris* and *Larix spp.* of Eurasia with climate-transfer models. Euras J Forest Res (6–2):3–23

Rehfeldt GE, Tchebakova NM, Parfenova EI (2004) Genetic responses to climate and climate-change in conifers of the temperate and boreal forests. Rec Res Dev Genet Breeding 1:113–130

Shumilova LV (1962) Botanical geography of Siberia. Tomsk University Press, Tomsk

Soja AJ, Tchebakova NM, French NF, Flannigan MD, Shugart HH, Stocks BJ, Sukhinin AI, Parfenova EI, Chapin FS III, Stauckhouse PW Jr (2007) Climate-induced boreal forest change: predictions versus current observations. Global Planet Change 56:274–96

Solomon AM, Prentice IC, Leemans R, Cramer W (1993) The interaction of climate and landuse in future terrestrial carbon storage and release. Water Air Soil Pollut 70:595–614

Tchebakova NM, Rehfeldt GE, Parfenova EI (2003) Redistribution of vegetation zones and populations of *Larix sibirica Ledeb.* and *Pinus sylvestris* L. in Central Siberia in a warming climate. Siberian Ecol J 6:677–686

Tchebakova NM, Rehfeldt GE, Parfenova EI (2006) Impacts of climate change on the distribution of *Larix spp.* and *Pinus sylvestris* and their climatypes and Siberia. Mitig Adapt Strat Glob Change 11:861–882

Tchebakova MN, Parfenova EI (2006) Predicting forest shifting in a changed climate by the end of the 20th century. Comput Technol 7(3):77–86 (in Russian)

Timoshok EE, Narozhny YuK, Dirks MN, Berezov AA (2003) Joint glaciological and botanical of primary vegetation successions on young moraines in Central Altai. Russ J Ecol 2:101–107

Monthly reference books on climate of the USSR (1961–1990). Krasnoyarsk Hydrometeorological Observatory Publishers

Reference books on climate of the USSR (1969–1970). Hydrometeoizdat Publishers

Velichko AA, Nechaev VP (1992) Evaluation of the permafrost zone dynamics in Northern Eurasia under global climate warming. Trans Russ Acad Sci Geogr 324:667–71

Walter H (1985) Vegetation of the Earth and ecological systems of the geo-biosphere, 3rd English edn. Springer-Verlag, New York

Zavelskaya NA, Zukert NV, Polyakova EY, Pryazhnikov AA (1993) Predictions of climate change impacts on the boreal forest of Russia. Russ J Forest Sci 3:16–23

Chapter 6
Wildfire Dynamics in Mid-Siberian Larch Dominated Forests

V.I. Kharuk, K.J. Ranson, and M.L. Dvinskaya

Abstract The long-term wildfire dynamics, including fire return interval (FRI), in the zone of larch dominance and the "larch-mixed taiga" ecotone were examined. A wildfire chronology encompassing the fifteenth through the twentieth centuries was developed by analyzing tree stem fire scars. Average FRI determined from stem fire scar dating was 82 ± 7 years in the zone of larch dominance. FRI was found to be dependent on site topography. FRI on north-east facing slopes in the zone of larch dominance was 86 ± 11 years. FRI was significantly less on south–west facing slopes at 61 ± 8 years and flat terrain at 68 ± 14 years. For bogs FRI was found to be much longer at 139 ± 17 years. The FRI decreased from 101 years in the nineteenth century to 65 years in the twentieth century. Connection of this phenomenon with natural and anthropogenic factors was analyzed. The relationship of extreme fire events with summer air temperature deviations at the regional and sub-continental levels was presented. Wildfire impact on permafrost thawing depth was analyzed. The implications of the observed trends on the larch community are discussed.

Keywords Wildfires • Fire return interval • Topography • Climate • Larch forests

6.1 Introduction

Larch (*Larix* spp.) dominated forests are an important component of the global circumpolar boreal forest. In Russia, larch is the widest-spread species and is found from the tundra zone in the north to the steppes in the south. The zone of larch

---

V.I. Kharuk (✉) and M.L. Dvinskaya
V.N. Sukachev Institute of Forest, SB RAS, 50 Academgorodok, Krasnoyarsk 660036, Russia
e-mail: kharuk@ksc.krasn.ru; mary_dvi@ksc.krasn.ru

K.J. Ranson
NASA Goddard Space Flight Center, Greenbelt, MD 20771, USA
e-mail: jon.ranson@nasa.gov

dominance spreads from Yenisei ridge in the west to the Pacific, and from Lake Baikal in the south to 73° north, where it forms the world's northward Ary-Mas stand. In Central Siberia larch on its southern and western margins is contacting with evergreen conifers (Siberian pine, *Pinus sibirica*, pine, *Pinus silvestris*, spruce, *Picea obovata*, fir, *Abies sibirica*) and hardwoods (birch, *Betula pendula, B. pubescens*, and aspen, *Populus tremula*). Larch forms high closure stands as well as open forests, and is found mainly over permafrost, where other tree species barely survive. Wildfires are typical for this territory with the majority occurring as ground fires due to low crown closure. Due to the surface root system (caused by permafrost) and dense lichen-moss cover, ground fires in their majority are stand-replacing fires. Our analysis of SPOT VEGETATION- derived map of burns (JRC 2003) estimated that wildfires burned about 0.25% of the forested territory annually from 1996 to 2000. Ranson et al. (2003) and Kovacs et al. (2004) obtained similar results from an analysis of Terra MODIS data in central Siberia in 2001 and 2003. According to Schimel et al. (2001) the vast area of larch forests including the forest tundra ecotone, is generally considered as a "carbon sink". However, positive long-term temperature trends at higher latitudes (Hansen et al. 2002) and the resulting increase of the fire frequency may convert this area to a source of greenhouse gases.

Regional climatic factors and topographic and environmental gradients are the fundamental processes that determine fires cycles (Bessie and Johnson 1995; Swetnam 1996). Topography (slope aspect and elevation) has an important role in the occurrence, frequency and extent of wildfire. Soil moisture levels are strongly dependent on the azimuth and steepness of slopes, since windward slopes receive higher precipitation and runoff accumulates in areas of lower relief. The drying of combustible materials within the forest is also dependent on the azimuth and slope steepness. The site elevation creates a vertical climatic gradient that also impacts on the risk of wildfires. The connection of wildfires and topography in central Siberia was discussed earlier by Kurbatsky (1962) and Kharuk et al. (2007a, 2008) who noted more fires on southwest facing slopes. Several authors have found similar patterns throughout forests in North America (e.g., Bergeron 1991, Rollins et al. 2002, Beaty and Taylor 2001) and in Europe (e.g., Vazquez and Moreno 2001). Knowledge of topographic-fire relations is important for understanding fire frequency at present and in the future for separating out local effects from the longer term trend in wildfire occurrence.

It is known that about 70% of permafrost areas in Siberia are occupied by larch dominated forests with the remaining 30% in the tundra. The maximal depth of permafrost is about 30–100 m in north-western Siberia and 500–1,500 m in the northern parts of central and eastern Siberia. Depth of thaw in the summer is typically 5 cm to >1.0 m. Litter decomposition in the larch communities is reduced by low summer temperatures resulting in increased litter thickness. This layer, together with moss and lichens, becomes a thermal insulator that promotes permafrost formation at decreasing depth. During low-precipitation years the ground cover layer dries and becomes a fire fuel source. The storage of fuels has been estimated at about 4–5 kg/m^2 in central Siberia, increasing up to 8 kg/m^2 in the north (Fedorov and Klimchenko 2000). This facilitates the spread of fires over tens to hundreds of kilometres, which causes the emission of greenhouse gases, and affects the permafrost depth. It is important to understand the dynamics of permafrost following fire.

The purpose of this work is (1) to investigate wildfire history and long-term trend in fire occurrence in the Siberian larch dominated communities, (2) to examine the relation between fires and topography, (3) to analyze fire impact on permafrost thawing depth, (4) to analyze the long-term impact of the summer air temperature deviations on the extreme fire events at regional, sub-continental and global levels.

6.2 Study Area

The focus of this study is the larch dominated communities within the Niznyaya (Lower) Tunguska river watershed in Evenkiya region, central Siberia (Fig. 6.1). The stand-replacing wildfires were studied (Fig. 6.2). The larch-dominated communities are composed mainly of *Gmelinii* larch (*Larix gmelinii*) with a mixture of birch (*Betula pendula*). The larch dominance area is within the Central Siberian

Fig. 6.1 Map of field measurement locations

Fig. 6.2 Larch stand killed by stand-replacing ground fire (*Color version available in Appendix*)

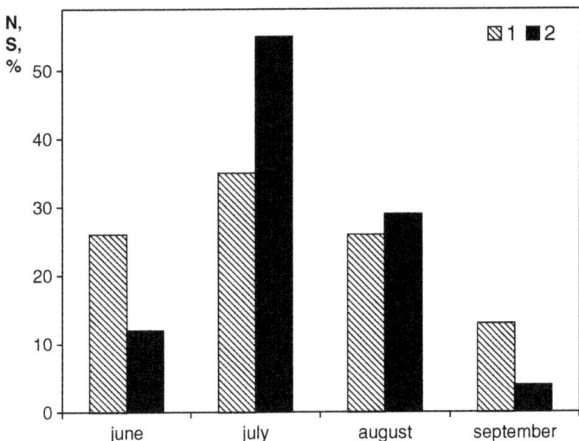

Fig. 6.3 The seasonal distribution of fires in the study area. 1 (N) – the proportion of fires by number, 2 (S) – the proportion of fires by area

plateau with maximum elevation of about 900 m. The climate is severe continental with average annual temperatures within −8 to −14°C. The precipitations are of about 250–400 mm. The seasonal fire distribution is uni-modal with its maximum number in July (Fig. 6.3). Note that late-spring and early-summer fires in its majority are not of stand-replacing type due to low dryness of on-ground "fuel".

6.3 Methods

Available maps and satellite imagery were used to aid in planning field measurements. A 1:1,000,000 digital topography map (NOAA 2003) and topography maps scaled 1:100,000 were used to identify areas of varying slope and aspect. Terra/MODIS data and a Eurasian forest map (JRC 2003) derived from SPOT VEGETATION data provided an overview of forest types in the study area. Individual Landsat-7 images were also used to plan routes and select stands for field sampling by identifying forest stands with apparent burn histories or for use as reference. The on-ground data were collected during the summers of 2001–2003 along the route, presented in Fig. 6.1. A total of 52 temporary test-sites (TS) were established. For analysis 38 sites with long fire history were selected (Fig. 6.4). TS radius was 9.8 m (0.03 ha). For each TS the following parameters were described: relief, vegetation cover type, forest stand structure, tree diameters and heights, regeneration structure, description of shrub and ground cover, and soil type. Each TS centre point coordinate was geo-referenced. For fire return interval (FRI) measurements, three to five trees in the vicinity (within 100 m) of each TS with fire scars were cut using a chainsaw to reveal trunk cross-sections. The samples were taken from between the base of the stump up to 1.0 m height. In addition to living trees, dead trees with apparent fire scars were also analyzed to extend the wildfire chronology by cross-dating the fire scars on the living and dead trees. For this purpose we looked for a match

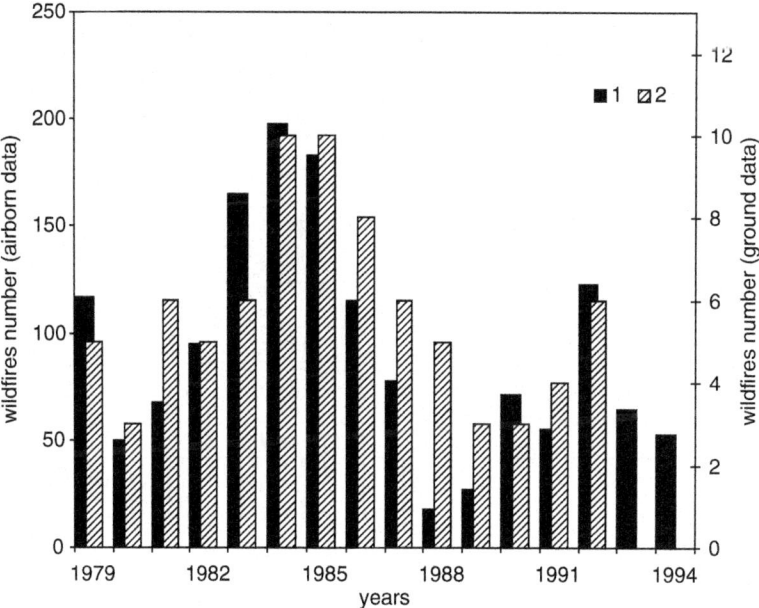

Fig. 6.4 The results of the fire dating with respect to landscape elements are shown for the test sites

of the two most recent fire scars present on both living and dead trees. In some cases (for example, if the tree's death was the result of the latest stand replacing fire) exact cross-dating of the fire-scar date was possible by matching a recent single fire scar on both trees.

In spite of periodic wildfires some trees were of considerable age (up to 400 years) in some of the TS. Typically, these TS yielded the most fire scars. The FRI was calculated as the number of tree rings between consecutive fire scars. In the temporal trend analysis the FRI values were referenced to the midpoint of an interval. For the purpose of estimating the accuracy of dating wildfires about 10% of the specimens (35 cross sections out of ~300) were analyzed using a "master" chronology method (Vaganov and Arbatskaya 1996). A combination of cross-correlation analysis and graphical cross-dating was used to detect double counted and missing rings (Holmes 1983, Rinn 1996). A good agreement between data obtained by both, "tree ring counting" and "master chronology" methods was found: the data difference was 0.97 +/- 0.4 year. To further check the accuracy estimation we compared the tree-ring derived fire dates with dates of fires derived from airborne observations (acquired between 1980 and 1993) in part of the study area. A high correlation ($r = 0.81$) was found between airborne observations and the tree ring observations (Fig. 6.5).

For analysis of the wildfire cycle and related fire danger periods the following technique was used. First, for each year of the analyzed period (1410–2002) the number of fires was plotted on the time axis. This data set was then filtered by

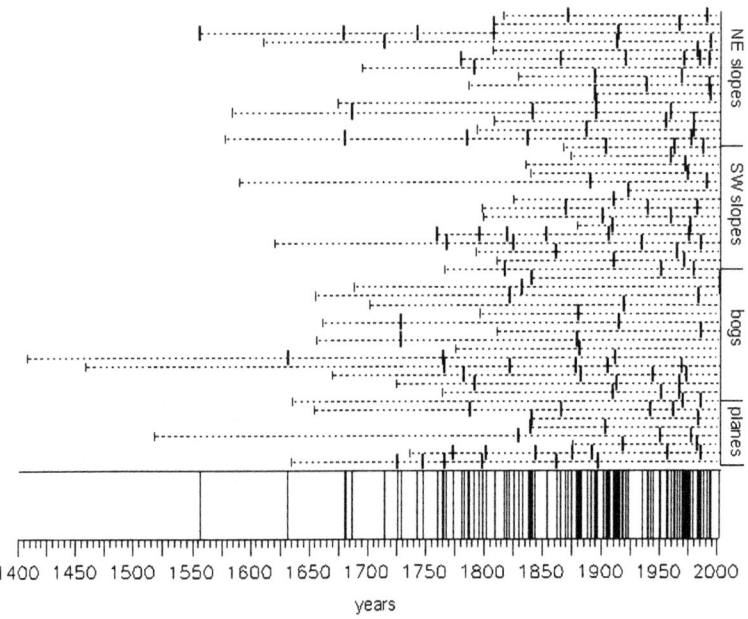

Fig. 6.5 Fire frequency according to airborne (1) and on-ground (2) data (determined by tree fire-scars)

using a "moving sum" filter of 11 years size to smooth fluctuations. The relationship between fire events and summer (June–July) temperature deviations was examined using reconstructed temperatures. The temperature datasets for the regional (northeast Siberia) and sub-continental (northern Eurasia) were obtained from the tree-ring analysis of Panyushkina et al. (2003), and Naurzbaev et al. (2004), respectively.

The analysis of the relationship between wildfires and topography was based on the ground-based measurements of slope and aspect measured with a clinometer and a compass, respectively, and elevation measured with a global positioning system (GPS). The distribution over the area was referenced to northern (azimuths 315°–45°), eastern (45°–135°), southern (135°–225°) and western (225°–315°), southwest (135°–315°) and north-east (315°–135°) aspects, and level bogs and plains. Sites with slopes 0° to 2° are referred to as plains and include mostly table lands and river terraces.

The normality of analyzed parameters before correlation analysis was checked, and then the Student's t-criterion was used. The non-parametric statistics, i.e. the Kendall tau, T [i.e., T = (number of agreements − number of disagreements)/total number of pairs)] also was used in the data analysis (StatSoft 2003).

6.4 Results

6.4.1 FRI and Landscape Characteristics

The on-ground data obtained with respect to the landscape type are presented in Fig. 6.4. Based on this, the FRI was calculated (Table 6.1). The analyzed time period covered 1410 to 2002. Table 6.1 shows that the wildfire return interval in the larch stand depends on terrain characteristics. Not surprisingly, bogs with persistent moisture have the longest FRI. Cool and wet northeast facing slopes have the longest FRI for upland forests. Southwest facing slopes with drier conditions have the shortest FRI which is only slightly longer for flat terrain. The differences between SW and NE slopes are significant at $p < 0.1$; bogs are different from all the other landscape elements at $p < 0.01$.

Table 6.1 The mean FRI values for various terrain aspects. SW = south-west facing slopes; NE = north-east facing slopes; s = standard deviation

Terrain aspect	FRI, years mean ±s	Number of test sites
SW	61 ± 8	11
NE		13
Bogs	139 ± 17	7
Plains	68 ± 14	7
All test sites	82 ± 7	38

6.4.2 Temporal Trends in the FRI

The data in Fig. 6.6 (open circles) shows that there is a trend of increasing number of annual fires in the twentieth century in comparison with the nineteenth century ($p < 0.05$). But this conclusion might be biased because the number of samples is less for older time periods (Fig. 6.4). This results in a variable number of tree samples for the different time periods ("fading effect"). To exclude the effect of decreasing sample size only the trees with age exceeding 200 years were examined for the FRI temporal trends analysis. This enabled balanced sampling for the both the nineteenth and twentieth century. The results of the analysis show an FRI decrease in the twentieth in comparison with the nineteenth century (Fig. 6.7): a reduction of mean FRI from 101±12 years to 65±6 years ($p < 0.01$), or approximately one third (Table 6.2). Although mean values are lower for other aspect classes for the twentieth century vs. the nineteenth century the differences were not statistically significant.

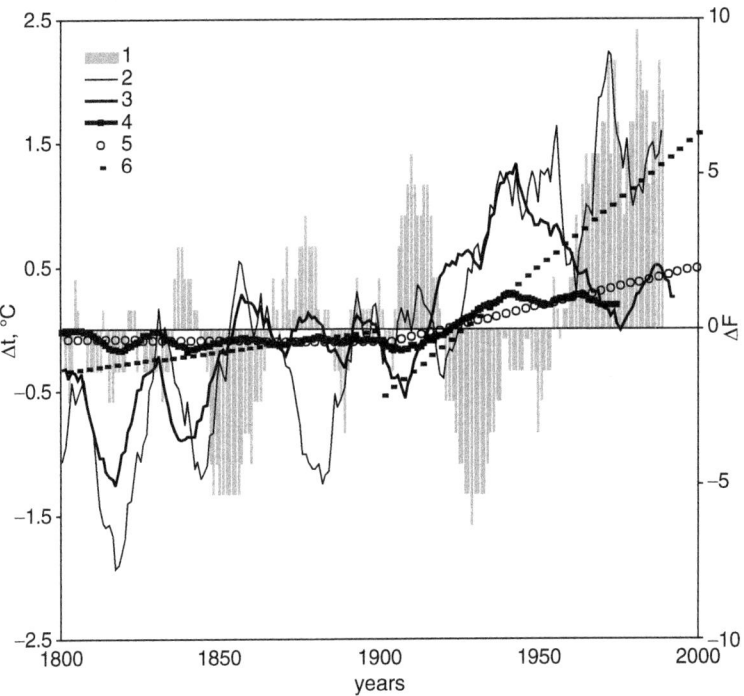

Fig. 6.6 The fire chronology in the zone of larch dominance and reconstructed summer air temperatures. 1 – deviations ($\Delta F, n$) from long-term (200 year) mean annual fire number distribution, 2, 3, 4 – temperature deviations ($\Delta t, °C$) from long-term (200 year) mean of northeast Siberia (Panyushkina et al. 2003) and northern Eurasia (Naurzbaev et al. 2004) temperature records. Straight lines indicate the temporal trends for the nineteenth and twentieth centuries for (symbol 5) northern hemispheric temperatures deviations (Mann and Jones 2003), and (symbol 6) for number of fires

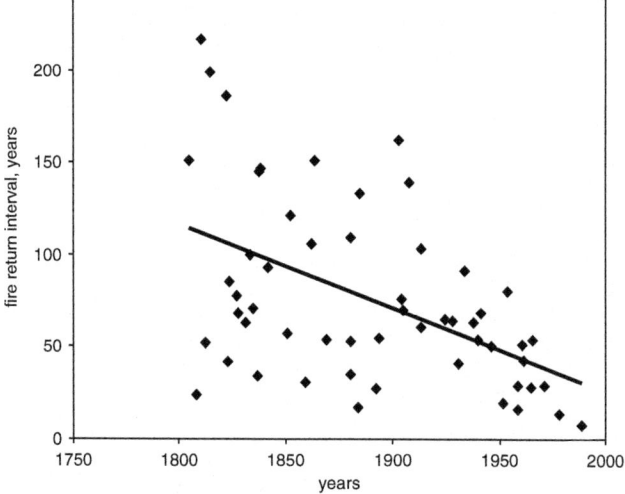

Fig. 6.7 The temporal trend in FRI in the nineteenth and twentieth centuries derived from cross-sections taken from trees that were more than 200 years old

Table 6.2 The mean values of FRI during the nineteenth and twentieth centuries. SW = south-west facing slopes; NE = north-east facing slopes, s = standard deviation

	Mean FRI ± s, yr	Number of sites
Nineteenth century		
SW	93 ± 29	3
NE	109 ± 25	6
Bogs	125 ± 20	6
All TS	101 ± 12	20
Twentieth century		
SW	58 ± 8	11
NE	74 ± 11	11
Bogs	81 ± 27	4
All test sites	65 ± 6	30

6.4.3 The FRI and Summer Air Temperature

The observed increase in fire frequency is occurring against the background of a significant ($p < 0.05$) upward summer temperature trend in the twentieth century (Fig. 6.6, symbol 5). Figure 6.8 represents reconstructed records of regional summer air temperature for northeast Siberia (Panyushkina et al. 2003), northern Eurasia (Naurzbaev et al. 2004), and the northern hemisphere (Mann and Jones 2003). All three data sets were processed by an 11-year filter, which made them

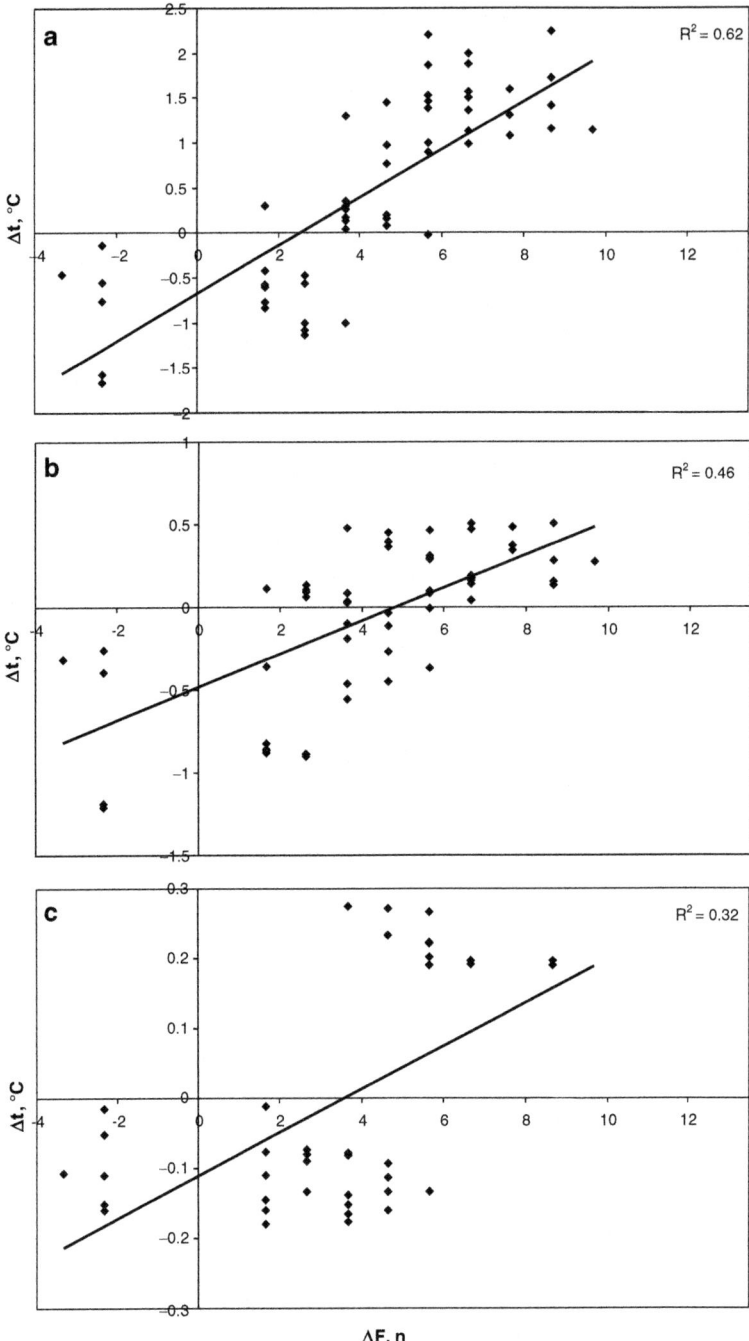

Fig. 6.8 The relationship of wildfire anomalies (ΔF,n – deviations from long-term mean; ΔF,n >1SD) with deviations of the mean reconstructed summer temperature record (Δt,°C – deviations from long-term mean) for (**a**) northeast Siberian (Panyushkina et al. 2003), (**b**) northern Eurasia (Naurzbaev et al. 2004), and (**c**) the northern hemisphere (Mann and Jones 2003)

temporally comparable with our wildfire data. For the analysis, temperature deviations (Δt, °C) were calculated as the difference with a mean summer temperatures for the 200 year period.

A significant ($p<0.01$) cross-correlation was found with the regional summer temperatures record ($\boldsymbol{R}_{ns} = 0.44$), where $\boldsymbol{R}_{ns}$ is a cross-correlation coefficient for northeast Siberia. The cross-correlation coefficients for northern Eurasia ($\boldsymbol{R}_{na}$), and northern hemisphere ($\boldsymbol{R}_{nh}$) are not significant. Similarly, the Kendall tau statistic for these variables was found significant ($p<0.01$) for regional summer temperatures ($\boldsymbol{T}_{ns} = 0.23$), and not significant for the other variables.

The next step in the analysis was based on the observation that most catastrophic wildfires were observed in the years 1914–1916, when wildfires were known to be present continuously from May until August in our study area. In those years the total affected area in Siberia was 160,000 km² (Krasnoyarsky region 1961; Levy et al. 2003). This event was caused by spring droughts and summer precipitation reduction to 30% of normal annual amounts. As a result, rivers, bogs, soils and especially the ground cover dried up. The 1914–1916 wildfires also covered huge expanses to the south, north and to an even greater degree to the west of the investigated area. The spike in the fire frequency distribution for this period is clearly visible in Fig. 6.6. After this catastrophic event an interval of reduced wildfires was observed, which was due to the fact that the majority of fire susceptible forests were burned in the preceding period. A similar fire event and following reduction of fires was observed in the nineteenth century (around 1846).

Thus, in spite of positive temperature trends at the regional and northern hemisphere levels, the number of fires during periods following large fires was minimal, because of a deficiency of combustible material. The consequences of such exceptional events do not match the regular "cause–effect" pattern, and should be analyzed separately. According to our observations, about 30 years are required for a severely burned area to accumulate enough combustible material to sustain another fire. Based on this, in the second step of analysis we excluded this "lag period" of 30 yr after the 1914–1916 and 1846 fire events. After this correction the cross-correlation becomes: $\boldsymbol{R}_{ns} = 0.67$, $\boldsymbol{R}_{na} = 0.43$, $\boldsymbol{R}_{nh} = 0.37$ ($p<0.01$). The Kendall tau statistic becomes: $\boldsymbol{T}_{ns} = 0.48$, $\boldsymbol{T}_{na} = 0.35$ ($p<0.05$) ($\boldsymbol{T}_{nh}$ is not significant).

The last step of the analysis was the study of the relationship of wildfire anomalies (>1 standard deviation) with temperature deviations (Fig. 6.8). Deviations from the mean in Fig. 6.7 were calculated separately for both the nineteenth and twentieth century. This minimized the impact of temporal trends in fires, which are significantly different in both centuries. The results show increased correlation with the reconstructed temperature record. All coefficients are significant ($p< 0.01$) with highest correlations for the northeast Siberia reconstructed temperature record ($\boldsymbol{R}_{ns}^{2}$). Thus, the extreme fire events showed higher correlations with the temperature deviations than the whole fire record pattern. A decrease of the relationship with temperatures from regional (northeast Siberia) to the northern Eurasia and northern hemisphere is observed (Fig. 6.8).

6.4.4 Wildfires and Species Migration into Zone of Larch Dominance

In the context of climate change wildfires may promote an invasion of "southern" conifer (Siberian pine, pine, spruce, fir,) and hardwood species (birch and aspen) into the zone of larch dominance: fire-scars represent a "starting place" for "southern species" invasion, due to better thermal and soil conditions, enriched biogenic compounds, increased soil thawing depth and drainage. This process eventually may cause substitution of larch in the upper canopy (Kharuk et al. 2007b). On the other hand, larch regenerates better, since after fire-induced sapling mortality, higher post-fire presence of larch trees may provides seeds for larch regeneration. Meanwhile the decrease of the fire return interval may interfere with the "southern species" invasion into the zone of larch dominance. Siberian pine regeneration is more abundant in old burns in comparison with fresh ones, whereas larch preferably occupies fresh fire scars, since there are more larch seed trees in the territories adjacent to fire-scars than "southern" conifers (Fig. 6.9). In the centre of its habitat larch dominates in the fire-scars, also birch participates in this process too. At favourable conditions (the wildfire coincides with the year of intensive cone production) the amount of larch regeneration on the fire-scars reaches up to 700,000 saplings/ha (Fig. 6.10). It should be noted that an alder (*Duschekia fruticosa*) is aggressively invading into fresh burns. On southern and western margins of the larch domination zone the fire-scars are intensively occupied by birch (up to ~1,000,000 saplings/ha).

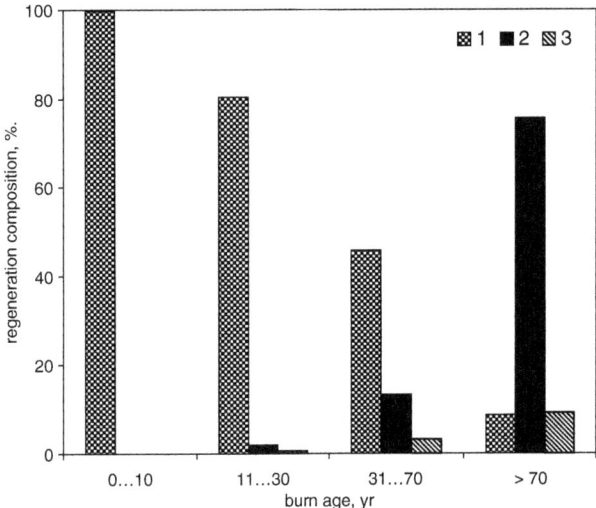

Fig. 6.9 Proportion of tree species vs burn age: 1 – larch, 2 – Siberian pine, 3 – spruce

6 Wildfire Dynamics in Mid-Siberian Larch Dominated Forests

Fig. 6.10 Larch regeneration after a ground fire happened in the late summer – begin of fall 1993. Fire event coincides with the year of high cone production. Number of saplings is about 700,000 thousand/ha (2001 year) (*Color version available in Appendix*)

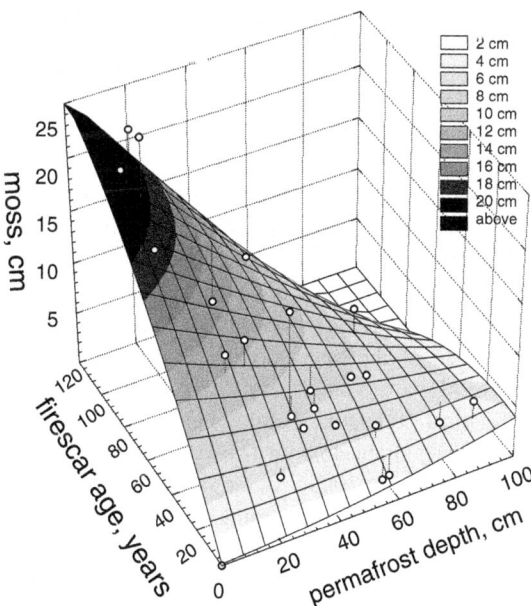

Fig. 6.11 Post-fire dynamic of permafrost seasonal thawing depth and moss cover height

6.4.5 Wildfires and Permafrost Thawing Depth

Figure 6.11 shows the patterns of soil thawing at fire sites in the season of the greatest thawing (the second half of August). As the period of time after the fire increased, the mean depth of thawing decreased from 80–100 to ~30 cm 180 years after the fire, or 0.3 cm/year. This was accounted for by an increased thickness of the moss–lichen cover serving as a heat insulator (Fig. 6.11). The rate of its growth was estimated at 0.1 cm per year. When the moss–lichen cover has dried, it becomes a combustible material that, in drought periods, creates favourable conditions for fires over areas as large as millions of hectares. Note that the increase in the depth of soil thawing after a fire is one of the factors accelerating the growth of surviving trees. We estimated this period at 10–30 years. The subsequent rise of the level of permafrost limits the development of the tree root system in the surface layer (30 cm or less). Therefore, although the stem in larch is protected with thick bark, damage to its roots from creeping fires causes a withering of tree stands.

6.5 Discussion

Fires are inherent to larch forests and all of the stands measured during the field investigations were of pyrogenic origin. Fires may result in uneven aged stands with two to four generations represented, or, with the complete destruction of the forest. An important result is the temporal trend of increased fire frequency in the twentieth century with a one-third reduction in average FRI (101 years in the nineteenth century to 65 years in the twentieth century). The results obtained in our study are possibly the result of two components: (1) the increase of anthropogenic impact, and (2) influence of a warmer climatic trend with resulting extreme temperature and drought conditions in the twentieth century. An increase of anthropogenic activity in the study area was observed between the 1940s and 1980s, and was associated with an increase of population and, especially, increases in mineral exploration. Note that in the area of investigation there is no mineral resource exploitation. Since 1990, changes in the Russia economy have led to a dramatic decrease of resource exploration (in some places complete cessation) in the study area, and, consequently, a decrease of the anthropogenic sources of fires. The population density within the Evenkiya region is very low (<0.03 people/km^2). According to aerial surveys, in the most populated Evenkiya area (which corresponds to ~8% of the total forest fund area); about 45% of fires were caused by lightning (Russian Ministry of Natural Resources 2001). Ivanova and Ivanov (2004) stated that in the low populated areas of the northern taiga about 90% of fires were caused by lightning. The number of fires typically reaches its maximum in July (Fig. 6.3), also the month with maximum lightning. It should be added that in the permafrost zone lightning causes ignition of the fires more "effectively" than in the non-permafrost zone. According to Sapozhnikov and Krechetov (1982) the effect is two times higher than in non-permafrost areas because lightning energy dissipates through the roots compressed in the narrow surface strata. Based on the above described observations, we consider that temperature increase in the twentieth century is a significant factor contributing to the observed fire frequency increase.

The relation of FRI and topography generally agrees with reported results from other forest regions (e.g., Beaty and Taylor 2001) and is useful information for ecosystem modelling (e.g., climatic trends impact on the FRI and fire- induced successions) and fire risk analysis (e.g. Maselli et al. 1996). The relationship of fires with topography could be associated not only with the fuel loading and thermal conditions on the different slope aspects, but also with different levels of fire ignition. For example lightning strikes increased by three to four times with an elevation increase of 400 to 600 m (Sapozhnikov and Krechetov 1982).

Other recent FRI studies show different results. For example, Bergeron (1991) in a North American boreal lake shore and island study showed FRI increasing after 1870 and attributed this to changing climate. Barrett et al. (1991) reported a large decrease in fires after 1935 in western USA forest mostly due to fire suppression. Wallenius and Vanha-Majamaa (2004) found an increase of the FRI in the Kola Peninsula (near the Russian border with Finland) in the nineteenth century, and related this to a decrease of anthropogenically ignited fires (as well as fire suppression and climatic trends). Buechling and Baker (2004) in studies of the fire history in the Rocky Mountains, USA, did not find a linkage with the topographical differences in the study area. They related low fire activity in the twentieth century with decreased severity and frequency of drought. Generally, the decrease of the fires in the end of nineteenth–twentieth century is attributed to fire suppression activity (Heyerdahl et al. 2001) and global climatic change (Flannigan et al. 1998, Weir et al. 2000).

A recent finding by Gillett et al. (2004) showed an increase of the area burned by forest fires in Canada the last 4 decades of the twentieth century and that the human-induced climate change has had a detectable influence on the area burned by forest fire in Canada over recent decades.

Our observed increase in fire rate, which is non-uniform with respect to the different landscape elements, could affect the biodiversity of the northern landscapes. According to Kharuk et al. (2007) there is evidence of evergreen conifers climate-driven migration into the larch dominated taiga. Burned areas offer a starting place for migration of other plant species into the larch dominated zone. Larch, in turn, is migrating into the tundra area (Kharuk and Fedotova 2003). An increase of fire frequency helps maintain the dominance of larch in the northern forests, since it is protected by thick bark. The main cause of larch forest mortality is damage of the root system. With an increase of summer thawing depth and, consequently, increased rooting depth and an overall increase of larch resistance to fires may be expected. It is known that in the non-permafrost zone, larch is the most fire-resistant species (followed by pine, Siberian pine, spruce, and fir). Importantly, larch regenerates poorly over a moss and lichen ground layer, and extremely well on mineral soil, especially if the year of fire coincides with seed production (Fig. 6.10).

Finally, it is necessary to emphasize a double role of fires in the larch-dominated forests. Its negative component includes greenhouse gas emissions during the burning and after the disturbance. The observed and predicted climate trends in high latitudes (Gordon et al. 2000) will cause increasing fire frequency. The "natural low limit" of the FRI should be about 30 years, as was mentioned above, due to fuel material accumulation. Another issue of fire frequency increase is the possible transformation of the extensive larch dominated area from a sink to a source of greenhouse gases. It's also important to note that in the case of severe large-area fires (>5–10 thousand ha) there may be a deficiency of seeds for regeneration, and the burned area will convert to grass, shrub, or birch for an extended period.

The positive role of fire includes the improvement of the eco-physiological conditions for post-fire growth (permafrost thawing depth increases, increased available nutrients, and enhanced canopy light regime). For example, the post-fire active layer depth increase leads to growth acceleration of surviving trees which lasts, according to our estimations, 10–30 years. The issues of larch dynamics in the face of changing climate and human activities are complex and require substantial further research.

Acknowledgments The work was supported in part by NASA's Science Mission Directorate and Russian RFFI grant No. 06-05-64939.

References

Barrett SW, Arno SF, Key CH (1991) Fire regimes of western larch – Lodgepole pine forests in Glacier National Park, Montana. Can J For Res 21(12):1711–1720

Beaty RM, Taylor AH (2001) Spatial and temporal variation of fire regimes in a mixed conifer forest landscape, Southern Cascades, California. USA J Biogeogr 28(8):955–966

Bergeron Y (1991) The influence of island and mainland lakeshore landscapes on boreal forest-fire regimes. Ecology 72(6):1980–1992

Bessie WC, Johnson EA (1995) The relative importance of fuels and weather on fire behavior in subalpine forests. Ecology 76:747–762

Buechling A, Baker WL (2004) A fire history from tree rings in a high-elevation forest of Rocky Mountain National Park. Can J For Res 34(6):1259–1273

Fedorov EM, Klimchenko AV (2000) The dynamics of combustible materials in the northern larch forests. Russ J Forest 2:48–49

Flannigan MD, Bergeron Y, Engelmark O, Wotton BM (1998) Future wildfire in circumboreal forests in relation to global warming. J Veg Sci 9:469–476

Gillett NP, Weaver AJ, Zwiers FW, Flannigan MD (2004) Detecting the effect of climate change on Canadian forest fires. Geophys Res Lett 31(18):Art. No. L18211. doi:10.1029/2004GL020876

Gordon C, Cooper C, Senior CA, Banks H, Gregory JM, Johns TC, Mitchell JFB, Wood RA (2000) The simulation of SST, sea-ice extents and ocean heat transport in a version of the Hadley Centre coupled model without flux adjustments. Climate Dynamics 16:147–168

Hansen J, Ruedy R, Sata MKI, Lo K (2002) Global warming continues. Science 295(5553):275–275

Heyerdahl EK, Beubaker LB, Agee JK (2001) Spatial controls of historical fire regimes: a multiscale example from the interior west, USA. Ecology 82:660–678

Holmes RL (1983) Computer-assisted quality control in tree-ring dating and measurement. Tree-ring Bull 44:69–75

Ivanova GA, Ivanov VA (2004) The fire regime in the forests of the Central Siberia. In: Furyaev VV (ed) Forest fire management at regional level, Alex, Moscow, pp 147–150

JRC (2003) Global 2000 Landcover map. Global Vegetation Monitoring Unit, Institute for Environment and Sustainability, Joint Research Center. http://geoserver.isciences.com:8080/geonetwork/grv/en/metadata.show, last accessed 02.03.2010

Kharuk VI, Fedotova EV (2003) Forest-tundra ecotone dynamics. In: Bobylev LP, Kondratyev KY, Johannessen OM (eds) Arctic environment variability in the context of global change. Springer-Practice, Heidelberg, pp 281–299

Kharuk VI, Ranson K, Dvinskaya M (2007a) Evidence of evergreen conifer invasion into larch dominated forests during recent decades in Central Siberia. Euras J Forest Res 10–2:163–171

Kharuk VI, Kasischke ES, Yakubailik OE (2007b) The spatial and temporal distribution of fires on Sakhalin Island, Russia. Int J Wildland Fire 16(5):556–562

Kharuk VI, Ranson KJ, Dvinskaya ML (2008) Wildfires dynamic in the larch dominance zone. Geophys Res Lett 35:L01402. doi:10.1029/2007GL032291

Kovacs K, Ranson KJ, Sun G, Kharuk VI (2004) The Relationship of the Terra MODIS Fire Product and Anthropogenic features in the Central Siberian Landscape. Earth Interact 8(18):1–25

Kirillov MV, Zcherbakova Yu A (eds) (1961) Krasnoyarsky region. Krasnoyarsk, Krasnoyarsk Publishing house, p. 404

Kurbatsky NP (1962) Technique and tactics of forest fire extinguishing. Nauka, Moscow

Levy GK, Zadonina NV, Berdnikova NE, Voronin VI, Glizin AV, Yazev SA, Baasandjav B, Ningbadgar S, Balginnyam M, Buddo VU (2003) Modern geodynamic and geliodinamic. Irkutsk University Publishing House, Irkutsk

Mann ME, Jones PD (2003) Global surface temperatures over the past two millennia. Geophys Res Lett 30(15):1820

Maselli F, Rodolfi A, Bottai L, Conese C (1996) Evaluation of forest fire risk by the analysis of environmental data and TM images. Int J Remote Sens 17(7):1417–1423

Naurzbaev MM, Hughes MK, Vaganov EA (2004) Tree-ring growth as sources of climatic information. Quatern Res 62:126–133

NOAA (2003) The Global Land One-km Base Elevation (GLOBE) Project. National Geophysical Data Center. http://www.ngdc.noaa.gov/mgg/topo/globe.html. Accessed 31 Jan 2008

Panyushkina IP, Huges MK, Vaganov EA, Munro MAR (2003) Summer temperature in northeastern Siberia since 1642 reconstructed from tracheid dimensions and cell numbers of *Larix cajanderi*. Can J For Res 33:1905–1914

Ranson KJ, Sun G, Kovacs K, Kharuk VI (2003) MODIS NDVI response following fires in Siberia. In: Proceedings of IGARSS03, 5, pp 3290-3292. Toulouse, France

Rinn F (1996) TSAP V 3.6 Reference manual: computer program for tree-ring analysis and presentation. Bierhelder weg 20, D-69126, Heidelberg, Germany

Rollins MG, Morgan P, Swetnam T (2002) Landscape-scale controls over 20(th) century fire occurrence in two large Rocky Mountain (USA) wilderness areas. Landscape Ecol 17(6):539–557

Russian Ministry of Natural Resources (2001) Chart of forest fires airborne observations in Evenkya: General report. Russian Ministry of Natural Resources, Moscow

Sapozhnikov VM, Krechetov AA (1982) Meteorological and geophysical aspects of underground cables lightning damage, Atmospheric Electricity, edited by A. Evteeva. Gidrometeoizdat Publishing, Leningrad pp. 256–258

Schimel DS, House JI, Hibbard KA, Bousquet P, Ciais P, Peylin P, Braswell BH, Apps MJ, Baker D, Bondeau A, Canadell J, Churkina G, Cramer W, Denning AS, Field CB, Friedlingstein P, Goodale C, Heimann M, Houghton RA, Melillo JM, Moore B, Murdiyarso D, Noble I, Pacala SW, Prentice IC, Raupach MR, Rayner PJ, Scholes RJ, Steffen WL, Wirth C (2001) Recent patterns and mechanisms of carbon exchange by terrestrial ecosystems. Nature 414(6860)):169–172

StatSoft Inc (2003) Nonparametric atatistics. http://www.statsoft.com/textbook/stnonpar.html. Accessed 31 Jan 2008

Swetnam TW (1996) Fire and climate history in the central Yenisey region, Siberia. In: Goldammer JG, Furyaev VV (eds) Fire in ecosystems of Boreal Eurasia, Kluwer Academic Publisher; Dodrecht, Boston, London, pp 90–104

Vaganov EA, Arbatskaya MK (1996) The climate history and wildfire frequency in the Mid of Krasnoyarsky Kray. I. Growing seasons' climatic conditions and seasonal wild fire-distribution. Siberian J Ecol 3(1):9–18

Vazquez A, Moreno JM (2001) Spatial distribution of forest fires in Sierra de Gredos (Central Spain). For Ecol Manag 147(1):55–65

Wallenius THTK, Vanha-Majamaa I (2004) Fire history in relation to site type and vegetation in Vienansalo wilderness in eastern Fennoscandia, Russia. Can J For Res 34(7):1400–1409

Weir JMH, Johnson EA, Miyanishi K (2000) Fire frequency and the spatial age mosaic of the mixed-wood boreal forest in western Canada. Ecol Appl 10:1162–1177

Chapter 7
Dendroclimatological Evidence of Climate Changes Across Siberia

V.V. Shishov and E.A. Vaganov

Abstract A major focus of the study described here is an attempt to reveal the nature of local and any widespread tree-growth responses to the recent warming seen in the instrumental observations. Namely, this chapter discusses spatial variation in the trends of radial tree-ring growth in Siberia and the Far East during different periods of the 18th to 20th centuries. That distribution of trends is compared with spatial NDVI trends and temperature changes in the northern hemisphere over the past 20 years.

A new classification approach is described to associate different spatial-temporal samples of tree-ring trends with recent tendencies of Siberian vegetation activity. Obtained shot-term tree-growth tendencies for the eighteenth–twentieth centuries are characterized by a common decrease of trend values from West to East with highest values for southern latitudes of Siberia and comparable to NDVI trends. But a recent spatial sample of tree-growth trend is described by highest variation for the last 250 years which can be explained by the most significant recent warming in the context of 2,000 years.

Keywords Climate change • Tree-ring trends • Normalized differentiated vegetation index (NDVI) • Siberia • Far East • Tree-ring chronology • Temperature

---

V.V. Shishov (✉)
IT and Math. Modelling Department, Krasnoyarsk State Trade-Economical Institute,
L. Prushinskoi Street, Krasnoyarsk, 660075, Russia
and
Dendroecology Department, Sukachev Institute of Forest,
Siberian Branch of Russian Academy of Sciences,
Akademgorodok Street, Krasnoyarsk, 660036, Russia
e-mail: shishov@forest.akadem.ru

E.A. Vaganov
Siberian Federal University, 79 Svobodnji Ave, Krasnoyarsk 660041, Russia
e-mail: institute@forest.akadem.ru

Recent global temperature changes in the Northern Hemisphere can be characterized by an average temperature increase of 0.37°C per decade for the last 40 years (Jones and Moberg 2003). This is approximately four times faster than the average for the century (Briffa et al. 2001; Jones and Moberg 2003; Mann et al. 1998). But there are a number of complexities concerning the regional character of such warming (Briffa et al. 1995; 1998b; Cook et al. 1994; Cook and Peters 1997; Liu et al. 2004; Shishov et al. 2002). Moreover, some areas show insignificant cooling trends in recent decades. So the start and end of the vegetation growing season, and the associated total degree day patterns of warming are variable since 1950 (Briffa et al. 2008).

There are a number of papers describing potential tree growth response to increasing temperature, CO_2 concentration, additional irrigation etc. (Briffa et al. 1998a, b; McDowell et al. 2002; Robertson et al. 1997; Waterhouse et al. 2004). There is a need to know how trees will respond in different forest stands, and different physiographic regions. This uncertainty arises for several reasons. First, most studies have involved small trees for short durations, inside greenhouse or field chambers that modify the environmental conditions (Nunez-Elisea et al. 1999; Ojeda et al. 2004; Thacker et al. 2003). Few used old trees in natural conditions (Graybill and Idso 1993; Karnosky 2003; LaMarche et al. 1984, 1986; Takeuchi et al. 2001). Secondly, the tree responses are significantly different between species, by season, nitrogen fertility level, co-occurring pollutant concentrations, etc. (Aasamaa et al. 2002; Naumburg et al. 2001; Takeuchi et al. 2001). A series of interesting studies were carried out in controlled conditions in Sweden and other regions of Europe (Bergh et al. 1998; Linder 1998; Linder et al. 1997; Linder and Ostlund 1998). These studies explored the young trees' response to a combination of a temperature increase and fertilization which could be considered as CO_2 fertilization. Additionally, spatial analysis of NDVI trends has revealed several regions with increasing primary productivity (Myneni et al. 1997; Zhou et al. 2001). However, there are no possibilities yet to classify which global factor is a cause of such changes.

It is well known, that climate signals are detectable in tree-ring data from ecologically sensitive sites (Briffa 2000; Briffa et al. 2008; Briffa and Osborn 1999; Briffa et al. 1998a, b; Hughes and Diaz 1994; Hughes et al. 1999; Vaganov et al. 1999). The spatial analysis of the long-term tree-ring chronologies near the Northern timberline has revealed significant differences in temperature changes (the principal factor limiting tree growth) and tree growth trends after 1960 (Briffa et al. 1998b, 2004 Shishov et al. 2002; Vaganov et al. 2000; Vaganov et al. 1999). There are a number of hypotheses relative to these differences (Briffa et al. 1998b; Vaganov et al. 1999). The reasons are likely complicated, combining interactions between limiting and growth-accelerating factors. So, for Northern Eurasia unprecedented concordance in tree growth was revealed in the twentieth century over the last 2 millennia (Briffa et al. 2008). It means that high-latitude territories become more homogeneous in their tree response to external forcing factors. Similar homogeneity but significantly lower amplitude is observed in the medieval warm period (Briffa et al. 2008)

7 Dendroclimatological Evidence of Climate Changes Across Siberia

This chapter discusses the spatial variation in the trends of radial tree-ring growth in Siberia and the Far East during different periods of the eighteenth–twentieth centuries. This distribution of trends is compared with spatial NDVI trends and temperature changes in the northern hemisphere over the past 20 years.

The Ural-Siberian dendrochronological database was used for calculation of tree-growth trends. In total, 154 long-term tree-ring chronologies were used in the analysis (Shishov et al. 2002) (Fig. 7.1).

All chronologies were obtained for conifer species by the same standardization method – using a negative exponential curve (Cook and Kairiukstis 1990). This method allows to illuminate age-depended components from individual tree-ring series and to retain low-frequency tree-ring components which could be associated with regional climatic fluctuations. It is known such standardization could distort variation of an individual tree-ring series especially for the last years of tree growth (Melvin 2004). So, those individual tree-ring series were included in the analysis which negative exponential curve was optimal (i.e. maximizing the determination coefficient) for the age trend approximation.

Two parameters were calculated for all consecutive 19-year moving time windows for the last 260 years (1740–2000); firstly, the trend (the slope coefficient of a least-squares fitted regression line) and secondly, the arithmetic mean, both obtained from each of 154 tree-ring chronologies. Values were assigned to the centre year of the moving window (i.e. the 1985 values refer to the window 1976–1994). As a result, 154 tree-ring chronologies were transformed to corresponding trend chronologies.

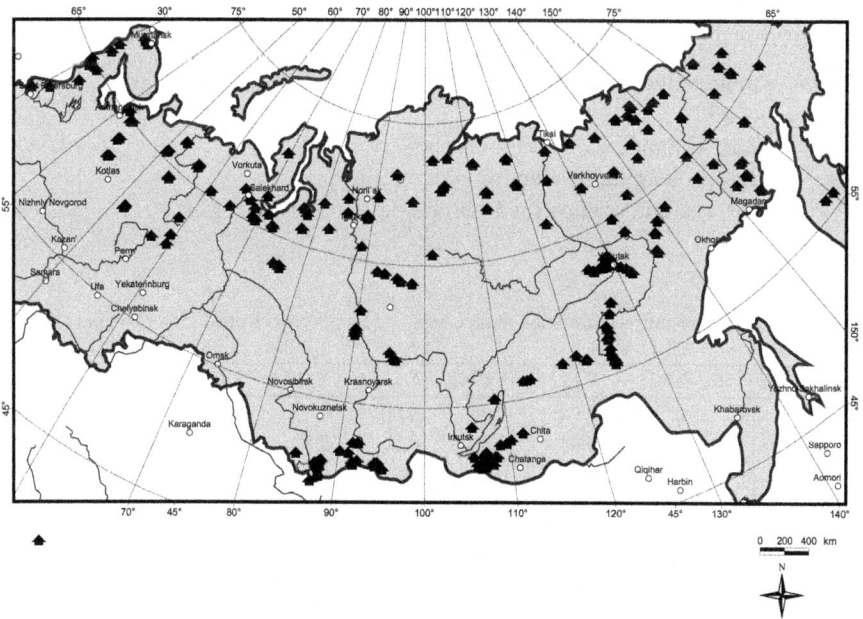

Fig. 7.1 Location of tree-ring chronologies using for calculation of tree-ring trends

Selection of a 260-year interval is based on the finding that 95% of analysed chronologies cover this time interval. The selection of a 19-year window was based on two reasons. Firstly, the NDVI trend map was obtained for 1981–2000 (centre year – 1990) (Zhou et al. 2001). The same map of trends was obtained for tree-ring data with a goal to compare the spatial distribution of these datasets. Secondly, it is well known that tree growth for the most part of Siberia is limited by summer temperatures (Cook and Kairiukstis 1990; Briffa et al. 1998b; Vaganov et al. 1999; Shishov et al. 2002), in which 20–30-years cooling and warming periods are obtained by preliminary analysis.

Each year from the 250-year interval is characterized by corresponding 154 trend values.

An analysis of the common 250 × 250 correlation matrix, obtained for all years from 1750–1990, has revealed a block structure of that matrix (Table 7.1). Each block consists from 6 × 6 cells beaded on the main diagonal with highly significant correlations close to 1. That result testifies a significant autocorrelation of sixth order between analysed years.

As a result, each fifth year from 250 was used in analysis of the spatial trend distribution, i.e. 1750 (corresponded interval – 1741–1759), 1755 (1746–1764), 1760 (1751–1769), 1765 (1756–1774), ..., 1900 (1891–1909), 1905 (1896–1914), 1910 (1901–1919), ..., 1990 (1981–1999). In total, 49 years were used for the spatial trend analysis.

The next task was to determine a nonlinear regression operator F which adequately described the spatial trend distribution for all selected years:

$$Trend_{tk} = F_t(Lat_k, Lon_k, Alt_k),$$

where

$Trend_{tk}$ – trend value for t-year ($t=1,750, ..., 2,000$) and k-chronology ($k=1, ..., 154$),
   Lat_k – latitude value for k-chronology,
   Lon_k – longitude value for k-chronology,
   Alt_k – altitude value for k-chronology.

Table 7.1 Part of common correlation matrix obtained for all years from 1750–1990 and all trend chronologies

	Y1750	Y1751	Y1752	Y1753	Y1754	Y1755	Y1756	Y1757	Y1758
Y1750	1.00	0.98	0.94	0.84	0.80	0.75	0.62	0.46	...
Y1751	0.98	1.00	0.98	0.88	0.86	0.80	0.67	0.48	...
Y1752	0.94	0.98	1.00	0.95	0.92	0.86	0.73	0.55	...
Y1753	0.84	0.88	0.95	1.00	0.97	0.93	0.83	0.67	...
Y1754	0.80	0.86	0.92	0.97	1.00	0.98	0.89	0.75	...
Y1755	0.75	0.80	0.86	0.93	0.98	1.00	0.95	0.84	...
Y1756	0.62	0.67	0.73	0.83	0.89	0.95	1.00	0.92	...
Y1757	0.46	0.48	0.55	0.67	0.75	0.84	0.92	1.00	...
Y1758	...	...	...	...	...	...	...	...	...

7 Dendroclimatological Evidence of Climate Changes Across Siberia

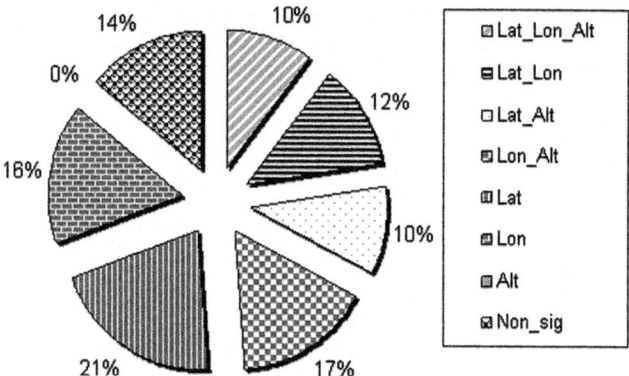

Fig. 7.2 Distribution of independent variables significance in time (100% is 49 selected years)

By nonlinear step-wise regression the sought operator F was estimated as:

$$Trend_{tk} = b_1 Lat_k + b_2 Lon_k + b_3 Alt_k + b_4 Lat_k^3 + b_5 Lon_k^3 + b_6 Alt_k^3 + b_7$$

Unknown parameters from the obtained equation were estimated for all 49 years with the goal to construct spatial maps of tree-growth trends for the last 250 years (Fig. 7.2).

An analysis of the partial correlation coefficients revealed temporal changes of the significance independent variables. So, latitude, longitude, altitude and their nonlinear transformation (Lat_Lon_Alt) in aggregated form are significant variables in 10% of cases, latitude and longitude (Lat_Lon) – in 12%, latitude and altitude (Lat_Alt) – in 10%, longitude and altitude (Lon_Alt) – in 17%, single latitude (Lat) – in 21%, longitude (Lon) – in 16% (Fig. 7.2). There was no equation where single altitude was a significant variable to describe the spatial variation of tree-growth trends. It is explained by linear changes of local temperature with increasing altitude. On the other hand, temperature is a limiting tree growth factor for most ecotopes of Siberia. Therefore, increasing increasing/decreasing altitude does not affect on tree growth trend.

The obtained distribution of the significance of geographical variables means a significant spatial-temporal redistribution of principal tree-growth tendencies for the last 250 years which are dependent in most cases on latitude and longitude.

Such regression equations, determined by seven b-coefficients and describing dependence between growth trends and geographical coordinates, generate similar spatial-temporal maps by data extrapolation. In that case each year (object) can be characterized by seven standardized b-coefficients. Cluster analysis was used to reveal homogeneous groups of selected years (Fig. 7.3). To define the number of clusters of objects (years) the tree clustering technique (Tryon 1939) was used along with Ward's linkage-rule method (Ward 1963).

The linkage rule determines when two clusters are sufficiently similar to be linked together and attempts to minimize the sum of squared distance between any two (hypothetical) clusters that can be formed at each step. Euclidean distance (or geometric distance in the multidimensional space) was used as linkage distance.

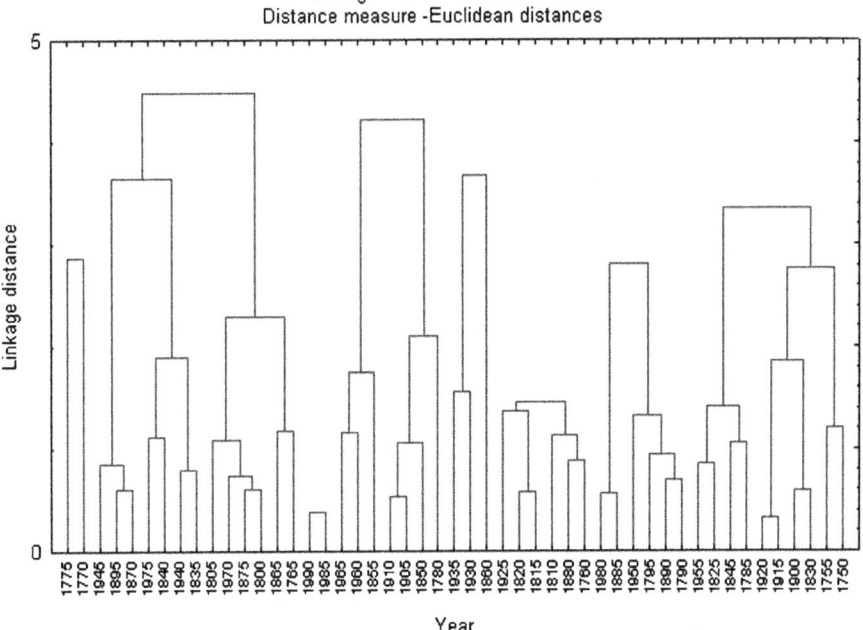

Fig. 7.3 Clustering tree obtained for 49 analysed objects (years)

To define which objects belong to each cluster the "k-means" clustering procedure (Hartigan 1975; Hartigan and Wong 1978) was used. This produces creates exactly k (in this case, $k=10$) different clusters of greatest possible distinction. This algorithm tries to move objects in and out of groups to get the most significant ANOVA results (http://www.statsoft.com/textbook/).

As a result of the K-means approach 49 objects (years) were separated into six groups. The First group consists of 6 years, second – 5, third – 2, fourth – 12, fifth – 15, sixth – 9 (Table 7.2).

There is a hypothesis of compactness as a basis of classification theory: objects from the same cluster should be located compactly (i.e. generate a group) in factor space. Many criteria of classification quality are based on that hypothesis. In our case, belonging of objects (years) to the same cluster means similarity of these objects in seven-dimensional space of regression b-coefficients.

The obtained classification by tree clustering and the k-means procedure was checked using discriminant analysis (DA) (Huberty 1994; McLachlan 2004; Shishov 2000). DA is used to classify objects into the values of a categorical dependent, usually a dichotomy (or obtained classification). DA shares all the usual assumptions of correlation, requiring linear and homoscedastic relationships, and untruncated interval or near interval data (Huberty 1994). DA is relatively robust even when there are modest violations of homogeneity of variances and multivariate normal distribution (Lachenbruch 1975), but Klecka (1980)

Table 7.2 K-means clustering of objects (years). D is Euclidean distance between group centroid and objects

Cluster	Objects											
1	1,780	1,850	1,905	1,910	1,985	1,990						
D	0.36	0.58	0.54	0.39	0.73	0.74						
2	1,855	1,925	1,930	1,960	1,965							
D	0.37	0.39	0.73	0.5	0.47							
3	1,770	1,775										
D	0.58	0.58										
4	1,760	1,765	1,800	1,805	1,810	1,815	1,860	1,865	1,875	1,880	1,935	1,970
D	0.39	0.54	0.33	0.33	0.37	0.5	0.99	0.59	0.15	0.39	0.69	0.28
5	1,750	1,755	1,785	1,790	1,795	1,820	1,825	1,830	1,845	1,890	1,900	1,915	1,920	1,950	1,955
D	0.69	0.55	0.46	0.57	0.44	0.48	0.2	0.48	0.45	0.43	0.45	0.28	0.3	054	0.38
6	1,835	1,840	1,870	1,885	1,895	1,940	1,945	1,975	1,980			
D	0.24	0.55	0.46	0.45	0.45	0.19	0.32	0.53	0.43			

Table 7.3 Classification matrix estimating an identification quality of objects from different clusters

Observed classification	Percent correct	Predicted classification					
		Cluster 1 $p=.12245$	Cluster 2 $p=.10204$	Cluster 3 $p=.04082$	Cluster 4 $p=.24490$	Cluster 5 $p=.30612$	Cluster 6 $p=.18367$
Cluster 1	100	6	0	0	0	0	0
Cluster 2	100	0	5	0	0	0	0
Cluster 3	100	0	0	2	0	0	0
Cluster 4	100	0	0	0	12	0	0
Cluster 5	100	0	0	0	0	15	0
Cluster 6	100	0	0	0	0	0	9
Total	100	6	5	2	12	15	9

points out that dichotomous variables, which often violate multivariate normality, are not likely to affect conclusions based on DA. We can visualize how the two different functions discriminate between groups (clusters) by plotting the individual scores for the two discriminant (canonical) functions on a plate (http://www.statsoft.com/textbook/).

F-criteria, Wilks' Lambda statistics and the classification matrix were used to estimate classification quality.

By using such a classification approach the results led to an optimal separation of objects into groups.

The mean quality of the correct identification in the training sample was 100%, i.e. the classification algorithm did not make any mistake in that sample (Table 7.3). Next the statistical significance values ($p < 0.0001$) confirmed a good quality of identification: Wilk's lambda $\lambda_w = 0.009$ and F-criterion $F = 11.4$.

The recent spatial tree growth tendency for 1981–2000 was described by trend values with centre year 1990. That year (object) belongs to the first cluster composed of 6 objects. Note that there are satellite-sensed vegetation index data which cover 1981–2000 and were analysed with the goal to study global vegetation activity (Myneni et al. 1997; Zhou et al. 2001). For the same period an NDVI trend map was obtained (Zhou et al. 2001).

ERDAS Imagine and ARC VIEW GIS software were used. The former was used for an extrapolation of tree growth trends as a function of geographic coordinates and the latter for map compilation. A digital topographic map (scale 1:2,500,000) and a digital map of Siberian forests were used to compile maps of tree-ring growth trends (Forest of the USSR 1990). As a result, maps of tree growth trends were obtained for all objects from the first cluster (Fig. 7.4).

There are no complete analogues of spatial temporal tree-growth variation for the last time interval (1981–2000) excluded adjacent one (1976–1994). In the last case, the Euclidean distance D between objects is 0.38 and the correlation between the corresponding maps is 0.90 ($p<0.001$).

Spatial-temporal maps corresponding to objects from the first cluster can be interpreted as relative analogues of the recent one. So, the correlation between the map corresponding to 1901–1919 and the recent map is 0.54 ($p<0.01$) and $D = 2.68$.

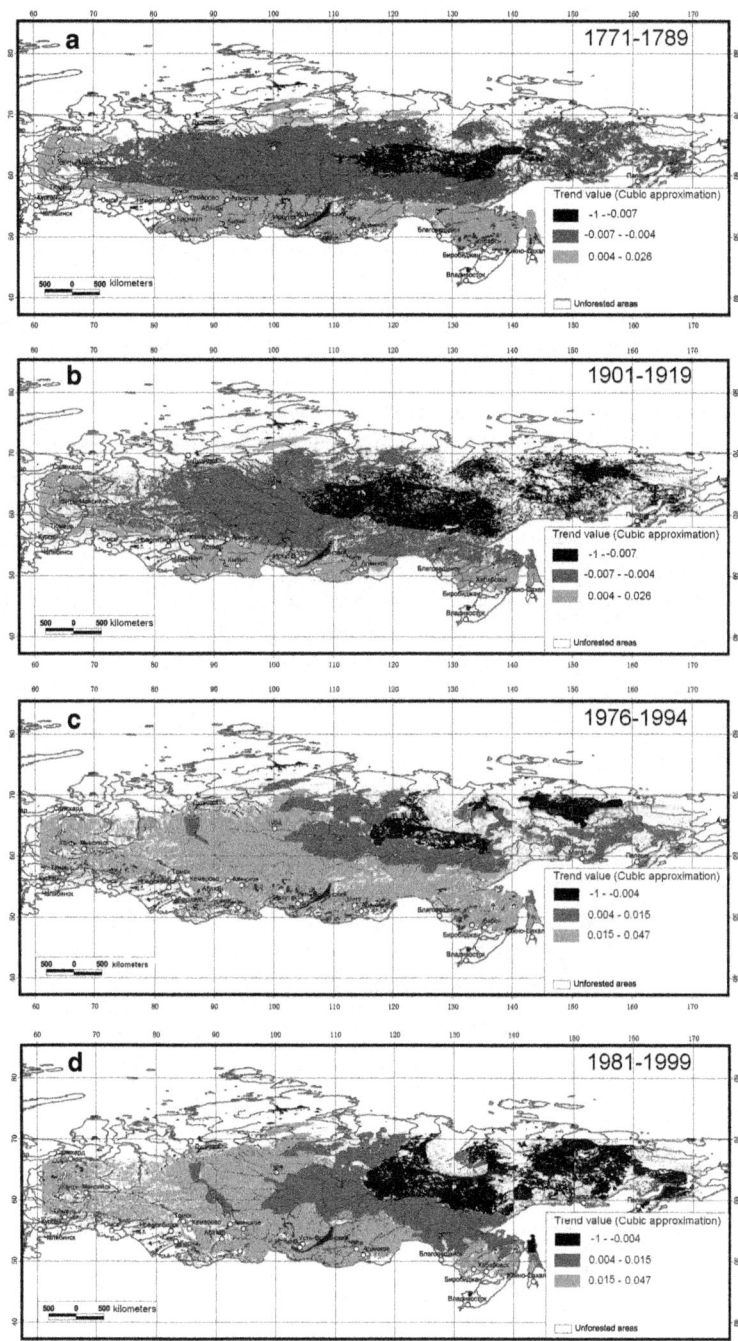

Fig. 7.4 Maps of tree-ring growth trends obtained for different time periods: (**a**) 1771–1789 (centre year is 1780); (**b**) 1901–1919 (1910); (**c**) 1976–1994 (1985); (**d**) 1981–1999 (1990). There is common tendency in trend values which decreases from West to East (*Color version available in Appendix*)

And, finally, corresponding statistics between 1771–1789 and recent spatial samples are r = 0.79 ($p<0.001$), D = 2.12.

Therefore, the first cluster is relatively homogeneous, i.e. all objects of the cluster retain similar relative spatial tendencies. These tendencies are characterized by a common decrease of trend values from West to East with highest values for southern latitudes of Siberia (Fig. 7.4). But the recent spatial sample of tree-growth trend is described by the highest variation for the last 250 years which can be explain by the most significant recent warming in context of 2,000 years (IPCC 2007; Briffa et al. 2008).

We analysed the spatial accordance between a map of NDVI (normalized differentiated vegetation index) trends (Zhou et al. 2001) and the spatial distribution of tree-growth trends.

Preliminarily, the map of tree-growth trend was transformed by grouping of trend values into six classes. The first class contained all values less then −0.015, second one – values located in [−0.015; −0.005), third – [−0.005; 0), fourth – [0; 0.005), fifth – [0.005; 0.02) and, finally, unselected values (more than 0.02) in the sixth class. This transformation was made to adjust the growth trend values with categorical values of NDVI trends.

To compare spatial samples a simple correlation analysis was used. A spatial mask containing 238 common points uniformly distributed for both samples was applied. As a result of that comparison a high concordance between spatial NDVI and the growth-trend series was obtained (Fig. 7.5).

The correlation coefficient is significant ($p<0.001$) and positive (r = 0.60).

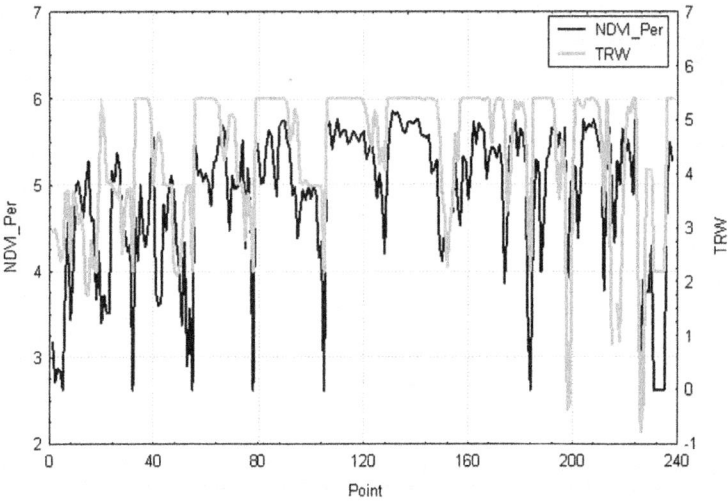

Fig. 7.5 Spatial variances of NDVI-trend series (NDVI_Per) and categorical values of tree-growth trends (TRW) obtaining for Siberia

7 Dendroclimatological Evidence of Climate Changes Across Siberia 111

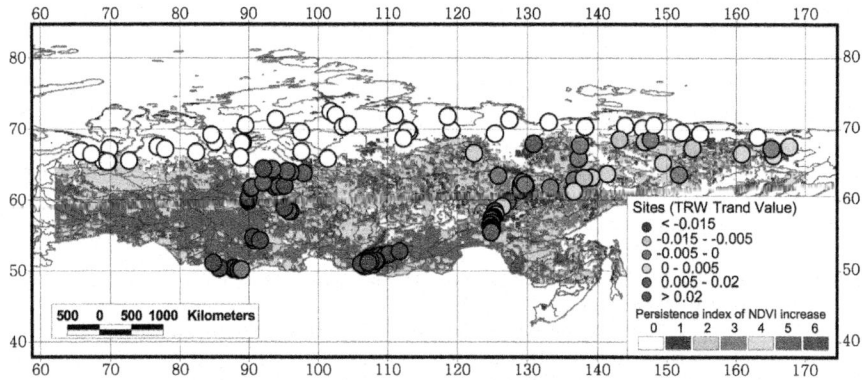

Fig. 7.6 The map of NDVI trends superposed with spatial values of tree-growth trends (*coloured circles*) for Siberia and Far East (*Color version available in Appendix*)

Moreover, by compilation of the NDVI-trend map with spatial growth-trend values spatial a high accordance was obtained (Fig. 7.6). These tendencies are resided to the first cluster (see text above) and characterized by a common decreasing of trend values from West to East with highest values for southern latitudes of Siberia.

Most tree-ring chronologies used in this chapter show significant correlations with summer temperature variations (Briffa et al. 1995, 1998b, 2008; Vaganov et al. 1999, 2000; Shishov et al. 2007). Therefore, the obtained tree-ring growth trends can be interpreted as regional temperature trends. That statement is consistent for Siberia, where strong dependence of total vegetation productivity with temperature changes is observed. In this case, an important peculiarity is revealed. Under scenarios of CO_2 doubling, forecasts by global climate models show the most significant warming in Subarctic areas (IPCC 2007). In our case, this result is partially correct partially. For the last decades radial growth increased and, therefore, significant temperature changes are observed for the western and southern parts of Siberia. There are two possible explanations of the spatial heterogeneity of tree-ring trend forcing by temperature changes. Firstly, theoretical estimation of the greenhouse-gas contribution are overestimated and recent temperature changes are related to natural phenomena. Secondly, global climate models are insufficiently adapted to the regional level even for large areas as, for example, Siberia.

Both interpretations require more adequate calibration and accurate verification of recent global models. Our obtained results show that developing a system of dendroclimatic monitoring can be useful to calibrate and verify climate models on regional and global scales.

Acknowledgments The authors are very grateful to Keith Briffa and Tom Melvin, whose competent advices were invaluable in preparing this work. This work was supported by the Royal Society (UK) (Royal Society No R14577), the Russian Foundation of Basic Researches (RFBR No 09-05-00900-a), the Grant of Russian Federation President (No MD-7845.2010.5).

References

Aasamaa K, Sober A, Hartung W, Niinemets U (2002) Rate of stomatal opening, shoot hydraulic conductance and photosynthetic characteristics in relation to leaf abscisic acid concentration in six temperate deciduous trees. Tree Physiol 22:267–276

Bergh J, McMurtrie RE, Linder S (1998) Climatic factors controlling the productivity of Norway spruce: a model-based analysis. For Ecol Manag 110:127–139

Briffa KR (2000) Annual climate variability in the Holocene: interpreting the message of ancient trees. Quat Sci Rev 19:87–105

Briffa KR, Osborn TJ (1999) Perspectives: climate warming – seeing the wood from the trees. Science 284:926–927

Briffa KR, Jones PD, Schweingruber FH, Shiyatov SG, Cook ER (1995) Unusual 20th-century summer warmth in a 1,000-year temperature record from Siberia. Nature 376:156–159

Briffa KR, Schweingruber FH, Jones PD, Osborn TJ, Harris IC, Shiyatov SG, Vaganov EA, Grudd H (1998a) Trees tell of past climates: but are they speaking less clearly today? Philos Trans R Soc B-Biol Sci 353:65–73

Briffa KR, Schweingruber FH, Jones PD, Osborn TJ, Shiyatov SG, Vaganov EA (1998b) Reduced sensitivity of recent tree-growth to temperature at high northern latitudes. Nature 391:678–682

Briffa KR, Osborn TJ, Schweingruber FH, Harris IC, Jones PD, Shiyatov SG, Vaganov EA (2001) Low-frequency temperature variations from a northern tree ring density network. J Geophys Res-Atmos 106:2929–2941

Briffa KR, Osborn TJ, Schweingruber FH (2004) Large-scale temperature inferences from tree rings: a review. Global Planet Change 40:11–26

Briffa KR, Shishov VV, Melvin TM, Vaganov EA, Grudd H, Hantemirov RM, Eronen M, Naurzbaev MM (2008) Trends in recent temperature and radial tree growth spanning 2000 years across Northwest Eurasia. Phil Trans R Soc B, Special issue. doi:10.1098/rstb.2007.2199. V. 363:2271–2284

Cook ER, Kairiukstis LA (1990) Methods of dendrochronology. Applications in the environmental sciences. Kluwer, Dordrecht/Boston/London, 394

Cook ER, Peters K (1997) Calculating unbiased tree-ring indices for the study of climatic and environmental change. Holocene 7:361–370

Cook ER, Briffa KR, Jones PD (1994) Spatial regression methods in dendroclimatology – a review and comparison of 2 techniques. Int J Climatol 14:379–402

Graybill DA, Idso SB (1993) Detecting the aerial fertilization effect of atmospheric CO_2 enrichment in tree-ring chronologies. Global Biogeochem Cycles 7:81–95

Hartigan JA (1975) Clustering algorithm. Wiley, New York

Hartigan JA, Wong MA (1978) Algorithm 136. A k-means clustering algorithm. Appl Stat 28:100

Huberty CJ (1994) Applied discriminant analysis. Wiley-Interscience (Wiley Series in Probability and Statistics), New York

Hughes MK, Diaz HF (1994) Was there a Medieval Warm Period, and if so, where and when. Climatic Change 26:109–142

Hughes MK, Vaganov EA, Shiyatov SG, Touchan R, Funkhouser G (1999) Twentieth-century summer warmth in northern Yakutia in a 600-year context. The Holocene 9:629–634

IPCC (2007) Climate change 2007: the physical science basis. Summary for Policymakers. Contribution of Working Group I to the Fourth Assessment Report of the Intergovernmental Panel on Climate Change. http://ipcc-wg1.ucar.edu/

Jones PD, Moberg A (2003) Hemispheric and large-scale surface air temperature variations: An extensive revision and an update to 2001. J Climate 16:206–223

Karnosky DF (2003) Impacts of elevated atmospheric CO_2 on forest trees and forest ecosystems: knowledge gaps. Environ Int 29:161–169

Klecka WR (1980) Discriminant analysis. In: Quantitative applications in the social sciences series, vol 19. Sage, Thousand Oaks, CA

Lachenbruch PA (1975) Discriminant analysis. Hafner, New York

LaMarche VC, Graybill DA, Fritts HC, Rose MR (1984) Increasing atmospheric carbon-dioxide – Tree-ring evidence for growth enhancement in natural vegetation. Science 225:1019–1021

LaMarche VC, Graybill DA, Fritts HC, Rose MR (1986) Carbon-dioxide enhancement of tree growth at high elevations. Science 231:860–860

Linder P (1998) Structural changes in two virgin boreal forest stands in central Sweden over 72 years. Scand J Forest Res 13:451–461

Linder P, Ostlund L (1998) Structural changes in three mid-boreal Swedish forest landscapes, 1885-1996. Biol Conserv 85:9–19

Linder P, Elfving B, Zackrisson O (1997) Stand structure and successional trends in virgin boreal forest reserves in Sweden. For Ecol Manag 98:17–33

Liu Y, Shi JF, Shishov V, Vaganov E, Yang YK, Cai QF, Sun JY, Wang L, Djanseitov I (2004) Reconstruction of May-July precipitation in the north Helan Mountain, Inner Mongolia since AD 1726 from tree-ring late-wood widths. Chinese Sci Bull 49:405–409

Mann ME, Bradley RS, Hughes MK (1998) Global-scale temperature patterns and climate forcing over the past six centuries. Nature 392:779–787

McDowell N, Barnard H, Bond BJ, Hinckley T, Hubbard RM, Ishii H, Kostner B, Magnani F, Marshall JD, Meinzer FC, Phillips N, Ryan MG, Whitehead D (2002) The relationship between tree height and leaf area: sapwood area ratio. Oecologia 132:12–20

McLachlan GJ (2004) Discriminant analysis and statistical pattern recognition. Wiley-Interscience (Wiley Series in Probability and Statistics), New York

Melvin T (2004) Historical growth rates and changing climatic sensitivity of boreal conifers. Ph.D. thesis. Climatic Research Unit, University of East Anglia, Norwich, UK, 220 p

Myneni RB, Keeling CD, Tucker CJ, Asrar G, Nemani RR (1997) Increased plant growth in the northern high latitudes from 1981 to 1991. Nature 386:698–702

Naumburg E, Ellsworth DS, Katul GG (2001) Modeling dynamic understory photosynthesis of contrasting species in ambient and elevated carbon dioxide. Oecologia 126:487–499

Nunez-Elisea R, Schaffer B, Fisher JB, Colls AM, Crane JH (1999) Influence of flooding on net CO_2 assimilation, growth and stem anatomy of Annona species. Ann Bot 84:771–780

Ojeda M, Schaffer B, Davies FS (2004) Flooding, root temperature, physiology and growth of two Annona species. Tree Physiol 24:1019–1025

Robertson I, Rolfe J, Switsur VR, Carter AHC, Hall MA, Barker AC, Waterhouse JS (1997) Signal strength and climate relationships in C-13/C-12 ratios of tree ring cellulose from oak in southwest Finland. Geophys Res Lett 24:1487–1490

Shishov V (2000) Statistical relationship between El Nino Intensity and summer temperature in the subarctic region of Siberia. Dokl Earth Sci 375a:1450–1453

Shishov VV, Vaganov EA, Hughes MK, Koretz MA (2002) Spatial variations in the annual tree-ring growth in Siberia in the past century. Dokl Earth Sci 387A:1088–1091

Shishov VV, Naurzbaev MM, Vaganov EA, Ivanovsky AB, Koretz MA (2007) The analyses of tree-ring growth on Eurasian north at the last decades. Izvestiya RAS. Geographical series, V3, 49–59

Takeuchi Y, Kubiske ME, Isebrands JG, Pregtizer KS, Hendrey G, Karnosky DF (2001) Photosynthesis, light and nitrogen relationships in a, young deciduous forest canopy under open-air CO_2 enrichment. Plant Cell Environ 24:1257–1268

Thacker JRM, Bryan WJ, McGinley C, Heritage S, Strang RHC (2003) Field and laboratory studies on the effects of neem (Azadirachta indica) oil on the feeding activity of the large pine weevil (Hylobius abietis L.) and implications for pest control in commercial conifer plantations. Crop Prot 22:753–760

Tryon RC (1939) Cluster analysis. Ann Arbor: Edwards Brothers

Forest of the USSR (1990) Scale 1:2,500,000. In: Garsia MG (ed.) Gos Upr Geol Kartogr, Moscow

Vaganov EA, Hughes MK, Kirdyanov AV, Schweingruber FH, Silkin PP (1999) Influence of snowfall and melt timing on tree growth in subarctic Eurasia. Nature 400:149–151

Vaganov EA, Briffa KR, Naurzbaev MM, Schweingruber FH, Shiyatov SG, Shishov VV (2000) Long-term climatic changes in the Arctic region of the Northern Hemisphere. Dokl Earth Sci 375:1314–1317

Ward JH (1963) Hierarchical grouping to optimize an objective function. J Am Stat Assoc 58:236–244

Waterhouse JS, Switsur VR, Barker AC, Carter AHC, Hemming DL, Loader NJ, Robertson I (2004) Northern European trees show a progressively diminishing response to increasing atmospheric carbon dioxide concentrations. Quat Sci Rev 23:803–810

Zhou LM, Tucker CJ, Kaufmann RK, Slayback D, Shabanov NV, Myneni RB (2001) Variations in northern vegetation activity inferred from satellite data of vegetation index during 1981 to 1999. J Geophys Res-Atmos 106:20069–20083

Chapter 8
Siberian Pine and Larch Response to Climate Warming in the Southern Siberian Mountain Forest: Tundra Ecotone

V.I. Kharuk, K.J. Ranson, M.L. Dvinskaya, and S.T. Im

Abstract The tree response to climate trends is most likely observable in the forest-tundra ecotone, where temperature limits tree growth. Here we show that trees in the forest-tundra ecotone of the mid of the south Siberian Mountains responded strongly to warmer temperatures during the past two decades. There was a growth increment increase, stand densification, regeneration propagation into the alpine tundra, and transformation of prostrate Siberian pine, larch and fir into arboreal forms. A temperature increase of 1°C allows regeneration to occupy areas ~40–100 m higher in elevation, depending on the site. Siberian pine and larch regeneration and arboreal forms now occur at elevations up to 200 m higher in comparison with the known location of the former tree line. These species surpass their upper historical boundary of 10–80 m elevation. Regeneration is propagating into the alpine tundra with the rate of 0.5–2.0 m/year. The observed winter temperature increase is significant for regeneration survival. Measurements of the radial and apical growth increments indicates an acceleration of krummholz transforming into arboreal forms in the mid-1980s. Larch surpasses Siberian pine in cold resistance, and has an arboreal growth form where Siberian pine is in krummholz form. Improving climate provides competitive advantages to Siberian pine in the areas with sufficient precipitation amount. Larch, as a leader in harsh environment resistance, received an advantage at the upper front tree line, and in the areas with low precipitation. Observed tree migration into the alpine stony tundra will decrease albedo, providing a positive feedback to global warming at the regional level.

Keywords Climate trends • Mountain forest-tundra ecotone • Pinus sibirica • Larix sibirica • Upward plant migration

---

V.I. Kharuk (✉), M.L. Dvinskaya, and S.T. Im
V.N. Sukachev Institute of Forest, SB RAS, 50, Akademgorodok, Krasnoyarsk, 660036, Russia
e-mail: kharuk@ksc.krasn.ru; mary_dvi@ksc.krasn.ru; stim@ksc.krasn.ru

K.J. Ranson
NASA Goddard Space Flight Center, Greenbelt, MD 20771, USA
e-mail: jon.ranson@nasa.gov

8.1 Introduction

Last century the global mean air temperature increase was 0.06°/decade; since 1976 this rate accelerates, and reaches 0.18°/decade. Since 1987 the ten warmest years since beginning of meteorological measurements have been observed (WMO 2002). The observed warming now covers a period of more than 20 years, long enough to expect a vegetation response. The tree response to climate warming should be stronger where temperature mainly limits tree growth, i.e. in northern and mountain forest-tundra ecotones, and this response should differ for trees with different temperature sensitivities. Indeed, there are now numerous reports of tree and stand responses to warming, including species invasion into the tundra, increase of stand density and tree radial increment along the northern tree line in the last decades of the twentieth century (e.g., Suarez et al. 1999; Payette et al. 2001; Lloyd and Fastie 2002). In the mountain (upper) tree line border data obtained are more controversial. Several publications have reported an upward shift of the upper tree line (Kullmann 2002; Shiyatov 2003; Kharuk and Fedotova 2003). However, other investigations reported the stability of the tree line during the last decades (Masek 2001; Klasner and Fagre, 2002). Mountain forest-tundra ecotone studies of tree response to climate warming have an advantage in comparison with the northern ecotone since stronger a temperature gradient exists in the mountains. Consequently, vegetation changes are "compressed" at an altitudinal scale of 10–100 m, whereas in the northern ecotone similar changes are expected over kilometres. With temperature increasing, in the mountain forest-tundra ecotone the following responses are expected: (1) regeneration increase and migration along the altitude gradient, (2) tree line shift to higher altitude, (3) tree radial and apical increment increase, (4) fast stand density increases, (5) migration of "warm –adapted" species into traditional range of "cold-adapted" species. Although the most significant temperature changes are observed and predicted in Siberia, there are few published reports on the Siberian species response to climate warming (Shiyatov 2003; Kharuk et al. 2005, 2006). Especially poorly investigated is the vast region of the South Siberian Altai – Sayan Mountain system.

The purpose of this study is an analysis of climate-driven changes in the mountain forest-tundra ecotone formed by Siberian pine and larch. The study parameters were tree apical and radial increments, tree morphology, regeneration abundance and age structure, and the upper limit of tree and regeneration line.

8.2 Materials and Methods

The studies were conducted on two sites: the first was within the optimum of Siberian pine habitat (northern part of West Sayan Mountains); the second was within the Siberian pine and larch contact area, the Tannu-Ola ridge (Fig. 8.1).

The western Sayan Mountains are a system of ridges divided by a dense drainage network. Northern slopes of ridges between 800 and 1,800 m in elevation are

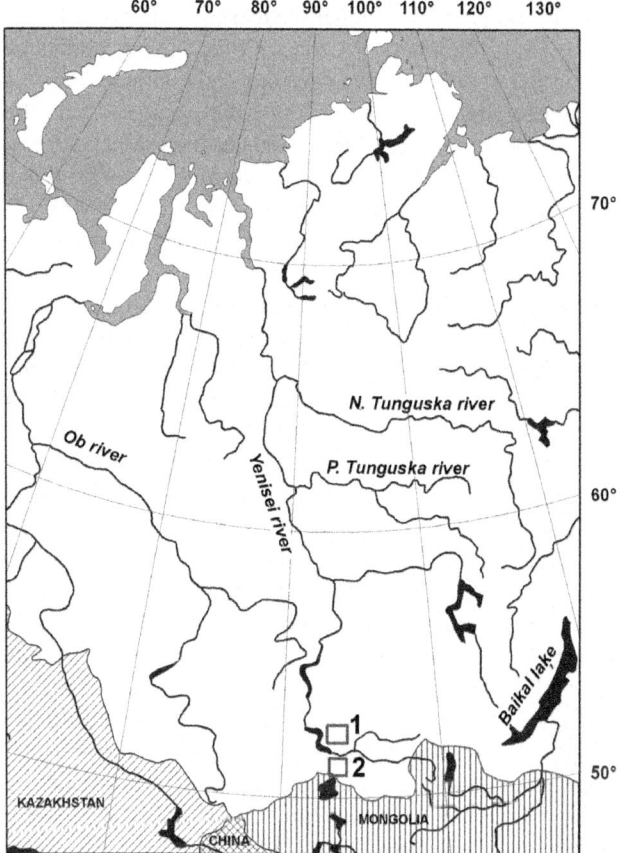

Fig. 8.1 The location of the study area. 1 – North Sayan, and 2 – Tannu – Ola sites (*Color version available in Appendix*)

occupied by so-called dark-needle conifers including Siberian pine (*Pinus sibirica*), fir (*Abies sibirica*) and spruce (*Picea obovata*). Altitudinal belts include forest-steppe (up to 250–300 m.), a narrow Scotch pine and mixed forest belt (up to 400–450 m), a "dark-needle" conifer belt (up to 1,600–1,700 m), and sparse Siberian pine –fir forest belt (1,600–1,800 m.). At the upper elevation limit forests are composed of sparse Siberian pine stands (on northern slopes) and larch (on southern slopes). The climate is severe continental. Mean monthly temperatures in the mountains range from −28°C to −34°C during January, and from 10°C to 12°C during July. Precipitation falls mainly during the summer. The amount strongly depends on altitude and slope orientation, varying from 400–500 mm a year (northern foothills) up to 1,000–1,500 mm (northern windward slopes). On southern slopes the amount of precipitation is 400–500 mm. Northern slopes are covered by moisture-dependent Siberian pine and fir stands, whereas southern facing slopes are covered mostly by larch, a drought-resistant species. The study site is within northern windward slopes.

The Tannu-Ola ridge is in southern Siberia, near the Mongolian border (Fig. 8.1). This is a transition between zone of Siberian pine and larch dominated areas. The vegetation patterns on the northern and southern slopes are different due to different precipitation amount (700–1,000 mm/year, and 400–500 mm/year, respectively). The upper forest belt on the northern slopes is formed by mixed Siberian pine (*Pinus sibirica*) and larch (*Larix sibirica*) stands; this is the southern border of the Siberian pine area. At higher elevations (>1,900–2,000 m) these stands transition into the forest-tundra ecotone. South facing, rain – shadowed slopes are covered by larch forests with patches of mountain meadows and steppes. The climate is severe continental with mean January temperature of −28°C to −35°C and +13°C to +15°C in July. For both sites temperature records showed an increase of air temperature for all seasons since the middle of the Eighties, whereas a positive precipitation trend is observable only for winter time. Climate variables considered in this study includes seasonal (DJF, MAM, JJA, SON), growing season (MJJAS), and mean annual temperatures and precipitation. Meteorological data were acquired until the year 2002 (http://www.cru.uea.ac.uk/~timm/grid/CRU_TS_2_1.html).

On-ground studies were made along the elevation transects oriented from sparse stands into the alpine tundra. Trees and regeneration (trees with $h < 2.5$ m) density measurements were made on the sample plot size (10×10 m). Seedling and regeneration age structure was determined by counting annual rings at the root neck level. Current year seedlings were recognized based on morphology features (Fig. 8.5). Regeneration was also divided into viable, nonviable and dead groups. The viable regeneration includes healthy samples (without signs of suppression or damage), damaged, but viable (the damaged or substituted top with stable or progressing apical increment). The North Sayan site includes five transects (within an elevation range from 1,570 to 1,790 m). The sample size was: for trees – 50 specimens, for regeneration – 1,881 specimen (all – Siberia pine). Mean regeneration density was ~24,000/ha (with max up to 70,000/ha). The Tannu-Ola site includes 4 transects (within an elevation range 1,900–2,450 m). The sample size was: trees – 87 specimen, regeneration – 404 specimen (95% Siberian pine, the rest were larch). Mean regeneration density was 1,400/ha (with max ~5,000/ha). Trees were sampled on its upper limit; disks for radial increment analysis were cut off at root neck level. Apical increments were measured as the distance between whorls, and covered the whole measurable period (15–25 year, depending on sampled tree). Increment analysis was based on the dendrochronology methods described by Shiyatov (2003). Since apical and radial increment values for different trees were unequal, the following normalization was done for comparison purposes. The increment values of a given sample were normalized by summing for the all years, and made equal to 1.0. Then, the increment value for a given year was taken equal to its proportion to the total sum. Samples of dead tree fossils in the forest-tundra ecotone were also collected. Tree mortality dates were determined by cross-analysis with a master-chronology (Shiyatov 2003).

8.3 Results

8.3.1 Tree Increment Dynamics

During the last decades (since 1987) a strong increase of larch and Siberian pine apical and radial increments was observed (Fig. 8.2). For Siberian pine this increase strongly correlates with summer temperatures ($R^2 = 0.89 - 0.93$); for larch the relationship is weaker ($R^2 = 0.43$–0.56) (Fig. 8.3). For the whole "pre-warming" period (1950–1986) all correlations are not significant. Larch, in comparison with Siberian pine, showed a weaker response to summer temperature increase. Correlation with precipitation is non-significant (excluding spring precipitation: $R = 0.43, \ldots, 0.49; p < 0.05$).

8.3.2 Regeneration Age Structure

For both sites age distribution showed that regeneration appeared mainly during the last two decades (Fig. 8.4). The maximum mortality (>20%) was observed in the youngest age group, the maximal proportion of damaged seedlings (with substituted

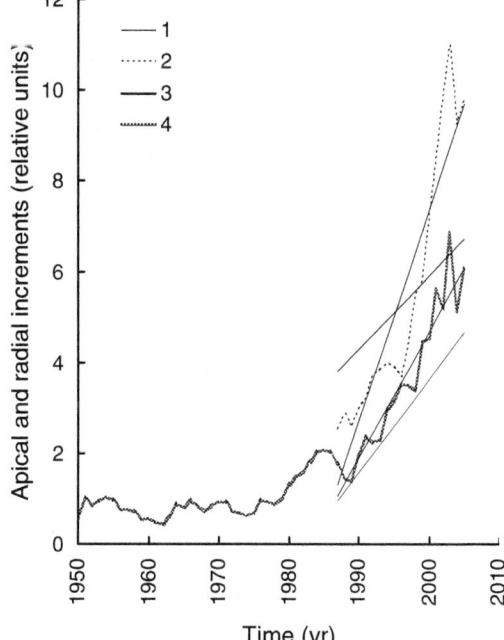

Fig. 8.2 Dynamics of apical and radial increments of Siberian pine and larch. Larch: 1 – apical, 3 – radial increment; Siberian pine: 2 – apical increment, 4 – radial increment; *Trees* (A ~ 55 year) were sampled at elevations 2,035–2,100 m. All trends (*straight lines*) are significant (p < 0.05)

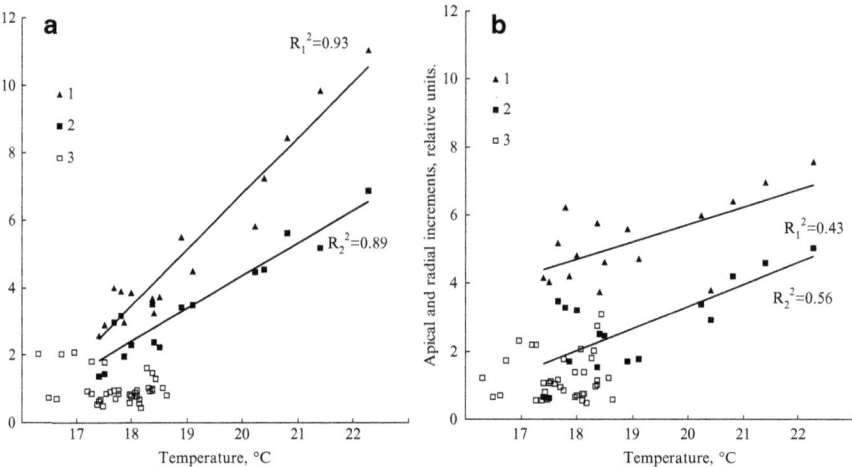

Fig. 8.3 Siberian pine and larch increments dependence on summer temperature. (**a**) Siberian pine. (**b**) Larch. 1, 2 – apical and radial increments for period 1987–2005; 3 – radial increments for period 1950–1986

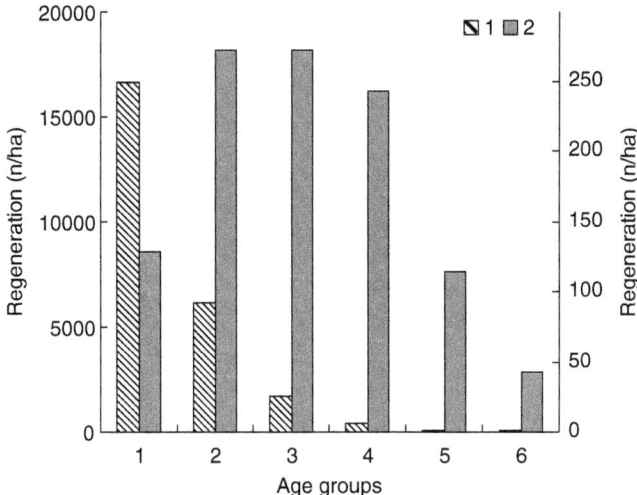

Fig. 8.4 Regeneration age structure in the forest-tundra ecotone. 1 – North Sayan site (*left ordinate*), 2 – Tannu-Ola site (*right ordinate*)

or dead tops) was found in the older group. Notably the maximal regeneration number at the Sayan site reached high values – up to 70,000 saplings/ha; more than 90% of the regeneration appeared during last two decades.

In the Tannu-Ola site regeneration density was considerably lower (mean values about 1,400/ha). These differences could be attributed to higher (optimal

Fig. 8.5 Regeneration is often sheltered by felled predecessors. *Inset*: current year sapling with nut remains on the needles (*Color version available in Appendix*)

for Siberian pine) precipitation amounts in the Sayan Mountains (>100 mm/ year). The majority of regeneration (95%) was Siberian pine; this proportion does not correspond to the ratio of species in the upper canopy of the Tannu-Ola site, where the larch proportion is about 40%; this difference may indicate a substitution of larch by Siberian pine.

Current regeneration within the forest-tundra ecotone is often sheltered by felled "predecessor" trees (Fig. 8.5). The remains of fallen trees were present within the current forest-tundra ecotone indicating warmer conditions at this site in the past. The fallen "predecessor" trees were found to have died in the eighteenth and early nineteenth century, which corresponds to a period of cooling. Current regeneration now grows among these fallen trees and 10–80 m higher in elevation, depending on site. But still its height is regularly lower than its predecessor's height (Figs. 8.5 and 8.6).

With respect to the location of the "mother stand", regeneration propagates upward by about 30–200 m, depending on the site.

Regeneration density is correlated with warming (Fig. 8.7). Data shown in Fig. 8.8 illustrate the regeneration density relationship with seasonal mean temperatures (DJF, MAM, JJA, SON), mean annual temperature, and growing season temperatures; the last one was approximated by MJJAS temperatures. The correlation pattern for regeneration (both overwinter and total) is different for the Sayan and the Tannu-Ola site (Fig. 8.8); this difference could be attributed to lower snow cover in the Tannu-Ola site, which leads to a stronger dependence on the winter temperatures. The data on Fig. 8.8 also show that winter temperatures are important for regeneration density, since low temperatures cause regeneration damage and mortality. For comparison, radial increments of Siberian pine and larch are presented; larch, as a cold-adapted species, shows less correlation with temperature.

Fig. 8.6 Siberian pine and larch regeneration reached and surpass the former tee line; also the height is still lower than its predecessors (*Color version available in Appendix*)

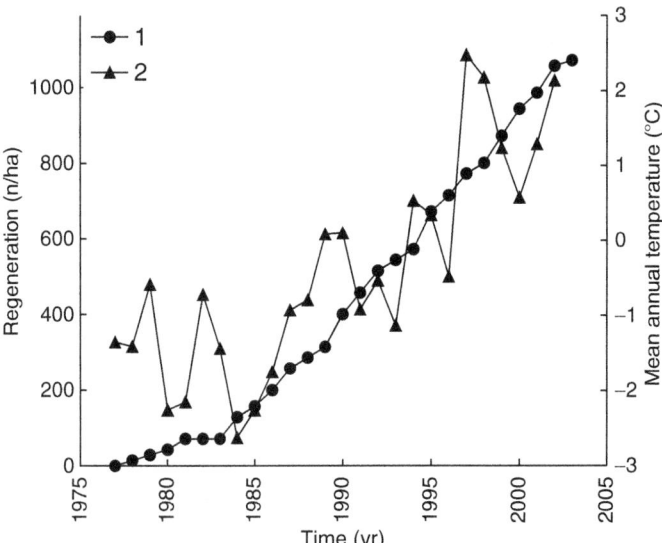

Fig. 8.7 Regeneration (cumulative) versus mean annual temperature: 1 – cumulative regeneration density, number/ha, 2 – mean annual temperature. Cycling in regeneration amount is due to Siberian pine seed production cycles

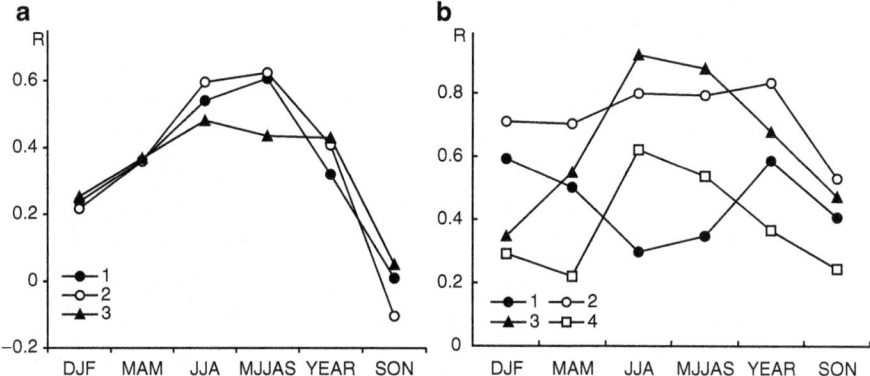

Fig. 8.8 "Regeneration amount–air temperature" regression coefficient values for different year seasons: 1 – regeneration (overwinter) number; 2 – regeneration (cumulative) number; 3, 4 – Siberian pine larch radial increments, respectively

8.3.3 Estimation of Regeneration Response to Warming

The regeneration response to warming, i.e. the propagation along the elevation gradient, could be estimated based on the comparison of the regeneration upper line position referred to the temperature changes for a given period. The upper line position for a given period could be found based on a regeneration age structure analysis. Calculations showed that for the north Sayan site a rise in temperature of 1°C allows regeneration to advance 100–150 m; for the Tannu-Ola site these values are lower (10–40 m). The difference could be attributed to harsher conditions in the Tannu-Ola site (less amount of precipitation, including winter, seedling-protecting snowfall), in conjunction with wintertime winds. The estimated vertical speed of regeneration propagation along the elevation gradient is about 0.5–2.0 m/year.

8.3.4 Prostrate Versus Arboreal

A transformation of the prostrate (or krummholz) forms of Siberian pine into arboreal forms in the mountain forest-tundra ecotone has been observed (Figs. 8.9, 8.10, and 8.11). Radial increment analysis showed that the beginning of krummholz transforming into arboreal form was observed in 1987 (Fig. 8.9). In the same year existing arboreal forms showed a similar strong increase of radial increment (Figs. 8.2 and 8.3). Note also the increase in the Forties which corresponds to climate warming in the Thirties and Forties of the last century (Fig. 8.9). Larch is much less (>10 times) likely than Siberian pine to be found in krummholz forms. This species surpasses Siberian pine in frost resistance, and grows in arboreal forms where pine is prostrate (Fig. 8.10). It should be noted that in the Sayan site prostrate fir is also transforming into arboreal forms.

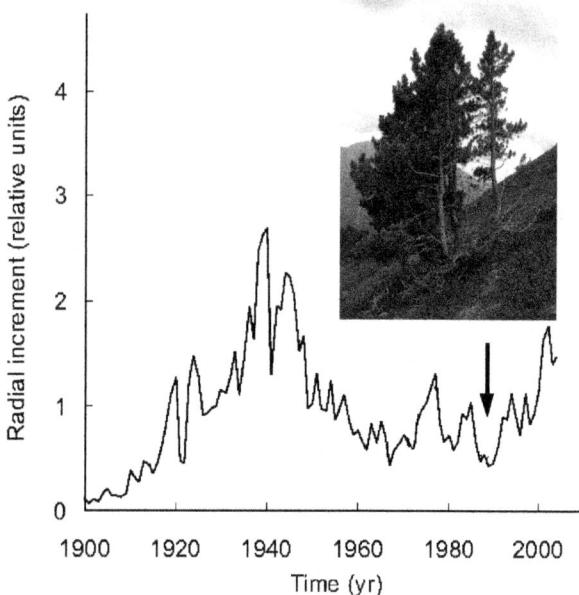

Fig. 8.9 Radial increment dynamics of "post-prostrate" form of Siberian pine. An arrow indicates beginning of prostates transforming into arboreal forms. *Inset*: Siberian pines transformed from krummholz to arboreal form (elevation is about ~2,000 m) (*Color version available in Appendix*)

8.4 Discussion

During the last two decades in the mid of the south Siberian Mountains system, in the forest-tundra ecotone trees are strongly responding to warming by increase of growth increments, stand densification, regeneration density, advancing tree line position, and transformation of krummholz to arboreal forms. A rapid increase of apical and radial increment for Siberian pine and larch was observed since the middle of the 1980s, and it strongly correlates with mean summer and vegetation period temperatures (Figs. 8.3, 8.7, and 8.8). Notably for the former period correlations were not significant (Fig. 8.3). The other notable observation is that winter and spring temperatures are significant for regeneration density (Fig. 8.8). Similar observations were reported for other regions (Kharuk et al. 2005; Kullmann 2007). The winter temperature increase promotes regeneration survival due to desiccation reduction. Increased winter time precipitation and higher snow cover protects young seedlings against desiccation and snow abrasion. The critical period for seedling survival is a moment when seedlings exceed snow cover height, and face winter desiccation and snow abrasion. This causes apical meristem damage, seedling mortality, or krummholz formation. Seedlings that effectively exceed this barrier have a typical "tree-in-skirt" shape (Fig. 8.11). On the other hand, an excess of winter precipitation may negatively affect both regeneration density and growth increments, since snow accumulation leads to a decrease of the vegetation period. Indeed, in the

Fig. 8.10 Larch surpasses Siberian pine in winter desiccation resistance, and growing arboreal where pine is prostrate (*Color version available in Appendix*)

Fig. 8.11 "Tree-in-skirt" Siberian pine. *Lower brunches* corresponds snow level (*Color version available in Appendix*)

Fig. 8.12 Snow accumulation behind propagating trees and regeneration caused a zone of tree vegetation depression. Picture was taken 04 July 2006 (*Color version available in Appendix*)

Altai Mountains we found a negative correlation between apical increment and snow precipitation. Regeneration itself could lead to a "snow-depression" effect: once regeneration appears, it works as a "snow-accumulation strip". Snowdrifts behind this strips caused a prolongation of the snow covered period, and consequently regeneration depression (Fig. 8.12). This phenomenon explains the "wave-propagation" pattern of forest migration to the tundra, which is often observable in the vicinity of a "mother stand". The period of these waves corresponds to the snowdrift length behind the "front wave", and is equal to ~2–3 tree heights (Fig. 8.12).

Alongside with temperature and precipitation, winter winds are essential for tree survival. It's known that climate severity for biological organisms is a function of temperature and wind speed. There is an empirical rule that 2 m/s winter wind speed is equal to a 1°C decrease. A classical example of wind impact is winter desiccation and snow abrasion, which often limits seedling survival (Shiyatov 2003; Kullmann 2002; Kharuk and Fedotova 2003). At the upper tree limit regeneration often grows in sheltered zones behind felled trees and rocks (Figs. 8.5, 8.6).

Wind – resistance is increasing by tree-clustering in the forest-tundra ecotone (Fig. 8.13). These clusters have a wind-protecting geometry: wind circumvents the cluster, rather than penetrating through, thus reducing desiccation and snow abrasion. Clusters often have a colonna-shape, and are oriented along the prevailing winds. A leader of a column is a mature or even dead tree, the successor's protector. A new tree generation around the clusters indicates improving growth conditions (Fig. 8.13).

Fig. 8.13 Tree clusters are oriented according prevailing winds. The *cluster's shape* is reducing wind impact (*Color version available in Appendix*)

Unfortunately, data on wind speed are poor, and more investigations are necessary for understanding combined wind-temperature impacts on regeneration survival.

The observation of a transformation of prostrate to arboreal forms, especially abundant for Siberian pine, is further evidence of climate warming (Figs. 8.9–8.11). Notably some proportion of "post-prostrate trees (about 5–15%) have a dead (or substituted by a lower branch) apical meristem, indicating that observing warming is not a straightforward process. Larch, the species which forms the northern tree line in Asia, is a leader in resistance to harsh growth conditions among Siberian conifers. At the edge of the mountain forest-tundra ecotone, where Siberian pine is presented by krummholz, larch is growing in arboreal form (Fig. 8.10). Larch can survive in extreme conditions because of its deciduous leaf habit and dense bark protecting the stem from winter desiccation and snow abrasion (Kharuk and Fedotova 2003). For Siberian pine, on the other hand, high proportions of living tissues are covered by thin bark (Fig. 8.14). These chlorophyll containing tissues provide additional assimilates which could be significant for surviving within the tree line zone. This is true also for larch, whose young stems also contain chlorophyll. The photosynthetic rate of bark photosynthesis may reach 30–50% that of leaves (Kharouk et al. 1995). Due to this, some species (a willow, *Salix* spp.) are vegetating in the tundra zone without leaves (Fig. 8.15).

Larch is also known as a drought-resistant species: it survives in areas with semi-desert precipitation levels (Siberian north-east, <250 mm/year; Kharuk et al. 2005). Siberian pine, on the contrary, is a "the tree-of fogs" growing optimally with precipitation

Fig. 8.14 Chlorophyll-contain tissues in Siberian pine bark (*Color version available in Appendix*)

Fig. 8.15 Willow (Salix spp.) is vegetating in the upper tree limit without leaves. Picture was taken 03 July 2006 (*Color version available in Appendix*)

Fig. 8.16 Trees are "diffusing" along elevation gradient. Peripheral trees have a krummholz form (*Color version available in Appendix*)

higher 1,000 mm/year. A precipitation increase, which is observable in the larch-dominated communities in Central Siberia and in the Altai-Sayan Mountain system, is evidently favourable for Siberian pine.

Larch and Siberian pine have a different strategy of propagation. Larch, as a "wind-dependent" species, is "diffusing" from the "mother stand wall", forming a "tree concentration" gradient along the elevation gradient (Fig. 8.16). In mountain tundra we found larch regeneration in a distance of 200–300 m from the seed trees. Evidently, ground winds and snowstorms allow light larch seeds travelling a long distance over snow- covered tundra. Numerous single trees (a "diffusion") are a sign of improving growing conditions (Fig. 8.17). For larch propagation solitary trees growing in a distance of up to 1–2 km from the "mother stand" are also important, "refugia" of former climate warming (Fig. 8.18). A similar observation was made for pine in Swedish Scandinavia (Kullmann 2007). Siberian pine is less dependent on solitary trees or tree clusters for its propagation. For Siberian pine the primary role in propagation is played by the so-called "cedar-bird" (*Nucifraga caryocatactes*): because of this bird regeneration appeared in a radius of 1–2 km from "mother stands". Moreover, this bird preferably "hides" Siberian pine nuts in the parcels with better edaphic and microclimate conditions for seedling germination (local depressions, wind-protected areas behind rocks and felled trees). This behaviour pattern is explained by co-evolution of both species.

Fig. 8.17 Trees are "diffusing" into tundra zone (*Color version available in Appendix*)

Fig. 8.18 "Mother-larch" during centuries is providing seeds for new regeneration waves (*Color version available in Appendix*)

The higher response of Siberia pine to temperature increases indicates competitive advantages of this species with temperature increases. With observing stand density increase (Kharuk et al. 2006), Siberian pine received an additional advantage since larch is not shadow tolerant. So, the current climate change will lead to a substitution of larch by Siberian pine in the area with medium to high precipitation. Species substitution, larch for pine, in concert with species migration to stony tundra (Figs. 8.10, 8.15, 8.17) will lead to an albedo decrease, i.e. it will cause a positive global warming feedback. The other expected consequence of substitution of larch by pine is an increase of biodiversity since Siberian pine dominated communities provide a better food source for animals and birds. Larch, due to high drought- and wind-resistance, has advantages in areas with low precipitation, and in the extreme area of the upper tree line (Figs. 8.10, 3.10.17).

The mountain tree line advancing and retreating is a periodic process (Kullmann 2007; Shiyatov 2007). Analysis of fissile wood showed that within the tree line zone trees died mainly during the eighteenth and early nineteenth century, a period of cooling. The actual tree-line is not corresponding to its potential position, since vegetation follows the climatic changes with a lag: during a warming period the tree line is behind its potential climatic border due to the limitation in the tree regeneration (germination capacity, seeds spreading, seedlings survival). Similarly, during cooling periods the tree line retreats with a lag since mature trees have a higher resistance to the environmental conditions.

The described findings in the south Siberian Mountains (a rapid increase in increments and regeneration number, stand densification, upper tree line shift, and transformation of krummholz to arboreal forms) is typical for the Sayan-Altai Mountain system. Moreover, the beginning of this phenomenon coincides with similar events on the western end of Eurasia, in Swedish Scandinavia (Kullman 2004). This is evidence of the Eurasian scale of the phenomena, caused by a temperature threshold "trigger" effect.

Acknowledgments This research was supported by the NASA Science Mission Directorate, Terrestrial Ecology Program, and Russian Fund for Fundamental Investigations # 06-05-64939.

References

CRU TS 2.1: grids of climate observations, Dr. Mitchell TD. http://www.cru.uea.ac.uk/~timm/grid/CRU_TS_2_1.html. Accessed 29 April 2008

Kharouk VI, Middleton EM, Spencer SL, Rock BN, Willams DL (1995) Aspen bark photosynthesis and its significance to remote sensing and carbon budget estimates in the boreal ecosystem. J Water Air Soil Pol 82:483–497

Kharuk VI, Fedotova EV (2003) In: Bobylev LP, Kondratyev KY, Johannessen OM (eds) Forest-tundra ecotone dynamics. Arctic environment variability in the context of global change. Springer-Practice, Heidelberg, pp 281–299

Kharuk VI, Dvinskaya ML, Ranson KJ, Im ST (2005) Expansion of evergreen conifers to the larch-dominated zone and climatic trends. Russ J Ecol 36:164–170

Kharuk VI, Ranson KJ, Im ST, Naurzbaev MM (2006) Forest-tundra larch forests and climatic trends. Russ J Ecol 37:323–331

Klasner FL, Fagre DB (2002) A half century of change in alpine treeline patterns at Glacier National Park, Montana, USA. Arct Antart Alp Res 34:49–56

Kullman L (2004) The changing face of the alpine world. Glob Change Newslett 57:12–14

Kullmann L (2002) Rapid recent range-margins rise of tree and shrub species in the Swedish Scandes. J Ecol 90:68–77

Kullmann L (2007) Tree line population monitoring of Pinus sylvestris in the Swedish Scandes, 1973–2005: implications for tree line theory and climate change ecology. J Ecol 95:41–52

Lloyd AH, Fastie CL (2002) Spatial and temporal variability in the growth and climate response of treeline trees in Alaska. Climatic Change 52:481–509

Masek JG (2001) Stability of boreal forest stands during recent climate change: evidence from Landsat satellite imagery. J Biogeogr 28:967–976

Payette S, Fortin M, Gamache I (2001) The subarctic forest-tundra: the structure of a biome in a changing climate. BioScience 51(9):709–718

Shiyatov SG (2003) Rates of change in the upper tree line ecotone in the Polar Ural Mountains. Pages News 11:8–10

Suarez F, Binkley D, Kaye MW (1999) Expansion of forest stands into tundra in the Noatak National Preserve, northwest Alaska. Ecoscience 6:465–470

WMO (World Meteorological Organization) (2002) WMO statement on the status of the global climate in 2002, vol 684, WMO Press Release. WMO, Geneva

Part II
Hydrosphere

Chapter 9
Remote Sensing of Spring Snowmelt in Siberia

A. Bartsch, W. Wagner, and R. Kidd

Abstract Active as well as passive spaceborne sensors can be used to monitor spring snowmelt on regional to continental scale. Change detection methods are used to determine dates related to the thaw period. They comprise initial thaw, primary thaw, start of diurnal thaw/refreeze period, mean date of thaw, end of thaw and start of greening-up of vegetation. Only the latter is determined by use of passive optical sensors and combines measurements of visible and infrared radiation. All other approaches use microwave data. Some instruments such as the scatterometer Seawinds on QuikScat and the radiometer AMSR-E on Aqua make several measurements per day allowing the detection of diurnal thaw and refreeze, which is characteristic of the spring snowmelt period in northern latitudes. A specific, diurnal difference approach developed for QuikScat allows the determination of the length of the final period of diurnal thaw/refreeze. This duration and the spatial dynamics are closely linked to surface hydrology and ecosystem processes.

Keywords Snowmelt • Active microwaves • Satellite data

9.1 Introduction

Snowmelt dynamics play an essential role in the hydrological cycle of northern latitudes. The entire Siberia is seasonally covered by snow. Due to its vastness the timing varies by several months within this region. It instantaneously impacts not only surface hydrology and the energy budget but also terrestrial biota and thus the carbon cycle. Global and climate variability can be monitored by analyses of temporal evolution of snow cover (Scherer et al. 2005).

Land surface phenology is the integral of a range of processes including snowmelt, changes in soil wetness, and vegetation development patterns (White and Nemani 2006). An evolution towards earlier snowmelt in the northern hemisphere has been determined from different satellite data records starting in the 1980s (Dye and Tucker 2003; Smith et al. 2004).

Remote sensing can deliver spatially continuous information which is especially important in regions with sparse ground measurement networks and which are prone to climate change. The extent of snow cover can be monitored with all optical sensors over various scales as long as the cloud coverage is limited. Microwave systems overcome the problem of both cloud cover and requirement for daylight. They have been used with increasing confidence for snow cover monitoring over the past decade. Compared to optical systems some microwave sensors even allow several acquisitions per day. Diurnal changes within the snow pack can be observed. This allows detailed capture of spring freeze/thaw over large areas such as Siberia. Additional parameters of interest are the snow water equivalent (SWE) and wet/dry state (Sokol et al. 2003).

Depending upon the application type and on the temporal resolution of the satellite data specific melt related events are selected and termed. Snow coverage extent and state is usually investigated for hydrological purposes. Vegetation re-growth after snowmelt, on the other hand, is chosen for phenological studies. As optical datasets are more suitable for vegetation mapping they are used for the latter purpose.

Within this chapter a range of sensors and approaches suitable for snowmelt monitoring on regional to continental scale are discussed. Special emphasis is given to the scatterometer, which is an active microwave system. In Section 9.4 an approach based upon adaptive thresholding of QuikScat data, using diurnal differences, is detailed followed by a comparison of the results of three different thaw detection methods. Some impacts of thaw timing on the environment are revealed using this kind of data. River runoff patterns are directly related to basin area that undergoes melt at a certain time (Section 9.6.1). During the melt period itself, CO_2 is released due to heterotrophic soil respiration. Uptake only exceeds release when snowmelt has ceased and plant growth starts (Section 9.6.2). The drainage of melt water from some peat lands in the permafrost transition zone is relatively slow. They can be identified by the use of coarse and medium scale active microwave data (Section 9.6.3).

9.2 Sensors

Medium to coarse resolution sensors (300–50 km) can provide the necessary temporal coverage for mapping on regional to continental scale. Both passive and active systems can be used for the purpose of final snowmelt detection. They cover wavelengths from 0.4 µm to 5.6 cm (visible light, near infrared and microwaves, Table 9.1).

Table 9.1 Selected frequencies/wavelengths employed for detection of end of spring snowmelt on regional to continental scale

Sensor type	Example studies	Wave length	Example instruments	Band names
Scatterometer (active)	Wismann (2000)	5.3 cm	ERS1/ERS2	C
	Nghiem et al. (2001)		Metop ASCAT	
	Kimball et al. (2001)	2.5 cm	Seawinds QuikScat	K_u
	Bartsch et al. (2007)			
Radiometer (passive)	Smith et al. (2004)	1.5 cm	AMSR-E (Aqua)	(19V/H)
	Grippa et al. (2005a, b)		SSM/I (DMSP)	
	Ashcraft and Long (2006)		SSMR (Nimbus 7)	
	Tedesco (2007)	0.8 cm	AMSR-E (Aqua)	(37V/H)
	Ramage et al. (2007)		SSM/I (DMSP)	
			SSMR (Nimbus 7)	
	Delbart et al. (2005)	1.63–1.67 µm	MODIS (Terra)	B6
	Hall and Riggs (2007)	1.58–1.75 µm	SPOT-VGT	SWIR-B4
		0.78–0.89 µm	SPOT-VGT	NIR-B3
		0.62–0.67 µm	MODIS (Terra, Aqua)	B1
		0.61–0.68 µm	SPOT-VGT	Red-B2
		0.54–0.56 µm	MODIS (Terra, Aqua)	B4
		0.46–0.48 µm	MODIS (Terra, Aqua)	B3
		0.43–0.47 µm	SPOT-VGT	Blue-B0

Optical multispectral instruments are commonly used across all scales of resolution. They record surface emissions in the visible and near infrared part of the electromagnetic spectrum. A major disadvantage is the cloud cover dependency. Therefore microwave techniques are preferred although they are of coarser spatial resolution. Passive microwave sensors measure the naturally emitted microwave radiation. Scatterometers are active instruments which emit radiation and record the received amount of backscatter from the earth surface. The electrical properties of snow are an important parameter considering interaction with microwaves and vary considerably between dry and wet conditions, making these sensors especially suitable for monitoring snow dynamics.

In the category of optical multispectral instruments, in particular AVHRR (Advanced Very High Resolution Radiometer) sensors have been used in the past to monitor vegetation re-growth patterns over long time periods (since 1982) and large regions (Myneni et al. 1997; Delbart et al. 2006). The spatial resolution of 1 km is improved with MODIS (Moderate Resolution Imaging Spectroradiometer, on Terra and Aqua platforms, 500 m, since 2000/2002) and similarly MERIS (ENVISAT, 300 m, launched in 2002). The suite of SPOT platforms and sensors also offer long term monitoring possibilities. A series of satellites have been launched starting in 1986. SPOT-VEGETATION (VGT, 10 × 10 km) products are available since 1998 on a regular and frequent basis.

Commonly used passive microwave sensors are the Special Spectral Microwave Imager (SSM/I) and the Advanced Microwave Scanning Radiometer (AMSR-E). They measure the brightness temperature T_b emitted from the ground in Kelvin. T_b increases as a result of snowmelt. These sensors each cover several wavelengths each. The SSM/I altogether provides seven different frequency polarization combinations. Snow melt can be derived by either solely using the 19 V (19.35 GHz) channel (e.g. Ashcraft and Long 2006) or a combination of 19 H and 37 V (e.g. Abdalati and Steffen 1997; Smith et al. 2004; Ashcraft and Long 2005; Grippa et al. 2005a, b). Similar channels are available with the Scanning Multichannel Microwave Radiometer (SMMR, Nimbus-7) and AMSR-E (Aqua platform).

Scatterometers are active microwave systems recording measurements of the backscatter usually expressed as the backscatter coefficient σ^0 in dB. Whereas T_b increases with snowmelt the backscatter coefficient σ^0 often strongly decreases if an originally frozen dry surface starts to thaw. The ERS1/2 scatterometer (50 km) and the ASCAT instrument (25 km) onboard the recently launched METOP satellite (2006) operate in C-Band and the Seawinds QuikScat (25 km) in K_u-band. Frozen surface conditions can be captured with C-band (5.3 cm; Wismann 2000) as well as K_u-band (2.1 cm). The latter however has been determined most suitable for the detection of snowmelt (e.g. Kimball et al. 2004a, b) due to good coverage and sensitivity to snow grain size. C-band is less sensitive to liquid water in snow than K_u-Band (Ashcraft and Long 2006; Ulaby et al. 1986).

SAR (Synthetic Aperture Radar) systems offer higher spatial resolution acquisitions than scatterometer ranging from tens to hundreds of meters. SARs typically operate at 1–12 GHz (P, L, C, X-bands). They are suitable for monitoring at local scale (Nagler and Rott 2000) but have mostly only monthly repeat cycles. ScanSAR

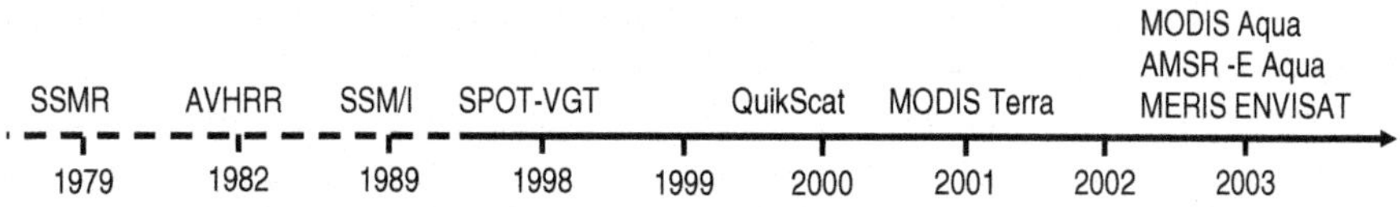

9 Remote Sensing of Spring Snowmelt in Siberia

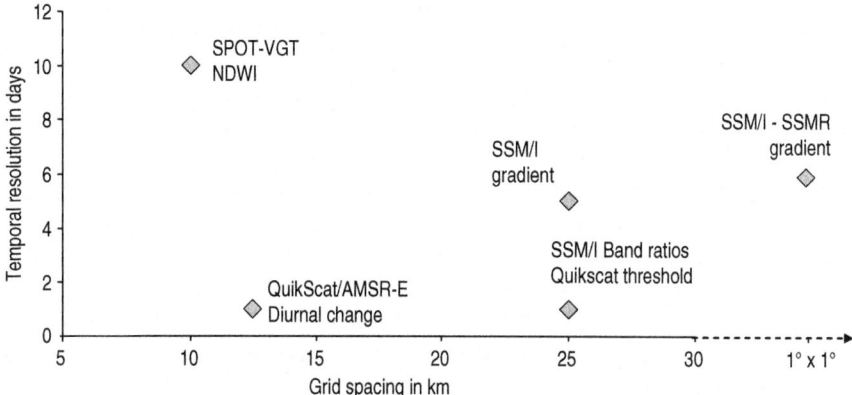

Fig. 9.1 Availability of regional spring snowmelt/greening up products from active and passive instruments

Fig. 9.2 Spatial and temporal resolution of regional to continental 'end of snowmelt' products

modes offer better than weekly sampling with medium resolution and larger swaths (e.g. ENVISAT ASAR WS with 1,510 m). They could therefore be used on regional scale for estimation of snow covered area (Luojus et al. 2007).

The longest records are available from SSM/I (Fig. 9.1) but regional snowmelt products are of coarse spatial and temporal resolution. The end of snowmelt is captured in higher detail (daily, 12.5 km) by sensors which only became available after 1999 (Fig. 9.2).

9.3 Methods for Detection of Seasonal Snow Cover

The general method for detection of snowmelt or start of vegetation re-growth respectively is change detection. Most methods are based on thresholds which are applied on the direct measurements of a specific band, on combinations of different frequencies, or on differences between acquisitions which are independent over time. For the best results local conditions need to be considered, although global thresholds are used within most methods.

The date of onset of greening up can be derived from indices such as NDWI (Normalized Difference Water Index) time series based on SPOT-VGT data (Delbart et al. 2005) and also NDVI (Normalized Difference Vegetation Index) or

NDSI (Normalized Difference Snow Index). The NDWI is based on a combination of near infrared and shortwave infrared (Gao 1996) compared to the NDVI which uses near infrared and visible red (Tucker 1979). The NDSI, which was developed for the detection of snow, is the normalized difference between visible blue and shortwave infrared (Hall et al. 1998). Time series of NDVI derived from AVHRR have already revealed that greening up dates occur earlier since those in the 1980s (e.g. Myneni et al. 1997), lengthening the active growing season. Most green-up studies use 10 day average data which may cause large uncertainties.

Snow parameters themselves are mostly derived from passive microwave data such as SSM/I. Changes in the combination of 19 H and 37 V allow for analysis of snow thaw on 5-day (Grippa et al. 2005a, Section 4.1.5) to 6-day time steps (Smith et al. 2004) on regional scale over non-glaciated terrain. Snow depth (Grippa et al. 2005a, b; Ashcraft and Long 2006, Biancamaria et al. in press) and to some extent the Snow Water Equivalent, SWE (Rawlins et al. 2007) can be estimated along with a snowmelt date. Both SSM/I derived snowmelt data and snow depth can be correlated with river runoff observed at certain time periods (Grippa et al. 2005b). Daily observations with SSM/I are used over glaciated terrain (Tedesco 2007). QuikScat time series have also been investigated on a daily basis (Kimball et al. 2001, 2004a, b, Bartsch et al. 2007a) but for seasonal snow cover. Changes in backscatter are not only detectable at the beginning and end of snowmelt but also the primary or major thaw can be determined (Kimball et al. 2001). The snowmelt period can thus even be split up into specific stages. In Kimball et al. (2004a) the onset of the growing season (assumed to be equal to the end of snowmelt) is determined by a threshold based approach and reports a good correlation with the onset of the growing season as determined by the start of xylem sap flow in pine trees. Anomalies of mean primary thaw correspond to variation in date of the seasonal CO_2 source-sink switch over the pan-Artic drainage basin (McDonald et al. 2004).

Diurnal differences are investigated in a range of studies since they indicate exactly when melt water is released from the snow pack. These single days of thaw and refreeze are summed up to obtain the number of melt days, a method which has been specifically used with passive microwave systems over large ice caps such as the Greenland ice sheet (Ashcraft and Long 2005, Tedesco 2007) and for mass balance studies over smaller ice caps (Wang et al. 2005). Ashcraft and Long (2006) compare six different snow melt detection methods which use constant thresholds for both scatterometer and radiometer measurements over the Greenland ice sheet. Results from different radiometer methods agreed well with each other as compared to ERS scatterometer observations and the diurnal difference method proposed by Nghiem et al. (2001) using Seawinds on QuikScat. By combination of ascending and descending orbits, thaw and refreeze can also be captured with SSM/I radiometer (Ramage and Isacks 2002; Smith et al. 2004; Tedesco 2007) and AMSR-E (Ramage et al. 2007) allowing the detection of incipient thaw timing as well as the number of melt days on glaciated terrain. SSMR provides diurnal acquisitions every other day. Data from this sensor are available from 1978 until 1987. Therefore long time series of spring thaw and autumn freeze ($1° \times 1°$, appr. 110 km at mid latitudes, Smith et al. 2004) can be derived in combination with SMM/I.

9 Remote Sensing of Spring Snowmelt in Siberia

The actual number of dates of snow thaw is of most interest for glacier mass balance studies but the final disappearance of snow together with the length of spring thaw is required in regions with seasonal snow cover. Clusters of consecutive days of diurnal cycling of freeze/thaw are characteristic for the final snowmelt period in boreal and tundra environments (Bartsch et al. 2007a). The start, end and duration of such periods give insight into spring CO_2 emissions and river runoff behaviour. The following sections provide details on an approach for the active microwave instrument Seawinds which uses adaptive thresholding in order to account for location specific noise. Results from central Siberia are compared with products from other snowmelt timing derivation methods.

9.4 Diurnal Differences over Siberia

SeaWinds Quikscat (K_u-band) measurements are available since 1999. The first entire snowmelt period over the northern hemisphere is covered in 2000. Large changes in backscatter between morning and evening acquisitions are characteristic for the snowmelt period, when freezing takes place over night and thawing of the surface during the day. A change from volume to surface scattering occurs in the case of melting. This may cause changes up to 6 dB (Kimball et al. 2004b). When significant changes due to freeze/thaw cycling cease, closed snow cover also disappears (Bartsch et al. 2007). For the identification of melt days over permanently snow or ice covered ground, only evening measurements are considered (Ashcraft and Long 2006). Diurnal differences (Bartsch et al. 2007) on the other hand are calculated for the delimitation of the final spring snowmelt period. The exact day of year of beginning and end of freeze/thaw cycling can be clearly determined with consideration of long-term noise (Bartsch et al. 2007a) in order to exclude unnatural effects, changes in soil moisture and snow pack characteristics. Therefore a location specific noise estimate s_σ is determined from analysis of long-term backscatter time series. Only then can significant diurnal differences be identified over a variety of environments.

Backscatter at each location in the northern latitudes is measured several times with spatially and temporarily varying footprints during a single day by QuikScat. As this occurs in irregular intervals the exact number of measurements in the morning and in the late afternoon/evening needs to be taken into account. The diurnal difference $\Delta\sigma^0$ is calculated by averaging all measurements acquired during morning and evening passes respectively and then calculating the difference (Bartsch et al. 2007a). Since the estimated noise of σ^0, s_σ is known, the standard deviation of $\Delta\sigma^0$, s_Δ, can be directly estimated. To identify diurnal changes which are significant at the 99 % confidence interval, $\Delta\sigma^0$ needs to exceed three times s_Δ. Figure 9.3a shows the diurnal differences $\Delta\sigma^0$ at Zotino. Significant values are highlighted.

A simple three-step approach is pursued to determine the most significant period of freeze/thaw indicators within a time series:

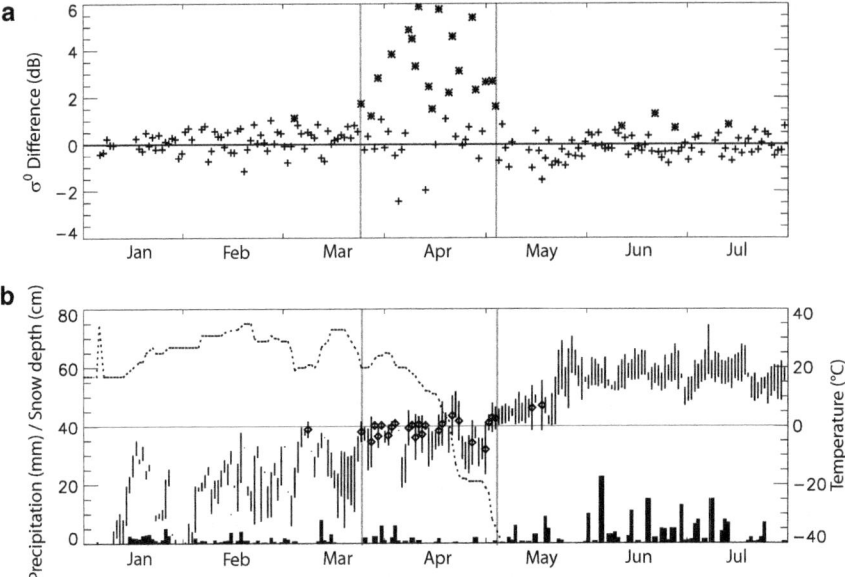

Fig. 9.3 Time series example for Zotino in 2000 (60.75° N, 89.38° E; adapted from Bartsch et al. 2007a): (**a**) backscatter difference between morning and evening measurements, significant values are indicated with*; (**b**) meteorological data including air temperature, precipitation and snow depth (WMO512); QuikScat determined start and end of thaw period is shown in both figures with *vertical grey lines*

1. For calculations on the northern hemisphere only periods between 1st of January and 1st of July are considered.
2. Indicators that are found initially within 10 days of each other are grouped into periods using temporal filtering.
3. The onset of the major and, usually final, snowmelt period is determined as the indicator period containing the greatest number of indicators. The onset of snowmelt is determined as the first day of this period.

Thus the duration of the snowmelt as detected from QuikScat corresponds to periods of significant diurnal change of surface conditions due to freezing and thawing. Example maps for Russia (2001) are shown in Fig. 9.4a. The number of days used for filtering (2) is crucial and therefore its sensitivity is assessed. A number of 20 days is used to assess where the results are reliable and especially at which locations multiple periods of significant diurnal thaw/refreeze occur (Fig. 9.4b). 87% of the investigated area has not been affected in 2001. No further diurnal thaw/refreeze cycling occurred for 92% of all grid points. The new maps allow a correction of the end of thaw maps if necessary.

Snowmelt trends have been derived over central Siberia for a time series covering 2000–2005 (corrected for multiple thaw periods) and are shown in Fig. 9.5. This region covers most of the Yenisei River Basin and the entire Siberia II project area.

There has been no trend in the end of snowmelt in the boreal continental and sub-arctic moderate regions. Start of thaw, however, seems to start earlier in the boreal continental zone lengthening the thaw duration. Contrary trends can be observed for the sub-arctic severe and arctic environments. The latter shows similar trends as in the severe boreal region where snow melt ceases earlier with at the same time shorter thaw duration. Thaw starts earlier and finishes later in the northern part of

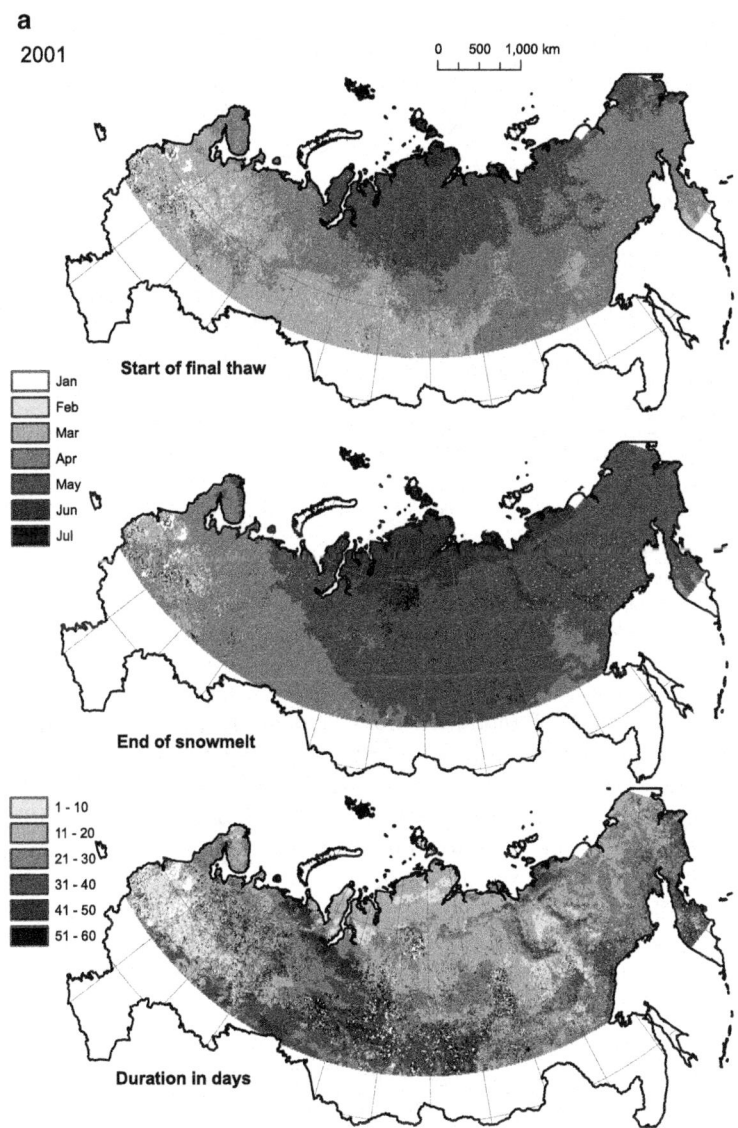

Fig. 9.4 (a) Spring freeze/thaw start, end and duration for Russia 2001 north from 55°N

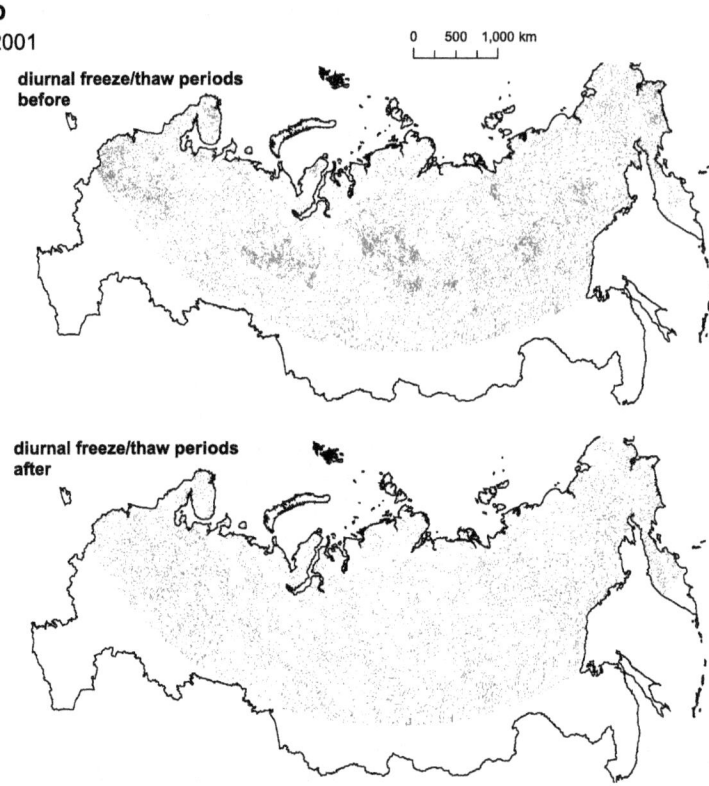

Fig. 9.4 (continued) (**b**) Sensitivity of temporal filtering parameter (10 days) regarding multiple thaw/refreeze periods: detected multiple freeze/thaw periods within 20 days before and after (*Color version available in Appendix*)

the sub-arctic severe zone (northeast from Putorana plateau). This may impact spring and early summer runoff, CO_2 fluxes and phenology (see Section 9.6). A trend of earlier mean thaw timing was detected by Smith et al. (2004) from SSM/I-SSMR for an analysis period from 1988–2002. The approach did neither differentiate between snow and soil nor between start and end of spring thaw. The results are therefore difficult to compare. The SSM/I derived primary thaw (McDonald et al. 2004) shows similar trends for 1988–2001. This indicator can, in cases, be seen similar to the start of diurnal thaw/refreeze. The Quikscat analyses would then confirm a continuation of this trend. Negative spring trends for NDVI in the sub-arctic southern tundra are observed with SPOT-VGT from seasonal mean mosaics 1998–2005 (Hüttich et al. 2007). This agrees with the later snow melt detected with QuikScat in the sub-arctic severe region only. Positive spring NDVI trends for taiga forest (~55–60°N) occur together with an earlier onset of thaw, a longer thaw period but not a change in end of snowmelt timing. Grippa et al. (2005b) found that in most areas of Siberia lower growing season NDVI (from AVHRR) corresponds

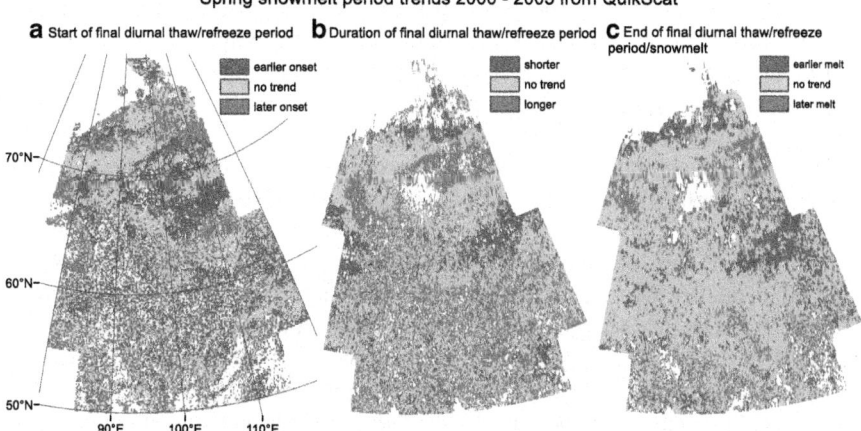

Fig. 9.5 Trends for the snowmelt period 2000–2005 from Seawinds QuikScat in central Siberia: (**a**) start, (**b**) duration, and (**c**) end of diurnal thaw/refreeze period/snowmelt with landscape groups (source: IIASA) within the Siberia II project area; 1 – arctic, 2 – sub-arctic severe, 3 – sub-arctic moderate, 4 – boreal severe, 5 – boreal continental, 6 – sub-boreal (steppe) (*Color version available in Appendix*)

to later snowmelt dates as derived from SSM/I. In the northern Taiga and Steppe however this is reversed. It is suggested that later snowmelt keeps soil water content at a suitable level for plant growth. The combination of AVHRR and SPOT-VGT allows investigations of green-up date for more than 20 years, starting in 1982 (Delbart et al. 2006). The greening up date advanced by 8 days over Northern Eurasia during the first 10 years and was delayed by almost 4 days during the following decade. For central Siberia, the QuikScat results suggest a continuation of delayed snowmelt in the severe sub-arctic (Tundra) only. Trend patterns coincide with landscape group boundaries in some cases. The landscape groups are based on climatology and vegetation. This suggests that there may be a relationship between the accuracy of the snowmelt detection and land cover.

9.5 Comparison of Different Methods over Central Siberia

Within the Siberia II project (Schmullius et al. 2003) spring timing was assessed using three different approaches and datasets. The Siberia II project dealt with multi-sensor concepts for greenhouse gas accounting in central Siberia (Chapter 1, this book). The investigated region stretches from the Taymir peninsula and Putorana Plateau in the North to the Sayan Mountains and Lake Baikal in the South. It covers approximately 3 Mio km² representing tundra, boreal forest and steppe grassland biomes. The exact boundaries are determined by administrative borders (Krasnoyarsk Kray, Irkutsk Oblast and Khakass Republic). The majority of the area belongs to the Yenisei basin. The eastern part experiences a more continental climate than the western part (Stolbovoi and McCallum 2002) which is also characterized by lower elevation a.s.l. The major bioclimatic zones (as used by Shvidenko et al. 2000) in the boreal forest biome are southern, middle and northern taiga (see Figure 1 in Bartsch et al. 2007a).

The different approaches followed within the project identified three different phases of the spring thaw, specially identifying the end of snow melt (Bartsch et al. 2007), free snow condition (Grippa et al. 2005a) and the onset of green up/budburst (Delbart et al. 2005). The end of snowmelt itself was derived from passive (SSM/I) and active microwave data (QuikScat). For the latter the method presented in Section 9.4 was developed. This provides daily values on 25 km resolution (re-gridded to 12.5 km). A spectral gradient has been derived from SSM/I channels (Grippa et al. 2005a) and a constant threshold applied in order to detect snow free conditions. Data have been averaged over 5-day periods and spatial resolution is 50 km (25 km samples). Where the difference between 19 and 37 GHz (both horizontally polarized) was below 3 K over a time period of at minimum three consecutive

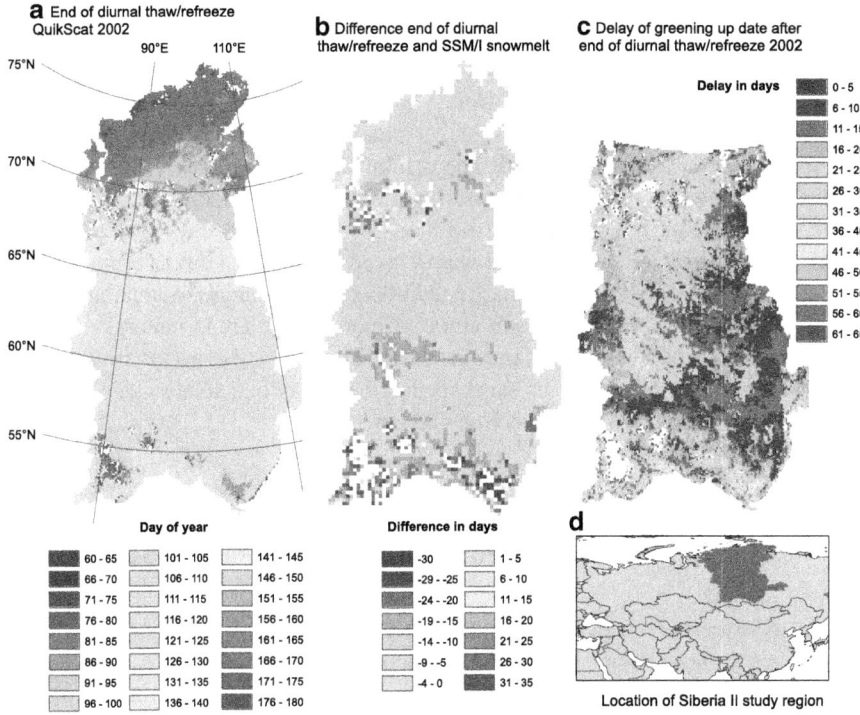

Fig. 9.6 Comparison of spring snowmelt and phenology products from the Siberia II project for 2002: (**a**) end of diurnal thaw/refreeze period from QuikScat (Bartsch et al. 2007a); (**b**) difference between end of diurnal thaw/refreeze period and end of snowmelt from SSM/I (Grippa et al. 2005a, b); (**c**) delay of greening up (Delbart et al. 2005) after end of diurnal thaw/refreeze period; (**d**) location map (Color version available in Appendix)

pentads it is assumed that snow has melted. The third approach is based on optical multispectral data from SPOT-VGT. The Normalized Difference Water Index (NDWI) was shown to be most suitable for the determination of greening up following the disappearance of snow (Delbart et al. 2005). The 10-day composites have 10 × 10 km pixel spacing.

Direct comparison of the three different snowmelt and phenology products, as shown in Fig. 9.6, is impeded by the varying temporal and spatial resolution. Allowing for a ± 10 day variation, 80 % of the Siberia II study region shows agreement between QuikScat and SSM/I end of snowmelt (Fig. 9.6a, b). Deviations beyond that threshold tend to be negative, indicating that SSM/I snowmelt is earlier than the end of daily freeze/thaw cycling. This occurs mostly in the southern mountainous region. Here, the offset (3–5 weeks) between freeze/thaw and budburst is also larger than average (1–2 weeks; Fig. 9.6c). Similar offsets can be observed for the Putorana Plateaux in between 65 and 70°N. As the greening up dates are derived from 10-day composites it can be concluded that the end of diurnal thaw and refreeze from QuikScat is also representative for budburst within Taiga forest with moderate terrain. Uncertainties occur in the steppe and agriculture areas (Bartsch et al. 2007a).

9.6 Impact of Thaw Timing and Duration on the Environment

The spatial and temporal patterns of thaw directly condition surface hydrology in northern latitudes. River runoff is largest in spring and early summer due to snowmelt. At the same time heterotrophic soil respiration starts and CO_2 is released to the atmosphere. Spring thaw also impacts the surface hydrology in boreal peat lands.

9.6.1 Runoff

The variability of runoff in Siberian rivers is important for the freshwater influx into the Arctic Ocean. River discharge has increased by 7% from 1936 to 1999 in Eurasian rivers (Peterson et al. 2002). Anomalies relate to the NAO (Northern Atlantic Oscillation) index, which has increased since the 1970s. Although there are uncertainties in the river discharge measurements of the investigated Russian rivers, this increase could be confirmed and quantified (Shiklomanov et al. 2006). An approximate increase of 2 ± 0.4 km^3/year has been estimated for the six largest river basins. The fresh water inflows from rivers which drain into to the Arctic Ocean are important for regulation of the global thermohaline circulation (Aagaard and Carmack 1989).

The three largest watersheds in Siberia are the Ob River (3 Mio km^2), Yenisei River (2.5 Mio km^2) and Lena River (2.5 Mio km^2). A high correlation of SSM/I derived average snowmelt date with river runoff in May has been determined for the

Ob basin (Grippa et al. 2005b). The later the snow has completely melted the lower the observed discharge in May (1989–2001). The correlation coefficient drops for June and July. June runoff is more related to the actual snow depth during the end of the winter. A shift in discharge patterns is recorded for the Yenisei River (Yang et al. 2004). Stream flow decreased in the early melt period (1935–1999) and increased in the late melt period. It is suggested that this is related to a delay in snow cover melt. The impact of permafrost thaw on the Lena River is discussed in Chap. 10.

Rawlins et al. (2005) used the QuikScat threshold method from Kimball et al. (2004ab) to detect the incipient thaw timing. It has been compared to modelled snow water increase over a number of North American basins. Correlations of up to 0.75 were found in Arctic basins. Almost 50% of the Pan-Arctic area shows a difference of less than a week between primary thaw from QuikScat and modelled snow water increase.

Northern basins which have been investigated with QuikScat using the diurnal differences are the Lena (station Kyusyur, 2.4 Mio km^2) and Mackenzie (station Arctic Red River, 1.65 Mio km^2). The comparison was made for the year 2000 (Bartsch et al. 2007b). Both rivers cross different environments and their northern catchment area is located within permafrost. For most of the Lena basin snowmelt starts in April and ceases in May (Bartsch et al. 2007b) which is similarly as can be observed for the Ob River basin (Grippa et al. 2005b). In the Mackenzie this period starts already in March but also finishes only in May in the north and northwest. Effects from topography and latitude are reflected in this pattern. The maximum runoff in the Lena-Kyusyur basin is clearly related to snowmelt as it occurs when snowmelt has finished over the entire basin. The runoff starts to increase only after 70% of the contributing area is mostly snow free. The temporal offset between maximum area of snowmelt and maximum runoff was more than a month in 2000. During the second half of April 80% of the basin underwent daily freeze/thaw cycling at the same time.

Similarly to the Lena at Kyusyur, the runoff at Mackenzie/Arctic-Red-River starts to increase above the winter level when 70% of the upstream area is snow-free (Fig. 9.7).

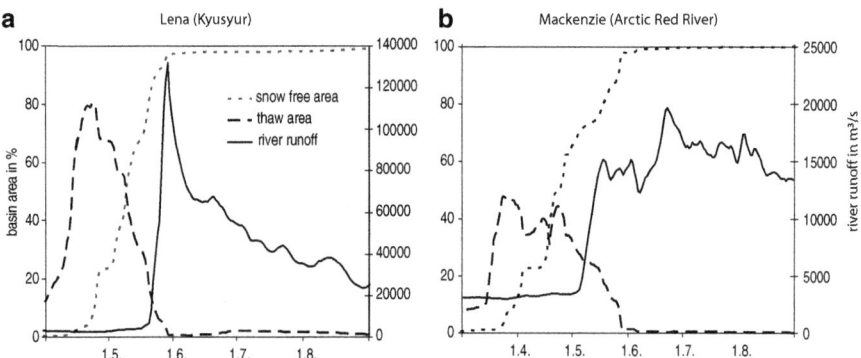

Fig. 9.7 Comparison of Lena (**a**) and Mackenzie (**b**) 2000 melt duration with river runoff (source: GRDC): percentage of snow free area (*short dashes*), the area that undergoes thaw in % (*long dashes*) and daily river runoff (*solid line*)

Apart from that, the Mackenzie basin shows different snowmelt and runoff behaviour throughout the spring and summer season. The area which undergoes snowmelt at the same time does not exceed 50% at any point in time due to topography and latitude. Snow melt area peaks in the end of March, which is 50 days before the first peak in runoff. A second peak can be observed 1 month later in the end of April, approximately 50 days before the second and major peak of runoff at the Arctic-Red-River confluence. For more than a month water from snowmelt is held back in the basin. The Quikscat detected end of diurnal thaw/refreeze cycling over the entire Mackenzie basin corresponds to the time period when the inundation within the wetland of the Peace Athabasca Delta (Temimi et al. 2007) is highest.

9.6.2 CO_2 Fluxes and Phenology

It is known that there is a relationship between soil thaw (Goulden et al. 1998), snowmelt (Aurela et al. 2004) and CO_2 fluxes from ground measurements. During spring two dates are important to carbon balance: firstly, when carbon starts to be released above winter baseline due to onset of heterotrophic soil respiration and secondly, the date at which the ecosystem switches from a net source to a net sink. This information can be retrieved from eddy flux tower data. CO_2 flux data have been collected as part of a measurement campaign by TCOS Siberia (Shibistova et al. 2002a, b, Chapter 5.1). The data set also contains air and soil temperature measurements. The site with the most complete CO_2 flux record within the central Siberia that also starts around onset of spring thaw is Zotino (60.75° N, 89.38° E). It is located within the boreal forest biome and also represents peat land and discontinuous permafrost (Stolbovoi and McCallum 2002). The comparison of daily cumulative eddy flux tower data for CO_2 with the onset and end of the freeze/thaw cycle period, calculated with the diurnal difference method, at Zotino is shown in Fig. 9.8 for 2000. For 2001 the diurnal difference freeze/thaw indicator determines onset of major thaw more than 2 weeks earlier than in 2000 (Bartsch et al. 2007b). The end of significant freeze/thaw cycling is much more stable during the analysed years and occurs on day of year 125 ± 1 day at Zotino. The switch from positive to negative accumulated fluxes in 2000 and 2001 was close to the end of thaw date (1 day before in 2000 and 6 days later in 2001). The onset of the spring thaw period matches with the first days with fluxes above the late winter base line in 2001. The increase phase of fluxes in 2000 is not distinct (Fig. 9.8). An early short thaw period already occurred before the final thaw. This event was captured by the diurnal difference method but was correctly not determined as start of the major thaw period.

For the years 2000–2005 a general trend towards earlier onset and a longer snowmelt period has been observed for the Taiga biome (Fig. 9.5). This would result in longer periods of CO_2 release.

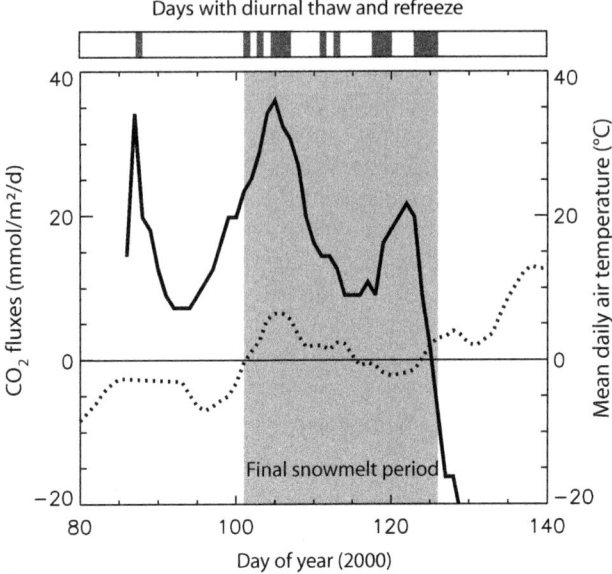

Fig. 9.8 Days with significant backscatter changes due to diurnal thaw and refreeze at Zotino (60.75° N, 89.38° E), smoothed (5 days) spring daily accumulated CO_2 fluxes (*solid line*) and mean daily air temperature (*dashed line*) in 2000 (source: TCOS Siberia). Duration of major and final melt period is indicated by *grey shading*. Adapted from Bartsch et al. (2007)

9.6.3 Peat Land Monitoring

Boreal peat lands are a type of wetlands and they show significant differences in surface properties during the spring thaw period compared to the summer depending on their morphology and the underlying permafrost. Temporary open water surfaces consisting of melt water can be especially found within the permafrost affected regions. If the timing of the melt period is known such wetlands can be identified and distinguished from other peat lands. This is important since boreal peat lands cover large areas in the northern hemisphere and thus play an important role as carbon storage and also for methane release. They feature as an important source of radiative forcing (Friborg et al. 2003). Northern peat lands also exert strong control on the hydrological regime (Pietroniro et al. 2005). The West Siberian lowland (WSL) is the largest wetland area of the world covering more than 2.7 Mio km² (Keddy and Fraser 2005). The actual size of wetland ecosystems accounts for 50% of the WSL and approximately 22.8 Mio tons of carbon are accumulated every year (Solomeshch 2005). Several of the largest connected peat land systems are part of the WSL. Most of the area is drained by the Ob and Yenisei River. Both rivers experienced continued increasing discharge in the recent past (Peterson et al. 2002)

By applying a threshold method, many boreal peat lands can be identified with data from SAR (Synthetic Aperture Radar) operating in C-Band due to their relatively high backscatter compared to the surroundings (Bartsch et al. 2007c). Soil water content, which contributes largely to the dielectric properties, is high and thus the backscatter is also high. The occurrence of abundant permanent and temporary open water surfaces below the spatial resolution of the SAR is characteristic of the permafrost transition zone and impedes this simple mapping approach (Bartsch et al. in press). Time series analysis, however, shows that backscatter values are lower than in homogeneous peat land during the summer period. During the snowmelt period backscatter remains relatively low compared to other peat land and importantly the typically surrounding forest areas. The location specific start and end of the melt period as derived from Seawinds Quikscat can in this case be used to select the correct dates for ENVISAT ASAR Global Mode acquisitions (1 km resolution) from this time period and allow for improved peat land identification. The duration of snowmelt varies between 1 and 2 months from year to year.

Figure 9.9 shows an example for a zone of raised string bogs for the year 2006. This wetland type represents almost half of all mires in the WSL (Solomeshch 2005). The area investigated covers a region from 58° N to 63° N and from 72° E to 75° E (Fig. 9.9 inset map). The Ob River traverses the area from East to West at approximately 61° N also passing the town Surgut. It covers the middle taiga zone south from the Ob River with sporadic permafrost as well as the discontinuous zone to the north (Stolbovoi and McCallum 2002) which is in the northern taiga vegetation zone (Solomeshch 2005). Basic information on the distribution of peat land exists as a GIS database (Sheng et al. 2004). It is compared to the peat land classifications results from the ENVISAT ASAR approach using temporal scene selection from QuikScat derived thaw dates.

Figure 9.9d shows the classification result without the use of temporal scene separation with QuikScat. The new result (Fig. 9.9c) provides the missing information from regions with abundant surface water (northern part). The ENVISAT ASAR map and the combined ASAR/QuikScat map together contain 70% of the peat lands reported in the GIS database (Sheng et al. 2004) for the study area. The new wetland map contains the actual wet peat land without forest cover in 2006 only. This may cause the discrepancies beside geolocation errors (Bartsch et al. 2007c). Although ENVISAT ASAR Global Mode only provides data with 1 km spatial resolution they can be used with the sensor combination to map the large and inhomogeneous boreal peat lands.

9.7 Summary

Regional snowmelt products are available from a range of sensors staring from 1979. More recently launched satellites have instruments which provide increased coverage and temporal sampling intervals. Microwave sensor products vary from 12.5 km to approximately 110 km grid spacing. Although optical multispectral

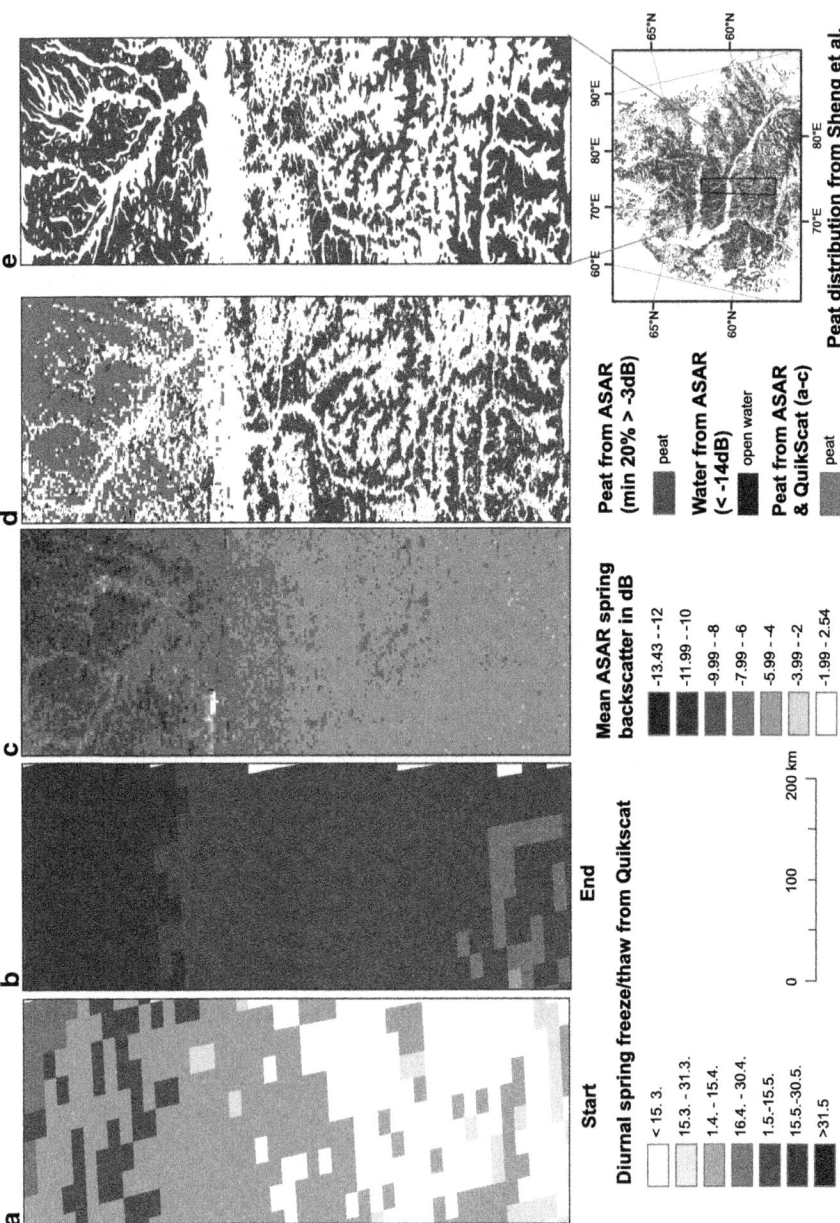

Fig. 9.9 Comparison of QuikScat freeze/thaw dates (2006) (**a**) onset, (**b**) end and (**c**) ENVISAT ASAR GM mean melt period backscatter (2006), (**d**) ASAR GM derived open peat land (July–Sept 2006, threshold method) and (**e**) peat distribution from the West Siberian Lowland database (Sheng et al. 2004). For location see inset map (*Color version available in Appendix*)

instruments are applicable for regional to global mapping with up to 300 m resolution (ENVISAT MERIS), continental scale snowmelt products are available so far of 10 km grid spacing only (SPOT-VGT). This is caused largely by the low number of revisit intervals and cloud cover dependency which demands temporal and spatial averaging.

K_u-band scatterometer and multi-band radiometers in the microwave spectrum can produce results which enable detailed investigation of the impact of snowmelt on the environment. The timing and specifically the temporal and spatial effects of snow thaw are related to hydrological and ecosystem processes. The monitoring of end of snowmelt alone provides insight only into a small part of the impacts of climate change on seasonal snow cover. Additional information is gained by extracting the entire duration of the final thaw period. The scatterometer Seawinds on Quikscat which is an active microwave instrument acquires the required data for the approach discussed in this chapter of detecting diurnal thaw/refreeze cycling.

References

Aagaard K, Carmack EC (1989) The role of sea ice and other fresh water in the arctic circulation. J Geophys Res 94(C10):14485–14498

Abdalati W, Steffen K (1997) Snowmelt on the Greenland ice sheet as derived from passive microwave satellite data. J Climate 10:165–175

Ashcraft IS, Long DG (2005) Differentiation between melt and freeze stages of the melt cycle using SSM/I channel ratios. IEEE Trans Geosci Remote Sens 43(6):1317–1323

Ashcraft IS, Long DG (2006) Comparison of methods for melt detection over Greenland using active and passive microwave measurements. Int J Remote Sens 27:2469–2488

Aurela M, Laurilla T, Tuovinen J-P (2004) The timing of snow melt controls the annual CO_2 balance in a subarctic fen. Geophys Res Lett 31:L16119

Bartsch A, Kidd R, Wagner W, Bartalis Z (2007a) Temporal and spatial variability of the beginning and end of daily spring freeze/thaw cycles derived from scatterometer data. Remote Sens Environ 1996:360–374

Bartsch A, Pathe C, Wagner W (2007c) Wetland mapping in the West Siberian Lowlands with ENVISAT ASAR global mode. Proceedings of the ENVISAT Symposium, Montreux 2007, ESA SP-636

Bartsch A, Wagner W, Rupp K, Kidd R (2007b) Application of C and Ku-band scatterometer data for catchment hydrology in northern latitudes. Proceedings of the 2007 IEEE International Geoscience and Remote Sensing Symposium. Barcelona, Spain, 23–27 July 2007

Bartsch A, Wagner W, Pathe C, Scipal K, Sabel D, Wolski P (2009) Global monitoring of wetlands – the value of ENVISAT ASAR global mode. J Environ Manage 90:2226–2233

Biancamaria S, Mognard NM, Boone A, Grippa M, Josberger EG (2008) A satellite snow depth multi-year average derived from SSM/I for the high latitude regions. Remote Sens Environ 112:2557–2568

Delbart N, Kergoat L, Le Toan T, Lhermitte J, Picard G (2005) Determination of phenological dates in boreal regions using normalized difference water index. Remote Sens Environ 97:26–38

Delbart N, Le Toan T, Kergoat L, Fedotova V (2006) Remote sensing of spring phenology in boreal regions: a free of snow-effect method using NOAA-AVHRR and SPOT-VGT data (1982–2004). Remote Sens Emviron 1001:52–62

Dye GD, Tucker CJ (2003) Seasonality and trends of snow cover, vegetation index and temperature in northern Eurasia. Geophys Res Lett 30(7):1405

Friborg T, Soegaard H, Christensen TR, Lloyd CR, Panikov NS (2003) Siberian wetland: where a sink is a source. Geophys Res Lett 30:2129–2132

Gao BC (1996) NDWI – a normalized difference water index for remote sensing of liquid water from space. Remote Sens Environ 58:257–266

Goulden ML, Wofsy SC, Harden JW, Trumbore SE, Crill PM, Gower ST, Fries T, Daube BC, Fan S-M, Sutton DJ, Bazzaz A, Munger JW (1998) Sensitivity of boreal forest carbon balance to soil thaw. Science 279:214–217

Grippa M, Kergoat L, Le Toan T, Mognard NM, Delbart N, L'Hermitte J, Vincente-Serrano SM (2005a) The impact of snow depth and snowmelt on the vegetation variability over Central Siberia. Geophys Res Lett 32:L21412

Grippa M, Mognard N, Le Toan T (2005b) Comparison between the interannual variability of snow parameters derived from SSM/I and the Ob river discharge. Remote Sens Environ 98:35–44

Hall DK, Riggs GA (2007) Accuracy assessment of the MODIS snow products. Hydrol Process 21:1534–1547

Hall DK, Foster JL, Verbyla DL, Klein AG, Benson CS (1998) Assessment of snow-cover mapping accuracy in a variety of vegetation-cover densities in central Alaska. Remote Sens Environ 66:129–137

Hüttich C, Herold M, Schmullius C, Egorov V, Bartalev SA (2007) Indicators of Northern Eurasia's land-cover change trends from SPOT-VEGETATION time-series analysis 1998–2005. Int J Remote Sens 28(18):4199–4206

Keddy PA, Fraser LH (2005) Introduction: big is beautiful. In: Fraser LH, Keddy PA (eds) The world's largest wetlands: ecology and conservation. Cambridge University Press, Cambridge, pp 1–10

Kimball J, McDonald K, Keyser A, Frolking S, Running S (2001) Application of the NASA Scatterometer (NSCAT) for Determining the Daily Frozen and Nonfrozen Landscape of Alaska. Remote Sens Environ 75(1):113–126

Kimball JS, McDonald KC, Frolking SE, Running SW (2004a) Radar remote sensing of the spring thaw transition across a boreal landscape. Remote Sens Environ 89:163–175

Kimball JS, McDonald KC, Running SW, Frolking SE (2004b) Satellite radar remote sensing of seasonal growing seasons for boreal and subalpine evergreen forests. Remote Sens Environ 90(2):243–258

Luojus KP, Pulliainen JT, Metsamaki SJ, Hallikainen MT (2007) Snow-covered area estimation using satellite radar wide-swath images. IEEE Trans Geosci Remote Sens 45(4):978–989

McDonald KC, Kimball JS, Njoku E, Zimmermann R, Zhao M (2004) Variability in springtime thaw in the terrestrial high latitudes: monitoring a major control on the biospheric assimilation of atmospheric CO_2 with spaceborne microwave remote sensing. Earth Interact 8(020):1–23

Myneni RB, Keeling CD, Tucker CJ, Asrar G, Nemani RR (1997) Increased plant growth in the northern high latitudes from 1981-1991. Nature 386:698–702

Nagler T, Rott H (2000) Retrieval of wet snow by means of multitemporal SAR data. IEEE Trans Geosci Remote Sens 38:754–765

Nghiem SV, Steffen K, Kwok R, Tsai W-Y (2001) Detection of snow melt regions on the Greenland ice sheet using diurnal backscatter change. J Glaciol 47(159):539–547

Peterson BJ, Holmes RM, McClelland JW, Vörösmarty CJ, Lammers RB, Shiklomanov AI, Shiklomanov IA, Rahmstorf S (2002) Increasing river discharge to the Arctic ocean. Science 298:2171–2173

Pietroniro A, Töyrä J, Loconte R, Kite G (2005) Remote sensing of surface water and soil moisture. In: Duguay CR, Pietroniro A (eds) Remote sensing in northern hydrology. AGU geophysical monograph, 163, 119–142

Ramage JM, Isacks BL (2002) Determination of melt onset and refreeze timing on southeast Alaskan Icefields using SSM/I diurnal amplitude variations. Ann Glaciol 34:391–398

Ramage JM, Apgar JD, McKenney RA, Hanna W (2007) Spatial variability of snowmelt timing from AMSR-E and SSM/I passive microwave sensors, Pelly River, Yukon Territory, Canada. Hydrol Process 21:1548–1560

Rawlins MA, McDonald KC, Frolking S, Lammers RB, Fahnestock M, Kimball JS, Vörösmarty CJ (2005) Remote sensing of snow thaw at the Pan-Arctic scale using the Seawinds scatterometer. J Hydrol 312:294–311

Rawlins MA, Fahnestock M, Frolking S, Vörösmarty CJ (2007) On the evaluation of snow water equivalent estimates over the terrestrial Arctic drainage basin. Hydrol Process 21:1616–1623

Scherer D, Hall DK, Hochschild V, König M, Winther J-G, Duguay CR, Pivot F, Mätzler C, Rau F, Seidel K, Solberg R, Walker AE (2005) Remote sensing of snow cover. In: Duguay CR, Pietroniro A (ed.) Remote sensing in northern hydrology. AGU Geophysical Monograph, 163, 7–38

Schmullius C, Hese S, Knorr D (2003) Siberia-II – a multi sensor approach for greenhouse gas accounting in northern Eurasia. Petermanns Geogr Mitt 147(6):4–5

Sheng Y, Smith LC, MacDonald GM, Kremenetski KV, Frey KE, Velichko AA, Lee M, Beilman DW, Dubinin P (2004) A high-resolution GIS-based inventory of the West Siberian peat carbon pool. Global Biogeochem Cycle 108:GB3004

Shibistova O, Lloyd J, Evgrafova S, Savushkina N, Zrazhevskaya G, Arneth A, Knohl A, Kolle O, Schulze ED (2002a) Seasonal and spatial variability in soil CO_2 efflux rates for a central Siberian Pinus sylvestris forest. Tellus B 54(5):552–567

Shibistova O, Lloyd J, Zrazhevskaya G, Arneth A, Kolle O, Knohl A, Asrekhantceva N, Shijneva I, Schmerler J (2002b) Annual ecosystem respiration budget for a Pinus sylvestris stand in central Siberia. Tellus B 54(5):568–589

Shiklomanov AI, Yakovleva TI, Lammers RB, Karasev IP, Vörösmarty CJ, Linder E (2006) Cold region river discharge uncertainty – estimates from large Russian rivers. J Hydrol 326:231–256

Shvidenko AZ, Nilsson S, Stolbovoi VS, Gluck M, Shchepashchenko DG, Rozhkov VA (2000) Aggregated estimation of the basic parameters of biological production and the carbon budget of Russian terrestrial ecosystems: 1. Stocks of plant organic mass. Russ J Ecol 31(6):371–378

Smith NV, Saatchi SS, Randerson JT (2004) Trends in high northern latitude soil freeze thaw cycles from 1988 to 2002. J Geophys Res 109:D12101

Sokol J, Pultz TJ, Walker AE (2003) Passive and active airborne microwave remote sensing of snow cover. Int J Remote Sens 24:5327–5344

Solomeshch AI (2005) The West Siberian lowland. In: Fraser LH, Keddy PA (eds) The world's largest wetlands: ecology and conservation. Cambridge University Press, Cambridge, pp 11–62

Stolbovoi V, McCallum I (2002) CD-ROM "Land Resources of Russia". International Institute for Applied Systems Analysis and the Russian Academy of Science. Laxenburg, Austria

Tedesco M (2007) Snowmelt detection over the Greenland ice sheet from SSM/I brightness temperature daily variations. Geophys Res Lett 34:L02504

Temimi M, Leconte R, Brissette F, Chaouch N (2007) Flood and soil wetness monitoring over the Mackenzie river basin using AMSR-E 37 GHz brightness temperature. J Hydrol 333:317–328

Tucker CJ (1979) Red and photographic infrared linear combinations for monitoring vegetation. Remote Sens Environ 8:127–150

Ulaby FT, Moore RK, Fung AK (1986) Microwave remote sensing: active and passive, vol III. From theory to applications. Artech House, Norwood

Wang L, Sharp MJ, Rivard B, Marshall S, Burgess D (2005) Melt season duration on Canadian Arctic ice caps, 2000–2004. Geophys Res Lett 32:L19502

White MA, Nemani RR (2006) Real-time monitoring and short-term forecasting of land surface phenology. Remote Sens Environ 104:43–49

Wismann V (2000) Monitoring of seasonal thawing in Siberia with ERS scatterometer data. IEEE Trans Geosci Remote Sens 38(4):1804–1809

Yang D, Ye B, Kane DL (2004) Streamflow changes over Siberian Yenisei river basin. J Hydrol 296:59–80

Chapter 10
Response of River Runoff in the Cryolithic Zone of Eastern Siberia (Lena River Basin) to Future Climate Warming

A.G. Georgiadi, I.P. Milyukova, and E.A. Kashutina

Abstract During the last several decades significant climate warming has been observed in the permafrost regions of Eastern Siberia. Observed environmental changes include increasing air temperature and to a lesser degree precipitation. Changes in regional climate are accompanied by changes in river runoff.

Seasonal and long-term changes of river runoff in different parts of the Lena river basin are characterized by distinct differences that can be explained by regional distinctions of climatic conditions, types and properties of permafrost, character of relief, hydrogeological conditions, etc. These factors determine the non-uniform response of river runoff amount and seasonal distribution to contemporary climate changes within the Lena river basin. Over the past 15–20 years river runoff has increased in different parts of the Lena river basin, but the scale of this increase over its territory is quite variable.

Hydrological responses to climate warming have been evaluated for the plain part of the Lena river basin based on a macroscale hydrological model featuring simplified descriptions of processes developed at the Institute of Geography of the Russian Academy of Sciences. Two atmosphere-ocean global circulation models used by the IPCC (ECHAM4/OPY3 and GFDL-R30) were used to model scenarios of future global climate. According to the results from hydrological modelling the expected anthropogenic climate warming in the twenty-first century can bring more significant river runoff changes in the Lena river basin compared to the twentieth century.

Keywords Cryolithic zone • Eastern Siberia • Scenarios of global climate warming • Model of monthly water balance • River runoff changes

---

A.G. Georgiadi (✉), I.P. Milyukova, and E.A. Kashutina
Institute of Geography, Russian Academy of Sciences, Staromonetny per., 29,
Moscow 119017, Russia
e-mail: galex50@gmail.com; mil-ira@list.ru; kategeo@mail.ru

10.1 Introduction

The flow of river water forming in the cryolithic zone of Eurasia, especially within the largest river basins (Lena, Yenisei, Ob'), has a strong impact upon regional climate and the surrounding seas, affecting the chemical composition of the water, formation of sea ice, and circulation of the Arctic and North Atlantic oceans. Changes of river runoff volume and seasonal distribution from watersheds within these basins can have considerable and in some cases critical impacts upon processes occurring within them.

During recent decades significant global climate warming has been observed. Climate warming has been largest in Northern Eurasia, particularly in the permafrost regions of Eastern Siberia (IPCC 2001a, b). Observed changes include rising winter air temperature and overall soil temperature (Varlamov et al. 2002), as well as increasing winter and autumn precipitation (Chapman and Walsh 1993; Razuvaev et al. 1996; Serreze et al. 2000; Wang and Cho 1997). These changes in regional climate have been accompanied by changes in river runoff within Eastern Siberia (Georgievsky et al. 1999; Savelieva et al. 2000; Yang et al. 2002; Simonov and Khristoforov 2005; Georgiadi et al. 2008).

Climate model projections of future climate related to increasing greenhouse gases in the atmosphere show that the permafrost regions of Eastern Siberia could be subject to one of the most notable changes (IPCC 2001). Regional climate warming will cause a considerable increase in soil temperature and consequent permafrost thawing (Anisimov et al. 1997; Demchenko et al. 2002; Malevsky-Malevich et al. 2001). These changes could alter the river runoff regime, particularly changes in the intra-annual distribution. Possible changes of the freshwater and heat inflow to Arctic have potentially important implications for ocean circulation and climate outside the region.

Until now, the processes regulating the water cycle in large basins have not been studied in sufficient detail, in particular their response to climate warming. The Lena River basin in Eastern Siberia was selected as one of the main regions of investigation. It is one of the world's largest basins (its area is 2,488,000 km^2, and the length is 4,400 km) and is almost completely covered by several metres of permafrost (with a thin seasonally-thawing active layer during the short summer). In addition, the Lena basin is characterized by low anthropogenic influence owing to low population density. The analysis of the Lena basin is also facilitated by a large intensive research programs carried out by Russian, bilateral, and international teams in the last 15 years. The most significant contribution to the improvement of our understanding of the processes in this region was achieved by field experiments that were conducted in 1996–2004 within the framework of the international program on studying of energy and water cycles (GEWEX) in Asia - Asian Monsoon Experiment (GAME). We have developed a model of the Lena river basin; exploiting data collected from GAME, and examined the changes in Lena runoff volume and seasonal distribution. We analyze changes in runoff in response to climate model projections forced by two scenarios of future climate, and project changes in runoff into the next century.

10.2 Methods

10.2.1 Model of Monthly Water Balance

The initial versions of the model and its application to the largest river basins of the Russian plain are considered in detail in Georgiadi and Milyukova (2002, 2006). In the model the basic processes of the hydrological cycle are described: infiltration and moisture accumulation in the soil, evaporation (on the basis of the modified Thornthwaite's method (Willmott et al. 1985), accumulation of water in the snow pack and snow melting on the basis of Komarov's method (Anon 1989), formation of surface, subsurface and groundwater flows in the rivers and full river runoff. The model can take into account macroscale heterogeneity of hydrometeorological fields and other territory characteristics, allowing a degree of reliability in the modelling of the river runoff changes. In the model of the monthly water balance the changes of the river runoff and other water balance elements are estimated in units of a regular grid, which facilitates the coupling of the model with climate model simulations.

In adapting the model for the conditions of permafrost soils it was assumed that the process of runoff formation in general can be presented as follows. In the cold season, precipitation arrives in solid form (snow) and accumulates. The top layers of soil at negative air temperatures freeze through, forming a practically waterproof layer. Positive air temperature signals the start of snow melt and soil thaw. Precipitation and melt water, getting to the surface of the catchment, partly evaporate from the surface, and the remainder seeps into the thawed active layer of the soil, the thickness of which gradually increases as the soil thaw progresses. At the same time, part of the water filters into the underlying horizons of underground water. If the soil layer becomes saturated to up to field capacity, a fast flow (mainly surface flow and flow from active layer of the supra-permafrost zone) is formed from the surplus moisture. The moisture held in underground water-bearing horizons is redistributed, with part forming surface flow and part infiltrating into deeper water-bearing horizons (i.e. groundwater) outside a zone of active water exchange.

The model is based on a conservation equation of average long-term monthly water balance of river catchments. In general it can be written down for each cell of a regular grid in the following way:

$$Q_s(t) + Q_{gr}(t) = P(t) - E(t) - I_d(t) - dW/dt \tag{10.1}$$

where $Q_s(t)$ is the total surface and subsurface (seasonal active layer) flow (mm); $Q_{gr}(t)$ is groundwater flow (mm). The sum of $Q_s(t)$ and $Q_{gr}(t)$ is the full river runoff. $P(t)$ is atmospheric precipitation (mm); $E(t)$ is evaporation (mm); $I_d(t)$ is infiltration of water to deep horizons of underground water outside of the active water exchange zone (mm), dW/dt is the change of the water amount in the active water exchange zone of the river basin, t is time.

For the water balance calculations in the Lena river basin a number of model blocks has been modified for the conditions of perennially frozen ground, and also blocks for calculations of seasonal soil thawing have been set up. In this version of the model two fundamentally different procedures are used: one allows to estimate the dynamics of seasonal thawing and freezing of the active layer and the other is used for areas where there is no perennially frozen ground, allowing to calculate the dynamics of the seasonally frozen layer depth and its influence on the flow formation.

For the calculation of the progression of the thawing (freezing) front, Eq. 10.2 is used (Pavlov 1979; Belchikov and Koren 1979), which has been established as a simplified solution of the classical single-front Stefan problem. The Stefan problem is a one-dimensional ordinary differential equation of heat conduction with a liquid/solid phase change at the boundary. In permafrost, the problem becomes a two-front problem since freezing does not only progress from above but also from below due to a store of cold in the deep frozen layer. The equation was defined assuming that the inflow of cold (heat) from below through the border of thawing (freezing) is constant (or a negligibly small quantity in comparison with the heat (cold) flux from the atmosphere). In this equation the height and density of snow cover are taken into account, and the initial temperature of soil thawing (freezing) may not be equal to 0°C.

$$Z^j = \pm St^j + \left\{ \left[St^j + Z^{j-1} \right]^2 \pm 2\lambda T_p^j \left(\Delta t / \rho_B L W_z^j \right) \right\}^{1/2} \quad (10.2)$$

where Z^j is the depth of the thawing front at day j and St^j is a variable which takes into account the availability and depth of the snow cover when calculating soil freezing or thawing. Note that when performing calculations for the changing boundary of soil freezing, $St^j = (\lambda / \lambda_c) H^j$, but for the changing boundary of soil thawing, $St^j = \lambda / A_k$. L is the specific heat of ice fusion, λ_c is the thermal conductivity of snow, λ is the thermal conductivity of frozen (at freezing) and thawed (at thawing) soil, H is the height of the snow layer, and T_p is the temperature of underlying ground, A_k is a parameter; W_z is volumetric soil moisture at the front of thawing (as a fraction of 1), ρ_B is the density of water, Δt is the time interval (days).

In the absence of a snow cover the dependence between air temperature and ground temperature was fixed, with the factor of permafrost being taken into account. Based on observation data, it is possible to set $T_p^j = T^j k_t$; ($k_t = 1.0 \approx 1.3$), where T is daily average air temperature and T_p is daily average soil temperature.

The thermal conductivity of thawed and frozen soil grounds was calculated by the relation

$$\lambda = k_p (j_{sk} + 0.1W - 1.1) - 0.1W \quad (10.3)$$

where j_{sk} is the relative density of the soil particle size fraction (g/cm³), $k_p = 1.5–1.7$ for sand, $k_p = 1.4–1.5$ for sandy loam, $k_p = 1.3–1.4$ for loamy sand and clay (the first value for thawing soil, the second – for frozen soils), W is the volumetric soil moisture in %. The thermal conductivity of snow was calculated as $\lambda_c = 4.32 + 1.64\rho_c + 518.4\rho_c^4$, where ρ_c is the snow density changing over time: $\rho_c = \rho_c^j + \rho_c^j \rho_c k_c, \rho_c^j, \rho_c^j = \rho_c^{j-1} + \rho_0 k_c$, ρ_0 is the initial snow density, K_c is a empirical parameter.

Infiltration to the underground water horizon of active water exchange (I_p) for day j was defined by the following relations:
in the absence of permafrost:

$$I_p^j = k_f (P^j + M^j - E^j) \, (W^j / W_{fc}) \tag{10.4}$$

in the presence of permafrost:

$$I_p^j = t_f k_f (P^j + M^j - E^j) \, (W^j / W_{fc}) \tag{10.5}$$

where M is daily snowmelt (mm), W_{fc} is the soil moisture reserve at field capacity (mm), t_f is a factor of vertical infiltration in ground water, t_f is a parameter dependent on the type of permafrost distribution in the given cell. Other symbols are given above.

The ground water runoff (mm of depth) was calculated as

$$Q_{gr}^j = k_{gr} (H_{gr}^j / H_{max}) \tag{10.6}$$

where H_{max} is the maximum capacity of the aquifer, k_{gr} is a factor of ground water runoff, and H_{gr} is the capacity of the aquifer on day j, calculated as

$$H_{gr}^j = H_{gr}^{j-1} + I_p^j - Q_{gr}^{j-1} - I_d^j \tag{10.7}$$

The moisture infiltration I_d into the deep horizons of groundwater outside of the active water exchange zone was set as a constant. The change of the groundwater capacity of active water exchange for day j was calculated by:

$$\Delta H_{gr}^j = H_{gr}^j - H_{gr}^{j-1} \tag{10.8}$$

Filtration of water into the aquifer of active water exchange (I_p) and groundwater flow (Q_{gr}) for different types of permafrost soil (continuous, discontinuous, sporadic and absent) were calculated using empirical factors derived from field observations.

10.2.2 General Scheme of Calculations

To initialise the model we used monthly average long-term data of air temperature and precipitation, soil moisture and physical soil characteristics (field capacity, soil volume and bulk density), types of permafrost lateral continuity (continuous, discontinuous, sporadic, absence of permafrost), the maximum depth of the active layer, and runoff for each grid cell.

Calculations were carried out on daily time steps. We used the long-term monthly average of the observational data. Interpolation of observational data was carried out per day. It was assumed that the monthly average air temperature falls at the 15th date of every month, the temperature on the 1st day of all months from January till July was the monthly average minus T_Δ^0, while the temperature of the last day of these months was set at the monthly average plus T_Δ^0. From August till December the reverse procedure was applied. Within a monthly interval, linear interpolation between the set air temperatures for the 1st, 15th and last day of each month were carried out. The deviation of the air temperature for the beginning and the end of the month from the long-term average of the month ($\pm T_\Delta^0$) was fitted empirically.

Calculations were carried out from 1st January till 31st December. However, for the calculations, besides initial information on climate data and soil characteristics, it is necessary to have data on soil moisture, snow water equivalent, and the aquifer capacity for the start date of calculations (the 1st of January). For the approximate initial conditions, calculations of these values were performed iteratively for the previous 4 months (September till December) and then used as initial conditions for the 1st January.

10.2.3 Methods of Parameter Calibration

The model includes 15 parameters. Seven of them (T_{cr}^0, t_m, k_t, t_f, $H_{gr_{\max}}$, T_{cl}^0, ρ_{cj}) were fitted empirically for homogeneous parts of the Lena river basin. For identification of the other parameters of the model, Rosenbrock's procedure of optimization (Rosenbrock and Story 1968) was used.

The optimization of the model was carried out on the basis of gridded average long-term data on monthly average air temperature, precipitation, evaporation and potential evaporation from the archive of the International Institute of Applied System Analysis (IIASA), Austria (Cramer and Leemans 1999), and also on monthly average river runoff of the plain part of the Lena river basin, obtained by interpolation of the instrumental stream flow measurements of representative medium river basins. For optimization of the model parameters, data of 50% of the grid cells, evenly distributed over the territory of this part of the Lena river basin, have been used. The remaining 50% of the data were used for model verification.

The optimization was carried out in several stages, separately for different blocks of the model. In the first stage the parameters α, α_1 and β were optimized using blocks for calculating the potential evaporation and evaporation from long-term data of monthly average evaporation (Georgiadi and Milyukova 2006). In the second stage, the parameters to calculate seasonal soil ground thawing (freezing), A_k and k_c, were optimized. Then, as an approximation to model runoff the melting factor k, the factor of vertical filtration into the groundwater k_f and the factor of a ground water runoff k_{gr} were optimized. In the fourth stage based on the data on seasonal soil thawing values the parameters A_k and k_c were specified. In the last stage the melting coefficient k, the factor of vertical filtration into the groundwater k_f and the factor of groundwater runoff k_{gr} were specified using runoff data. The least sensitive parameters whose magnitude could be set by approximation without loss of accuracy of the calculations were fit empirically.

The hydrological modelling shows that this multi-stage procedure for the identification of parameters, despite of all its complication, works very well and, as a rule, helps to increase the robustness of parameterization. In addition, according to the data on the spatial distribution of mean values of the maximal active layer depth (prepared by Fedorov from the Permafrost Institute of the Siberian Branch of the Russian Academy of Sciences) the optimization of the A_k and k_c parameters has been performed.

10.3 Description of the Data Used

10.3.1 Data on the Present-Day Climate

Mean monthly fields of air temperature, total precipitation and river runoff, soil moisture reserves were the initial information for model calculations. Mean monthly values of air temperature and precipitation from a 1° × 1° grid from supplemented global archive prepared at the International Institute of Applied System Analysis (Cramer and Leemans 1999) were used. These data as well as the data on river runoff and deviations of climatic elements were smoothed.

The data on spatial distribution of mean annual values of the maximal active layer for the conditions of modern climate have been prepared by A. Fedorov from Permafrost Institute of SB of RAS.

10.3.2 Data on River Runoff

Fields of mean monthly values of river runoff were calculated from monthly average long-term river runoff for medium rivers of the Lena river basin were used. The data on runoff referred to average weighted centre of basins, and then were interpolated into each cell. Data on river runoff averaged over time periods exceeding 30 years and referring mostly to time period between the 1930s–1940s and 1980s were used in calculations.

10.3.3 Data on Deviations of Climatic Elements from Their Present-Day Values

Results of numerical experiments on models of general circulation of the atmosphere and the ocean (AOGCMs) included in the IPCC program (MPIfM ECHAM4/

Table 10.1 Information about models of general circulation of the atmosphere and the ocean

No.	Organization where model was developed	Model	Grid cell	Description of model
1	Max Planck Meteorological Institute, Germany	ECHAM4/OPY3	$2.8° \times 2.8°$	Roeckner et al. (1996)
2	Geophysical Laboratory of Hydrodynamics of the Princeton University, USA	GFDL-R30	$2.2° \times 3.8°$	Wetherald and Manabe (1988); Broccoli and Manabe (1992)

OPY3 and GFDL-R30) were used as scenario estimations of climate change in the future (Table 10.1).

A2 and B2 families of scenarios for global socio-economic changes in the twenty-first century from the latest improved SRES series of such scenarios accepted in the IPCC program were used. Two periods (2010–2039, 2040–2069) were used for calculation of deviations of the mean monthly precipitation and air temperature from their present-day values (calculated for the period of 1961–1990).

For the mentioned above scenarios the results of independent calculations of corresponding mean annual values of active layer depth, carried out by Anisimov et al. (1997) have been used.

10.4 Discussion of Modelling Results

Climatic conditions. According to the above scenarios for the plain central part of the Lena River basin, the climate warming is expected to be particularly intensive in the middle of the twenty-first century (Figs. 10.1 and 10.2). According to the scenario of the Max Planck Meteorological Institute, however, climate warming is less notable. The both scenarios predict a more marked air temperature increase in the cold period of the year, which is likely to reduce this period. The character of changes in atmospheric precipitation is more complicated (Figs. 10.1 and 10.2). Only by the middle of this century, according to the both scenarios, the precipitation increase trend will be obvious. In this case, with the Max Planck Meteorological Institute scenario realized, a more perceptible precipitation increase can be expected in the warm season of the year, while, according to the scenario of the Geophysical Laboratory of Hydrodynamics, Princeton University, a more appreciable precipitation increase is to be observed in the cold season.

The character of the annual distribution of scenario changes in monthly average air temperatures and precipitation is rather similar to those observed in this region during the last decades of the current climate warming, but differs from the latter in scale.

Hydrological conditions. According to climatic scenarios of the Geophysical Laboratory of Hydrodynamics (Princeton University) and the Max Planck

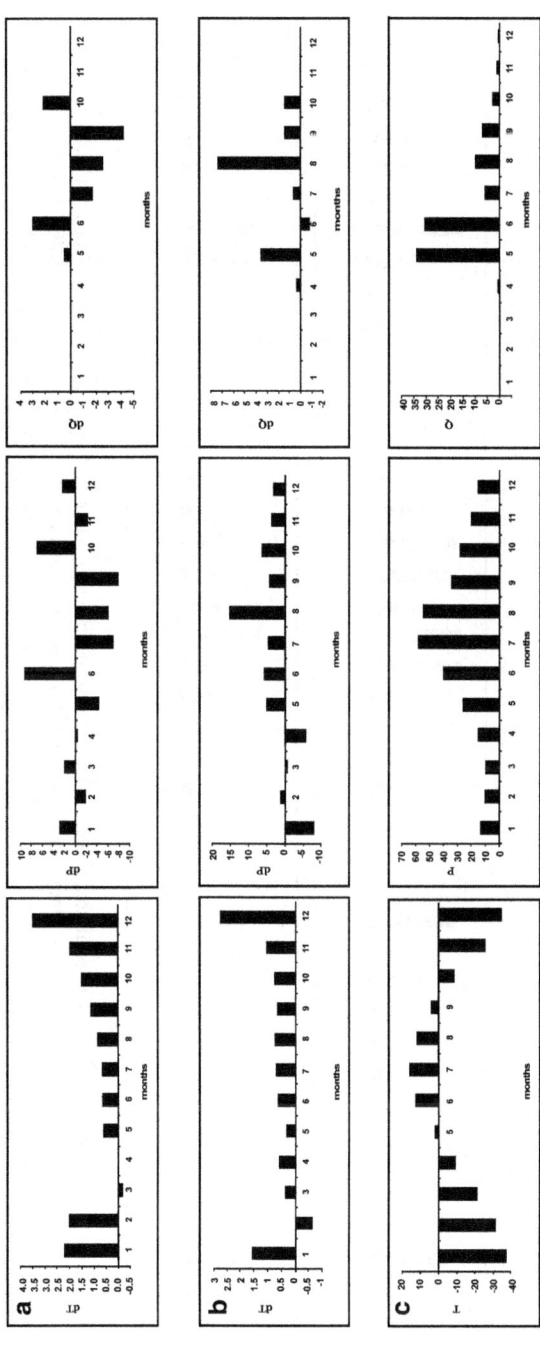

Fig. 10.1 Deviations of mean monthly values of air temperature (dT, °C), atmospheric precipitation (dP, mm) and river runoff (dQ, mm) from their recent ones, averaged in the cells of regular grid covering the plain part of the Lena basin, for the conditions of climate warming in 2010–2039. (**a**) – By Geophysical Laboratory of Hydrodynamics of the Princeton University (USA) scenario; (**b**) – by Max Planck Meteorological Institute (Germany) scenario; (**c**) – mean monthly air temperature (T, °C), atmospheric precipitation (P, mm), stream flow (Q, mm) of the modern climate conditions

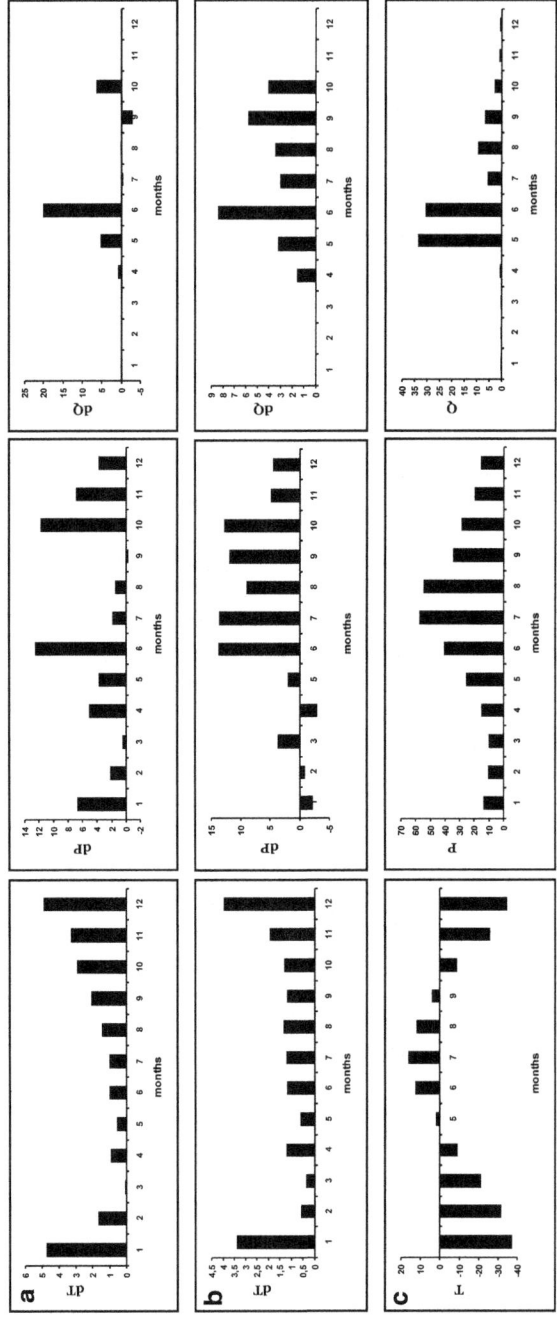

Fig. 10.2 Deviations of mean monthly values of air temperature (dT, °C), atmospheric precipitation (dP, mm) and river runoff (dQ, mm) from their recent ones, averaged in the cells of regular grid covering the plain part of the Lena basin, for the conditions of climate warming in 2040–2069. (**a**) – By Geophysical Laboratory of Hydrodynamics of the Princeton University (USA) scenario; (**b**) – by Max Planck Meteorological Institute (Germany) scenario; (**c**) – mean monthly air temperature (T, °C), atmospheric precipitation (P, mm), stream flow (Q, mm) of the modern climate conditions

Meteorological Institute for the plain part of the Lena River basin, during the first 30 years of the current century it is not likely to expect any notable increase in both annual river runoff and river runoff for the spring-summer flood period (Fig. 10.1). The distribution of the possible changes in the river runoff over the plain part of the basin is characterized by the essential spatial heterogeneity which is to be reduced appreciably by the middle of the century. According to the two climatic scenarios, a significant increase in annual and spring flood runoffs is expected to occur by the middle of the twenty-first century (Fig. 10.2).

In the first 30 years of the century, the both scenarios give insignificant changes in the wave of the spring–summer flood (Fig. 10.1), whereas river runoff changes following the main wave of high water may be of quite an opposite character (Fig. 10.1). By the middle of the century, the patterns of the changes in intra-annual distribution of stream-flow, which are derived from the two scenarios in question, differ from each other greatly (Fig. 10.2). However, it should be noted that the runoff changes are largely recorded in the warm season and the river runoff increases nearly throughout the season. According to the scenario of the Geophysical Laboratory of Hydrodynamics, Princeton University, the major changes occur during the passage of the flood wave, whereas according to the Max Planck Meteorological Institute scenario, these changes are distributed more uniformly throughout the warm season. In this case the flood onset shifts to earlier time, as compared with what is currently taking place. It should be recognized/mentioned that the character of the changes in intra-annual stream-flow distribution in the Lena River basin that are based on the scenario of the Geophysical Laboratory of Hydrodynamics, Princeton University, is alike the similar changes in the Volga River basin under the probable global climate warming in the current century.

10.5 Conclusions

1. According to the above scenarios for the plain central part of the Lena River basin, the climate warming is expected to be particularly intensive in the middle of the twenty-first century.
2. The character of the intra-annual distribution of scenario changes in monthly average air temperatures and precipitation is rather similar to those observed in this region during the last decades of the current climate warming, but differs from the latter in scale.
3. A significant increase, both in annual runoff and runoff for the period of spring-summer high water from the plain Lena River basin, is not likely to be expected during the first 30 years of the current century. Calculations based on the both climatic scenarios show a significant increase in annual and spring flood runoffs by the middle of the twenty-first century
4. A rather significant change in the character of the annual stream-flow distribution may take place by the middle of the twenty-first century. However, the above scenarios predict a different character of changes in the intra-annual stream-flow

distribution. According to the scenario of the Geophysical Laboratory of Hydrodynamics, Princeton University, the major changes occur during the passage of the flood wave, whereas according to the Max Planck Meteorological Institute scenario, these changes are distributed more uniformly throughout the warm season.

Acknowledgments The authors are grateful to Dr. R. Leemans, who contributed global database on the modern climate, prepared in the International Institute of Applied System Analysis, Laxenburg, Austria.

References

Anisimov OA, Shiklomanov NI, Nelson FE (1997) Effects of global warming on permafrost and active-layer thickness: results from transient general circulation models. Global Planet Change 15(2):61–77
Anon (1989) Manual on hydrological forecasts. Issue 1: Long-term forecasts of water regime of rivers, lakes and water reserves. Gidrometeoizdat, Leningrad, 358 pp (in Russian)
Belchikov VA, Koren VI (1979) Model of snow melting and rain runoff forming for forest watersheds. Proceedings of Hydrometeocenter of the USSR, Issue 218, pp 3–21 (in Russian)
Broccoli AJ, Manabe S (1992) The effects of orography on midlatitude Northern Hemisphere dry climates. J Climate 5(11):1181–1201
Chapman WL, Walsh JE (1993) Recent variations in sea ice and air temperature in high latitudes. Bull Am Meteorol Soc 74:33–47
Cramer WP, Leemans R (1999) Global 30-year mean monthly climatology, 1930–1960. V.2.1 (Cramer and Leemans). Available online [http://www.daac.ornl.gov/] from the ORNL Distributed Active Archive Center, Oak Ridge National Laboratory, Oak Ridge, Tennessee, USA
Demchenko PF, Velichko AA, Eliseev AV, Mokhov II, Nechaev VP (2002) Dependence of permafrost conditions on global warming: comparison of models, scenarios, and paleoclimatic reconstructions. Izvestiya, Atmos Ocean Phys 38(2):143–151. Translated from Izvestiya AN. Fizika Atmosfery i Okeana 38(2):165–174, 2002. English Translation Copyright © 2002 by AIK
Georgiadi AG, Milyukova IP (2002) Possible scales of hydrological changes in the Volga river basin during anthropogenic climate warming. Meteorol Hydrol 2:72–79 (in Russian)
Georgiadi AG, Milyukova IP (2006) Possible river runoff changes in the largest river basins of Russian Plain in XXI. Water management complex of Russia. No. 1, pp 62–77 (in Russian)
Georgiadi AG, Milyukova IP, Kashutina EA (2008) Recent and projected river runoff changes in permafrost regions of eastern siberia (Lena River Basin). Ninth International Conference on Permafrost. Kane DL and Hinkel KM (eds). Institute of Northern Engineering, University of Alaska Fairbanks 511–515
Georgievsky VYu, Ezhov AV, Shalygin AL, Shiklomanov IA, Shiklomanov AI (1999) Evaluation of possible climate change impact on hydrological regime and water resources of the former USSR rivers. "Meteorology and Hydrology", 1996, 11, pp 89–99. ISSN 0130-2906 (in Russian)
IPCC (2001a) Climate change 2001: the scientific basis. In: Houghton JT, Ding Y, Griggs DJ, Noguer M, van der Linden PJ, Dai X, Maskell K, Johnson CA (eds) Contribution of working Group I to the third assessment report of the intergovernmental panel on climate change . Cambridge University Press, Cambridge, UK and New York, NY, USA, 881 pp

IPCC II (2001b) Climate change 2001, impacts, adaptation, and vulnerability. In: McCarthy JJ, Canziani OF, Leary NA, Dokken DJ, White KS (eds). Cambridge University Press, 1032 p

Malevsky-Malevich SP, Molkentin EK, Nadyozhina ED, Shklyarevich OB (2001) Numerical simulation of permafrost parameters distribution in Russia. Cold Reg Sci Technol 32:1–11

Pavlov AV (1979) Thermophysics of landscapes. Nauka, Novosibirsk, 285 pp (in Russian)

Razuvaev VN, Apasova EG, Bulygina ON, Martuganov RA (1996) Assessment of natural variability of extreme temperatures over the former Soviet Union territory during the second half of 20th century. Trans RIHMI-WDC 162:3–13 (in Russian)

Roeckner E, Arpe K, Bengtsson L, Christoph M, Claussen M, Dümenil L, Esch M, Giorgetta M, Schlese U, Schulzweida U (1996) The atmospheric general circulation model ECHAM-4: model description and simulation of present-day climate. Max-Planck Institute for Meteorology. Report No. 218, Hamburg, Germany, 90 pp

Rosenbrock H, Story S (1968) Computational methods for chemist engineer. Translated from English. Moscow, Mir, 440 pp

Savelieva NI, Semiletov IP, Vasilevskaya LN, Pugach SP (2000) A climate shift in seasonal values of meteorological and hydrological parameters for Northeastern Asia. Prog Oceanogr 47(2–4):279–297

Serreze MC, Walsh JE, Chapin FS III, Osterkamp TE, Dyurgerov M, Romanovsky VE, Oechel WC, Morison J, Zhang T, Barry RG (2000) Observational evidence of recent change in the northern high-latitude environment. Climatic Change 46:159–207

Simonov Yu A, Khristoforov AV (2005) Analysis of long-term variability of stream flow of rivers of Arctic Ocean basin. Water Resour 32(6):645–652, in Russian

Varlamov SP, Skachkov Yu B, Skryabin PN (2002) Ground temperature regime of Central Yakutia permafrost landscapes. SB RAS, Yakutsk, 218 pp (in Russian)

Wang XL, Cho H-R (1997) Spatial-temporal structures of trend and oscillatory variabilities of precipitation over Northern Eurasia. J Clim 10:2285–2298

Wetherald RT, Manabe S (1988) Cloud feedback processes in a general circulation model. J Atmos Sci 45:1397–1415

Willmott CJ, Rowe CM, Mintz Y (1985) Climatology of the terrestrial seasonal water cycle. J Climatol 5:589–606

Yang D, Kane DL, Hinzman LD, Zhang X, Zhang T, Ye H (2002) Siberian Lena River hydrologic regime and recent change. J Geophys Res 107(d 23, 4694):14-4–14-10

Part III
Atmosphere

Chapter 11
Investigating Regional Scale Processes Using Remotely Sensed Atmospheric CO_2 Column Concentrations from SCIAMACHY

M.P. Barkley, A.J. Hewitt, and P.S. Monks

Abstract Satellite observations of atmospheric CO_2 are a rapidly emerging area of scientific research which have the potential to reduce the uncertainties in global carbon cycle fluxes and provide insight into surface sources and sinks. In this chapter, the potential of atmospheric CO_2 measurements, retrieved by the SCIAMACHY instrument on-board the ENVISAT satellite, to investigate regional carbon cycle processes is explored. The methodology for high precision measurements of the CO_2 total column retrievals from SCIAMACHY near infrared (NIR) spectral measurements are demonstrated using the *Full Spectral Initiation* (FSI) algorithm; which is based on the inclusion of suitable a priori information within the retrieval in order to minimize the errors on the retrieved CO_2 columns.

The monthly averaged CO_2 distributions over Siberia and also North America contain significant spatial features which correlate well with land vegetation type. Validation of the data from the FSI retrievals is also briefly discussed. Furthermore, the capability of the SCIAMACHY to observe lower tropospheric and surface CO_2 variability is then examined through comparisons to in situ aircraft observations over Siberia and additionally to surface CO_2 data over Eurasia. It is shown that strong similarities exist between the CO_2 anomalies measured by SCIAMACHY and those of the in situ instruments thus demonstrating the potential of SCIAMACHY to detect variations in lower tropospheric CO_2.

Keywords Carbon dioxide • SCIAMACHY • Satellite • Retrievals • FSI WFM-DOAS

M.P. Barkley (✉)
School of GeoSciences, University of Edinburgh, Crew Building, The King's Buildings,
West Mains Road, Edinburgh EH9 3JN, UK
e-mail: Michael.Barkley@ed.ac.uk

A.J. Hewitt and P.S. Monks
Earth Observation Science group, Departments of Physics and Chemistry,
University of Leicester, University Road, Leicester, LE1 7RH, UK
e-mail: ajh67@le.ac.uk; p.s.monks@le.ac.uk

11.1 Introduction

During the last 200 years there has been a rapid 30% increase in the amount of carbon dioxide (CO_2) within the atmosphere owing primarily to carbon emissions from the burning of fossil fuels, deforestation and from the industrial production of ammonia, lime and cement (Prentice et al. 2001). As CO_2 is the dominant anthropogenic greenhouse gas, this dramatic increase is likely to have a serious impact on the carbon cycle and climate, especially as present concentrations are now far greater than at any other time in the last 650,000 years (Siegenthaler et al. 2005). Of the anthropogenic carbon that is emitted into the atmosphere, only half remains there with the rest absorbed by two important sinks: the oceans and terrestrial biosphere (Sabine et al. 2004). Carbon dioxide that diffuses across the ocean surface is mixed to deep ocean waters via the solubility, biological and carbonates pumps where it is considered to be effectively removed from the climate system for several hundred years. This is in contrast to carbon sequestered by the terrestrial vegetation (for the creation and accumulation of plant biomass) which can be subsequently be released back to the atmosphere over much shorter timescales (e.g. through biomass burning events). Understanding how these sinks and the carbon cycle respond to escalating atmospheric CO_2 levels and global warming is essential for predicting future climate change, especially as feedback mechanisms within the cycle are still poorly quantified (Friedlingstein et al. 2003).

The oceanic and terrestrial fluxes are currently estimated, at continental and ocean basin scales, using inverse methods that employ transport models coupled with surface measurements of the atmospheric concentration provided by the sparse network of Earth System Research Laboratory Global Monitoring Division ERSL/GMD (http://www.esrl.noaa.gov/gmd/) ground stations (e.g., Gurney et al. 2002; Rödenbeck et al. 2003; Patra et al. 2006). In order to estimate regional carbon fluxes more observations of the atmospheric CO_2 distribution are needed so that tighter constraints can be placed on the model inversions. Satellite measurements can, in principle, provide the dense sampling that is needed and moreover give complete global coverage. However, the small spatial and temporal gradients associated with atmospheric CO_2 require satellite measurements to be made accurately and to a high precision to be of any value. A highly cited study by Rayner and O'Brien (2001) concluded that to improve over the existing ground network monthly averaged column data, at a precision of 1% (2.5 ppmv) or better, for an $8° \times 10°$ footprint are needed. However as shown by Houweling et al. (2004) this threshold can be relaxed for different geographical regions, highlighting that the precision cannot just be quantified by a single number.

During the last decade, preliminary CO_2 retrievals utilizing spectral information from either the thermal or the adjacent near infrared (NIR), have demonstrated that we are entering an era where satellite monitoring of atmospheric CO_2 concentrations is becoming more feasible (see, e.g., Chédin et al. 2003; Buchwitz et al. 2005, Barkley et al. 2006a; Engelen and McNally 2005) . Thermal infrared sounders, such as AIRS, have limited sensitivity to surface CO_2 since the light that the instrument detects originates from the mid-upper troposphere (Engelen and McNally 2005). In contrast,

NIR sensors like SCIAMACHY (the only currently operational NIR instrument) or the future Orbiting Carbon Observatory (OCO; http://oco.jpl.nasa.gov/) and the Greenhouse Gas Observing Satellite (GOSAT; http://www.jaxa.jp/projects/sat/gosat/index_e.html) missions*, are more sensitive to the lower troposphere since they detect sunlight that is reflected from the Earth's surface. The near surface sensitivity of the NIR makes it the ideal spectral region for observing surface fluxes (even though the absorption bands in the thermal IR are much stronger). Previous work by Buchwitz et al. (2005, 2006, 2007), Houweling et al. (2005) and Barkley et al. (2006a, b, c, 2007) have shown that CO_2 measurements from SCIAMACHY are possible with a precision that is approaching the 1% threshold requirement for inverse flux modelling. In this chapter, we will explore the potential of atmospheric CO_2 measurements, retrieved from SCIAMACHY NIR spectra using an algorithm called Full Spectral Initiation (FSI) (Barkley et al. 2006a), to investigate regional carbon cycle processes over areas such as Siberia.

11.2 The SCIAMACHY Instrument

Launched on board the ENVISAT satellite into a polar sun-synchronous orbit in March 2002, the Scanning Imaging Absorption spectrometer for Atmospheric CHartography (SCIAMACHY) is a passive hyper-spectral grating spectrometer whose scientific goal is to investigate atmospheric composition and processes (Bovensmann et al. 1999; Gottwald et al. 2006). SCIAMACHY detects sunlight that is either reflected from the surface and/or scattered by the atmosphere using eight separate grating spectrometers (or channels) that cover (non-continuously) the ultraviolet, visible and NIR regions (240–2,380 nm) of the electromagnetic spectrum. Though the spectral resolution of the channels differ, each consists of a linear array of 1,024 photo-diodes that sample a designated spectral range at about half the channel's full-width half maximum (0.2–1.4 nm). Atmospheric trace gases that have absorption features lying within the spectral intervals covered by SCIAMACHY contain information on their atmospheric abundance which can be derived using established retrieval techniques.

On the Earth's day side SCIAMACHY makes measurements in an alternating limb and nadir sequence. In nadir mode, the atmosphere directly below the sensor is viewed yielding observations with good spatial but poor vertical resolution; a typical set of observations consists of the nadir mirror scanning across track for 4 s followed by a fast 1 s back-scan, which is repeated for approximately 1 min. The ground swath viewed has fixed dimensions of 960 × 30 km² (across × along track). Atmospheric CO_2 distributions are mapped from space by retrieving the CO_2 vertical column densities (VCDs) (units of molecules cm^{-2}) from SCIAMACHY nadir observations made in the NIR using a small micro window within channel 6, centred on the CO_2 band at 1.57 µm. For channel 6, the nominal size of each ground

---

*OCO suffered launch failure on 24th February 2009. GOSAT successfully launched on 23rd January 2009.

pixel within the swath is 60 × 30 km² corresponding to a detector integration time of 0.25 s. Full (longitudinal) global coverage is achieved at the Equator within 6 days and more quickly at high latitudes.

11.3 Full Spectral Initiation (FSI) WFM-DOAS

The Full Spectral Initiation (FSI) retrieval algorithm has been developed explicitly to retrieve CO_2 from space using SCIAMACHY NIR spectral measurements (Barkley et al. 2006a; Barkley 2007). It is based on the WFM-DOAS technique first introduced by Buchwitz et al. (2000) which allows the trace gas VCD to be retrieved from a linear least squares fit of the logarithm of a model reference spectrum I_i^{ref} and its derivatives, plus a quadratic polynomial $P_i(a_m)$ to the logarithm of the sun normalized intensity I_i^{meas} measured at each ith detector pixel of centre wavelength λ_i:

$$\left\| \ln I_i^{meas}(\mathbf{V}^t) - \left[\ln I_i^{ref}(\mathbf{V}^m) + \sum_j \frac{\partial \ln I_i^{ref}}{\partial \bar{V}_j} \cdot (V_j^r - V_j^m) + P_i(a_m) \right] \right\|^2 \equiv \|RES_i\|^2 \quad (11.1)$$

The true and model vertical columns are represented by $V^t = (V^t_{CO_2}, V^t_{H_2O}, V^t_T)$ and $V^m = (V^m_{CO_2}, V^m_{H_2O}, V^m_T)$ and the retrieved columns denoted as V_j^r where j runs over the variables of CO_2, H_2O and temperature T (where V_T is not a vertical column but a scaling constant applied to the temperature profile). Each model derivative (or weighting function) expresses the change in radiance at the top of the atmosphere created by a relative scaling of the trace gas or temperature profile. The polynomial $P_i(a_m)$ which has coefficients a_0, a_1 and a_2 is included to compensate for broadband atmospheric scattering and the spectral continuum. Equation 11.1 can be solved, in a least squares sense, to minimize the fit residual RES_i with the fit parameters being the trace gas columns $V^r_{CO_2}$ and $V^r_{H_2O}$, the temperature scaling factor V^r_T and the polynomial coefficients. The error of each retrieved parameter σ_j is given by Eq. 11.2 where $(C_x)_{jj}$ is the jth diagonal element of the least squares fit covariance matrix, m is the number of spectral fitting points and n is the number of fit parameters.

$$\sigma_j = \sqrt{\frac{(C_x)_{jj} \times \sum_i RES_i}{(m-n)}} \quad (11.2)$$

In contrast to other applications of WFM-DOAS which perform global retrievals using only a finite number of reference spectra stored within a look-up table (Buchwitz et al. 2005, 2007), the FSI algorithm distinguishes itself by generating a reference spectrum for each individual SCIAMACHY measurement. This approach ensures each model spectrum closely approximates the observed radiance which increases the accuracy of the spectral fit and in turn minimizes the error on the retrieved CO_2 column. To create accurate reference spectra, synthetic radiances are computed using the radiative transfer model SCIATRAN (Rozanov et al. 2002) using known (a priori) information about the atmosphere and surface at the time of each SCIAMACHY observation. This a priori data includes a CO_2 profile (taken from a specially prepared climatology, Remedios et al. 2006); ECMWF water vapour, temperature and pressure profiles interpolated on to the centre of each SCIAMACHY pixel; a background aerosol scenario and an inferred a priori surface albedo. Once the model spectrum has been calculated a WFM-DOAS fit is performed and the vertical column retrieved.

The FSI algorithm is only applied to cloud free SCIAMACHY ground pixels, which are identified using a cloud detection method devised by Krijger et al. (2005), whose spectra have been calibrated in-house to overcome various instrumental issues that hamper NIR retrievals (see e.g., Gloudemans et al. 2005). Pixels over the oceans are also not processed owing to the very low albedo of the ocean surface (which results in very noisy spectra). To produce a column volume mixing ratio (VMR) in parts per million by volume (ppmv) each retrieved CO_2 VCD is normalized using the input a priori ECMWF surface pressure and only CO_2 VMRs where the retrieval fitting error is less than 5% and which lie in a range 340–400 ppmv are used for analyses.

11.4 SCIAMACHY Global CO_2 Distributions

The calculation of a reference spectrum for each SCIAMACHY observation is time consuming, a consequence of which is that the FSI algorithm is only applied to SCIAMACHY data over selected target regions of particular relevance for carbon cycle processes. Thus, global maps of SCIAMACHY CO_2 often have large areas with no or few observations, which reflect both this regional selection and also persistent cloud cover. The tri-monthly averages for 2003 are shown in Fig. 11.1 and illustrate that SCIAMACHY is capable of observing the evolving CO_2 distribution. For example, over North America and northern EurAsian landmasses, there are low CO_2 VMRs during June–August, uniform mixing ratios during September–November, and high CO_2 during the winter months. These distributions are the signature of strong near surface seasonal CO_2 uptake and release by surface vegetation i.e. natural terrestrial fluxes.

Other interesting features of note include persistent low CO_2 mixing ratios over the tropical Amazon, which despite deforestation within this region, is indicative of a terrestrial sink. Low CO_2 is also seen in the Congo basin over the tropical rainforest, during September–November which moves southward over Angola and Zambia in the southern hemisphere summer. However, great care must be taken when interpreting spatial features. For example, the high VMRs over sub-Saharan Africa or over north-east India during March–May are more likely to be the result of aerosol scattering enhancing the light path (which increases the absorption) rather than CO_2 surface emissions (Houweling et al. 2005; Barkley 2007).

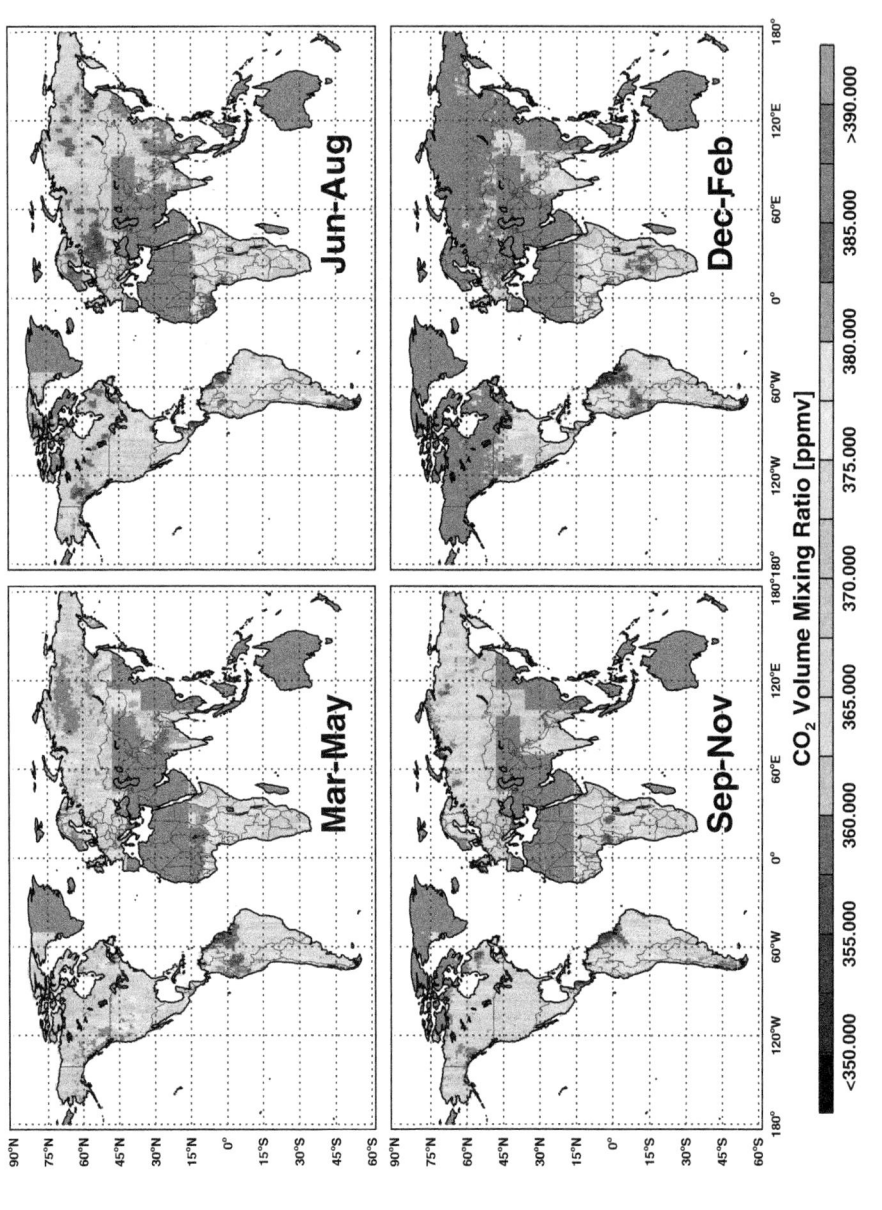

Fig. 11.1 Global SCIAMACHY tri-monthly CO_2 means for 2003 averaged on a $1° \times 1°$ grid and smoothed with a $3° \times 3°$ box car filter (*Color version available in Appendix*)

The CO_2 temporal behaviour, of each hemisphere and over target regions like Siberia, can be examined by plotting the time series of monthly means and anomalies, i.e. each scene's monthly mean minus its yearly mean. In the northern hemisphere, the time series is dominated by the uptake and release of CO_2 by vegetation with high mixing ratios in the winter and spring months, and much lower VMRs in the summer months. In the southern hemisphere, the time series has a 6 month phase difference relative to the north being more influence by inter-hemispheric mixing of air masses and the air-sea fluxes than by surface vegetation. As Fig. 11.2 shows, SCIAMACHY can capture this phasing and observes season cycle amplitudes of 11.2 and 9.2 ppmv for the northern and southern hemispheres respectively. Whilst the northern amplitude is credible, that observed in the south is rather high, amplitudes of ~3 ppmv being more realistic (Olsen and Randerson 2004). However as observation over land are only considered and regions like Australasia have not been processed by the FSI algorithm, this high figure may be biased by poor sampling. The seasonal cycle over Siberia (22.5 ppmv) is over twice that of the whole northern hemisphere, is far less smooth and is characteristic of highly variable CO_2 exchange between vegetation and the atmosphere.

The latitudinal gradient (north to south), which is created by more anthropogenic emissions in the northern hemisphere, is typically about 3–4 ppmv (Denning et al. 1995; Conway et al. 1994) at the surface and about 1–2 ppmv in the column integral (Olsen and Randerson 2004). The annual mean latitudinal gradient observed by SCIAMACHY is 1.57 ppmv, but this fluctuates over the course of the year and is even negative during the (northern) summer months owing to the large CO_2 uptake over regions like Siberia and North America.

11.5 Algorithm Validation

Before spatial maps of Siberia are examined, it is necessary to give a critical assessment of the SCIAMACHY/FSI CO_2 product. Analysis of the FSI retrievals reveal that the CO_2 column errors (given by Eq. 11.2) are typically 1–4%, with the root-mean-square (RMS) errors of the spectral fits themselves of order 0.1–0.3%, thus indicating a stable retrieval set-up (Barkley et al. 2006b). However, to assess the accuracy (bias) and precision of the retrievals it is necessary to validate SCIAMACHY CO_2 against a variety of sources, including both in situ observations and model simulations. In this section three validation comparisons, of particular relevance to Siberia, are briefly discussed. A concise summary of the complete SCIAMACHY validation exercises is given in Table 11.1.

The most useful and important technique of validating satellite observations of atmospheric CO_2 is through comparisons to ground based Fourier Transform Spectrometer (FTS) measurements of the CO_2 column integral (e.g. Dils et al. 2006). To this end SCIAMACHY/FSI CO_2 has been compared to data provide by three FTS instruments at Egbert (Canada), Bremen (Germany) and Park Falls (USA) (Barkley et al 2006b; Barkley 2007). Of these sites, Park Falls is the most

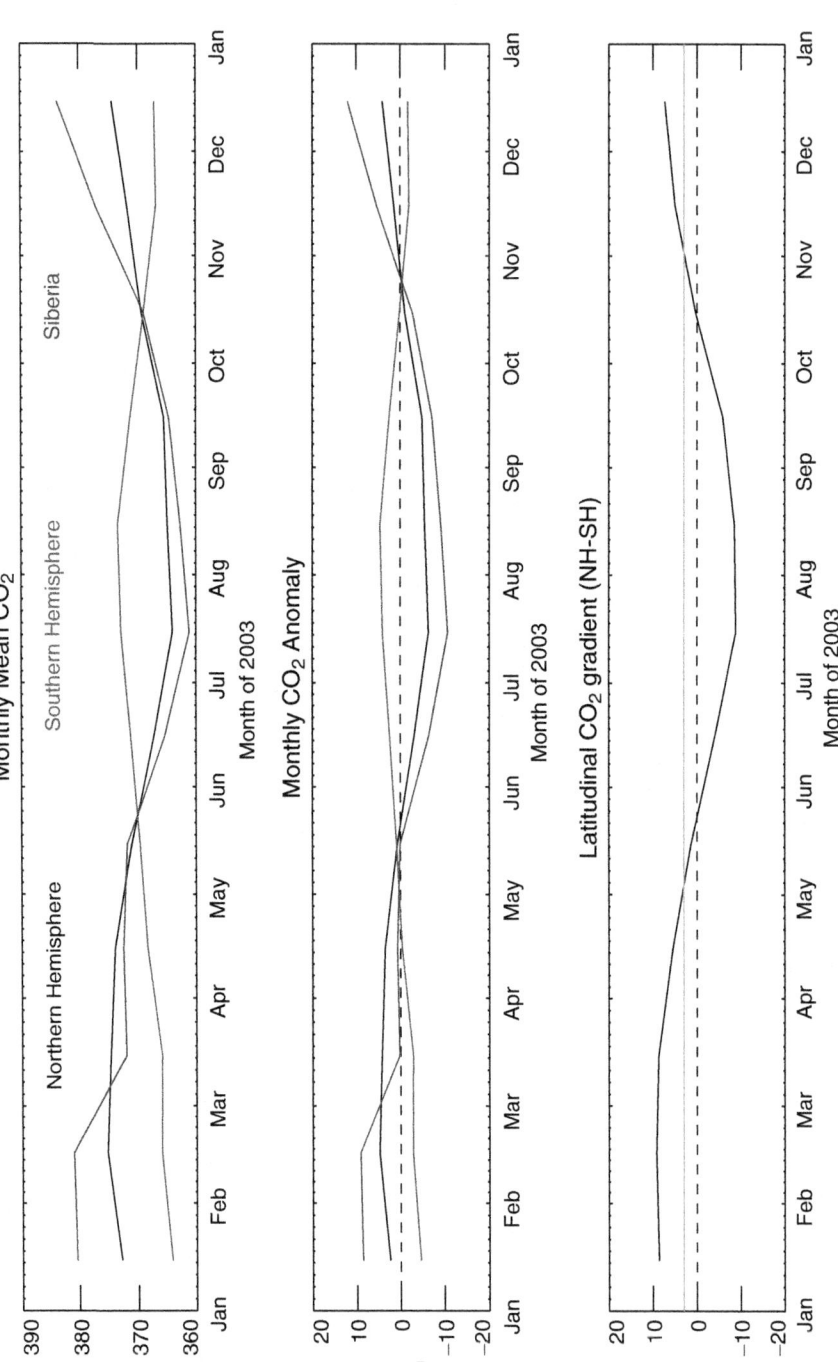

Fig. 1.2 *Top panel*: Time series for 2003 of the monthly means for the northern (*black*) and southern (*red*) hemispheres. The time series for Siberia (*blue*) is also shown. *Middle panel*: Time series of the CO_2 anomalies (i.e. each monthly mean minus the yearly mean). *Bottom panel*: The latitudinal inter-hemispheric gradient (north—south), the 3 ppmv gradient that is expected from surface observations (Conway et al. 1994) is shown as the *green line* (*Color version available in Appendix*)

Table 11.1 Summary of SCIAMACHY CO_2 validation exercises

Validation data	Summary	Reference
Egbert (IR) FTS	Possible contamination of FTIR columns measurements from nearby Toronto	Barkley et al. 2006c
	Bias ~−4%, 1σ error ~3%	
	Bias not constant, apparent seasonal trend	
Bremen (NIR) FTS	Few FTIR measurements available. Limited comparison	Barkley 2007
	Good agreement in January and December	
Park Falls (NIR) FTS	Comparison of daily averaged collocated SCIAMACHY measurements	Barkley et al. 2007
	Bias of ~−2%, 1σ error ~2%	
	No apparent seasonal trend in bias	
CO_2 retrieved from AIRS	Coincidental large scale features observed over North America during summer and autumn of 2003	Barkley et al. 2006c
CO_2 simulated by the TM3 Model	Annual model means detected to within ~2%	Barkley et al. 2006b
	Mean difference between SCIAMACHY and model 1–3%	
	Correlation between monthly time series >0.7	
	Seasonal cycle observed by SCIAMACHY is ~2–3 times larger than model	
	Good agreement in winter months but more CO_2 up take observed by SCIAMACHY in summer months	
Aircraft observations	Comparison over three Siberian locations	Barkley et al. 2007
	Correlation between time series >0.72	
	Mean difference between SCIAMACHY and aircraft ~2%	
Ground based surface and tower measurements	Consistent bias of ~−2%	Barkley et al. 2007
	Good agreement between monthly time series of CO_2 anomalies, correlations typically >0.7	
	SCIAMACHY able to track changes in tropospheric CO_2	
MODIS vegetation indices	Comparison limited to North America	Barkley et al. 2007
	Amount of CO_2 variability coincident with the amount of vegetation activity	

suitable since the Egbert FTS is in close proximity to Toronto, potentially suffering from regional pollution, and few measurements were available from the Bremen instrument during the time period of available SCIAMACHY retrievals. Comparison of the daily means retrieved by the FSI algorithm to those measured by the Park Falls FTS, reveal a negative bias in the SCIAMAHCY CO_2 that is a function of the collocation limits around the site tolerated for satellite overpasses. At close proximity (within $0.5° \times 0.5°$ of the FTS instrument) the mean bias is about −3.0% whereas at large distances (within $10.0° \times 10.0°$) the bias drops to −0.9%. This improvement is because (1) there are greater number of satellite and FTS match-ups, and (2) more SCIAMACHY observations are used to calculate each daily mean.

The standard SCIAMACHY/FSI product is often represented as a monthly average on a 1.0° × 1.0° grid; at these collocation limits the bias is −2.1% but the correlation between the SCIAMACHY and FTS daily means is quite low at 0.36. This indicates that SCIAMACHY cannot capture the day to day variability of the ground based measurements. It is only once the collocation boundaries are extended to 3.0° × 3.0° does the correlation become significant. However, if the 1.0° × 1.0° limits are selected but both sets of observations are assembled in to monthly averages, the correlation dramatically improves to 0.94 whilst the bias stays approximately the same. Thus, to detect monthly variability collocation limits of 1.0° × 1.0° are acceptable but to observe day to day variability a wider overpass criteria must be adopted. In either case the bias is about −2.0%.

In addition to Park Falls, another important validation exercise has been the comparison of SCIAMACHY CO_2 to spatially and temporally collocated aircraft CO_2 measurements over three Siberian locations: Novosibirsk, Surgut and Yakutsk. Flask samples were obtained during the aircraft flights during 2003, using the air-sampling method as outlined in Machida et al. (2001), with the CO_2 volume mixing ratios derived from the samples to an accuracy of ~0.10 ppmv, against standard gases, using a non-dispersive infrared analyzer (NDIR). To facilitate a meaningful comparison, SCIAMACHY observations occurring on the same day of each flight and collocated within ±10.0° longitude and ±8.0° latitude of each location, were averaged and compared to the mean of the aircraft measurements (convolved with a mean SCIAMACHY averaging kernel) over all sampling altitudes. Over Novosibirsk and Surgut aircraft measurements were made up to an altitude of 7 km but over Yakutsk only up to 3 km. The large spatial overpass limits are necessary as during some months there is little SCIAMACHY data owing to persistent cloud cover.

In general, SCIAMACHY CO_2 shows reasonable agreement to the aircraft data as in all instances the correlations between the data are greater than 0.7 (Table 11.2), suggesting that SCIAMACHY has the ability to track CO_2 variability over this region. For example over Novosibirsk (Fig. 11.3), the overall difference between the mean aircraft CO_2 and SCIAMACHY is approximately 2% with the correlation 0.77. In addition, the CO_2 anomalies measured over Novosibirsk show very good agreement between March–July and also similar behaviour between October–December. All the aircraft exhibit a large seasonal cycle amplitude (>20 ppmv) near the surface that decreases with altitude. Over Novosibirsk and Yakutsk, which are

Table 11.2 Summary of the aircraft and SCIAMACHY comparison over Siberia, using only coincidental observations (i.e. when both SCIAMACHY and aircraft observations occur on the same day)

Location	Vegetation	# of flights	Sampling Altitudes (km)	Seasonal cycle (ppmv)		Correlation r
				Aircraft	SCIAMACHY	
Novosibirsk	Forest	12	0.0–7.0	23.5	21.0	0.79
Surgut	Wetland	12	0.0–7.0	11.0	26.0	0.90
Yakutsk	Forest	20	0.0–3.0	25.0	17.5	0.72

11 Investigating Regional Scale Processes Using Remotely Sensed 183

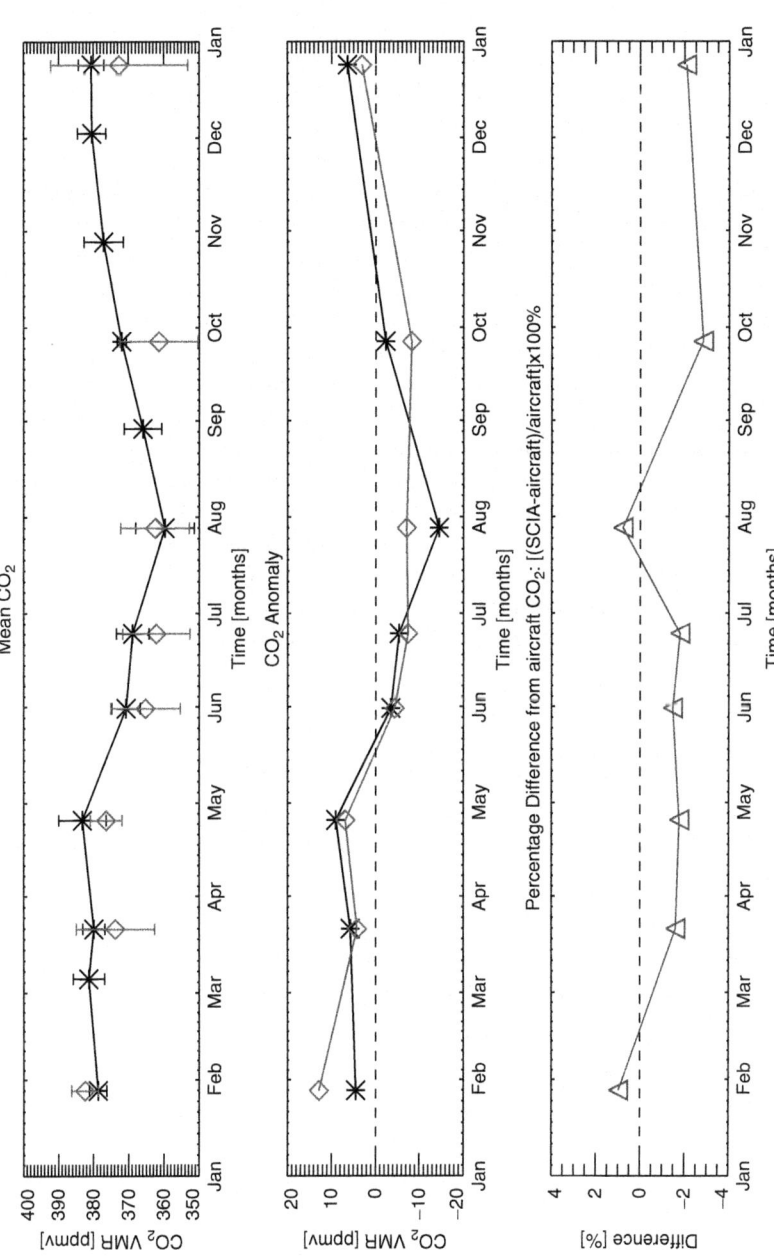

Fig. 11.3 SCIAMACHY (*red*) and aircraft (*black*) measurements over Novosibirsk. *Top panel*: The mean aircraft mixing ratio (over all altitudes) and the SCIAMACHY column VMRs. The *error bars* represent the 1 standard deviation uncertainty. *Middle panel*: The CO_2 anomaly (using only coincidental observations). *Bottom panel*: The percentage difference between SCIAMACHY and the mean aircraft mixing ratio (*Color version available in Appendix*)

surrounded by forests, the mean seasonal signal over all the aircraft sampling altitudes, are 23.5 and 25 ppmv respectively, which are larger than those observed by SCIAMACHY. Over Surgut, which is surrounded by wetlands, the seasonal amplitude measured by the aircraft is much smaller (11.0 ppmv), whilst SCIAMACHY observes more seasonal variation in the column VMR.

An important aspect of satellite CO_2 retrievals is the question of near surface sensitivity. Comparisons to CO_2 measurements made at the surface and on tall towers give an insight on the ability of SCIAMACHY to track changes in lower tropospheric CO_2. As shown by Barkley et al. (2007b), validation against surface measurements over Mongolia, Europe and North America revealed that out of 11 time series comparisons, eight had correlation coefficients greater than 0.7 and very similar seasonal cycle amplitudes, even over locations where smooth seasonal cycle did not exist. Unfortunately over Siberia there are not any surface stations that provide data; the nearest stations being located in Romania, Kyrgyzstan, Kazakhstan and Mongolia. In spite of this a comparison between the surface monthly anomalies, measured at these sites and those observed by SCIAMACHY are very similar. The SCIAMACHY anomalies are calculated as the mean of all surrounding grid boxes lying within 15° longitude and latitude of each site. These large collocation limits are selected so to be as close to or overlap the Siberia region. At each of these locations, the correlation between the time series are high (>0.8) with SCIAMACHY able detect a signal which is much alike to that measured at the surface. It is clearly not the case that retrievals are simply following the a priori CO_2 column (show in green in Fig. 11.4). Furthermore, the seasonal cycle amplitudes are also similar (see Table 11.3).

11.6 SCIAMACHY CO_2 Columns over Siberia

SCIAMACHY CO_2 column retrievals over Siberia (Fig. 11.5) are be characterized by three traits:

1. There are very few measurements during the winter months owing to persistent and widespread cloud coverage. The size of a SCIAMACHY ground pixel is quite coarse thus the probability of a cloud free scene is reduced which therefore

Table 11.3 Summary of the ground station and SCIAMACHY time series comparison (surface station data taken from the WDCGG network; http://gaw.kishou.go.jp/wdcgg/)

Location	Seasonal cycle amplitude (ppmv)		Correlation r
	Ground station	SCIAMACHY	
Black Sea (Romania)	16.1	20.7	0.81
Issyk-Kul (Krgystan)	12.3	13.5	0.74
Sary Taukum (Kazakhstan)	14.9	14.3	0.84
Plateau Assy (Kazakhstan)	13.8	13.4	0.87
Ulaan Uul (Mongolia)	11.5	13.6	0.95

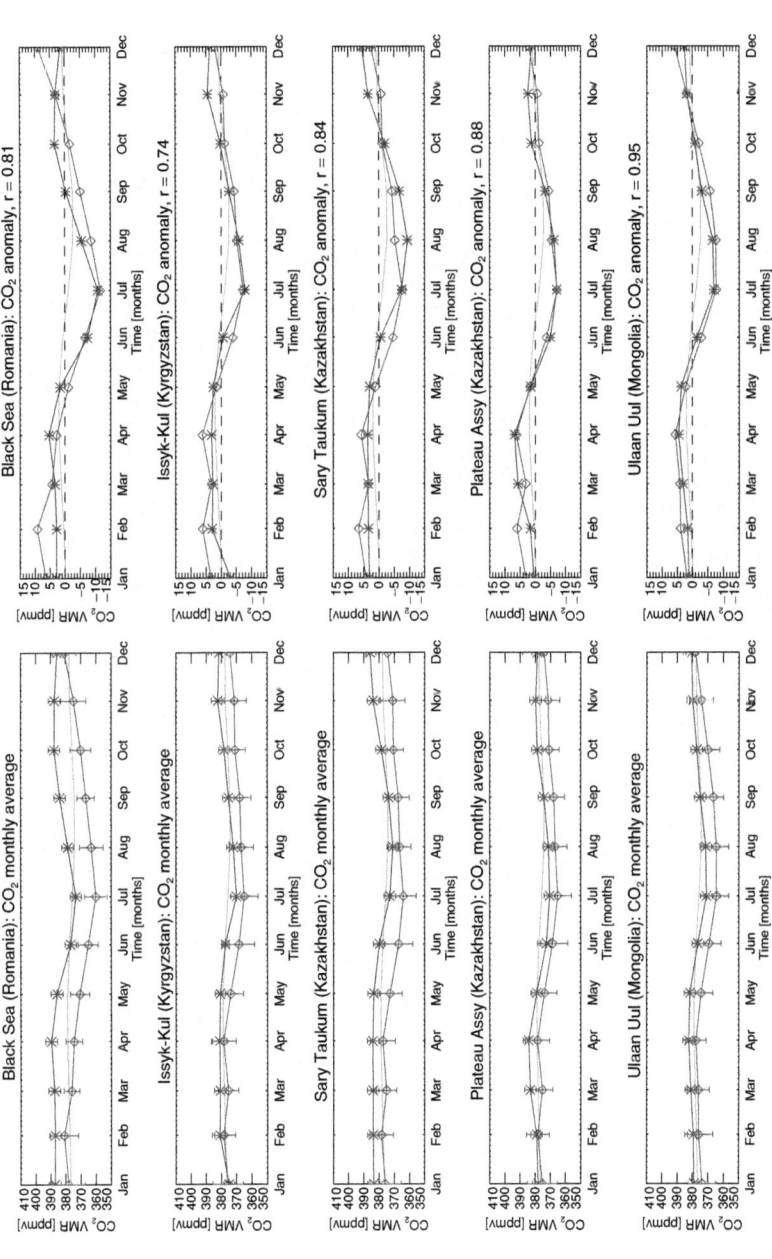

Fig. 11.4 Time series of ground based in situ observations (*blue*) versus SCIAMACHY CO$_2$ (*red*). *Left panels*: The CO$_2$ monthly averages with the 1 standard deviation error bars. *Right panels*: The corresponding CO$_2$ monthly anomaly. The a priori CO$_2$ column VMR is shown *in green*. The in situ data is from the World Data Centre for Greenhouse Gases (WDCGG) (http://gaw.kishou.go.jp/wdcgg/) (*Color version available in Appendix*)

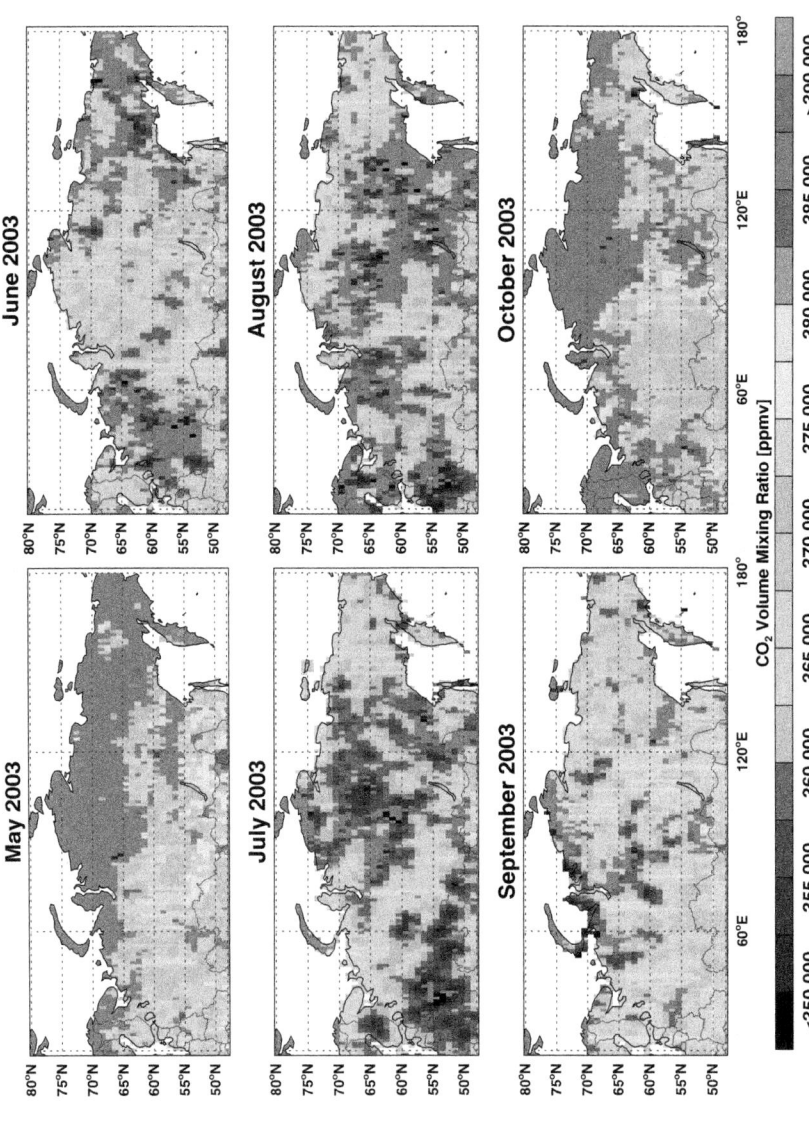

Fig. II.5 SCIAMACHY monthly CO_2 distributions over Siberia for May–September 2003 averaged on a $1° \times 1°$ grid and smoothed with a $3° \times 3°$ box car filter (*Color version available in Appendix*)

introduces a clear sky bias within the observations. In addition, snow and ice is very dark at NIR wavelengths hence, the retrieval over any cloud free scene is degraded by the lower signal to noise ratio of the recorded spectra, and is more likely to be discarded by the quality filtering.

2. Very low CO_2 VMRs during June, July and August are observed. This uptake is first noticeable in June over the West Siberian Plain and also on the along the coast of the Sea of Okhotsk and over East Siberian Mountains. By July the uptake is much more widespread with the lowest VMRs now over the Central Siberian Plateau. These features are still evident in August, although there is less observations owing to a de-icing phase of the SCIAMACHY detectors in this particular year. To the west of the Urals is massive uptake in July around regions of intensive agriculture.

3. There are sporadic localized enhancements, for example in October, SCIAMACHY sees an enhancement approximately along the Yenisey River (which splits the West Siberian Plain and the Central Siberian Plateau). Similarly in May, there are large CO_2 VMRs seen over the Yablonovyy mountain range (approximately 115° E 49° N). Such features are not discernable in model simulations (Barkley et al. 2006b). The enhancement in May is in the close vicinity of a cluster of wild fires as detected by the Along Track Scanning Radiometer (ATSR) instrument. It is difficult to clarify whether the high CO_2 VMRs are directly from biomass burning emissions or created falsely through aerosol scattering. Nevertheless, SCIAMACHY detects a signal that is a created by a surface process.

The question remains, are these features real, i.e. are they the signature of surface fluxes? Can the CO_2 enhancements and depletions be related to surface processes such as biomass burning or photosynthetic activity, or are they simply a residual trait of the retrieval itself, like a surface albedo effect? A (seasonal) surface reflectance bias can be discounted for two reasons. Firstly, an a priori albedo determined from the mean radiance of each SCIAMACHY measurement, is used within the FSI algorithm to generate each individual reference spectrum. Whilst the CO_2 distributions evolve considerably with time, the variation in surface albedo is small. Secondly, a comparison between AIRS and SCIAMACHY over North America, performed by Barkley et al. (2006c) demonstrated that both instruments essentially observe the same large scale features. AIRS is a thermal IR instrument which is insensitive to the surface albedo, thus if SCIAMACHY observes the same features, when accounting for surface reflectance within the retrieval, then it too must be observing the same CO_2 fluctuations.

The CO_2 drawdown that is widespread over Siberia is much more localized over North America (Fig. 11.6), demonstrating the power of satellite observations to, at least qualitatively, identify continental differences. For example, over North America during July, the transition from low CO_2 along the Canadian Shield and along the eastern US coast, to higher CO_2 over the arid south western states corresponds to a change from evergreen, deciduous broadleaf and mixed forest to land covered by crops and large grass plains (Fig. 11.7). It is therefore most likely that SCIAMACHY is detecting greater CO_2 uptake by the forest compared to the

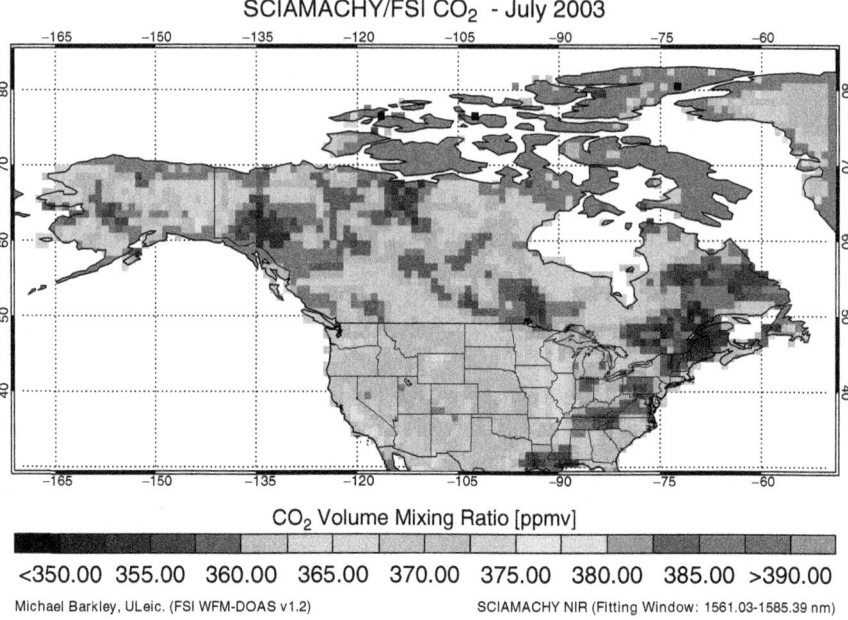

Fig. 11.6 SCIAMACHY CO_2 over North America for July 2003 (*Color version available in Appendix*)

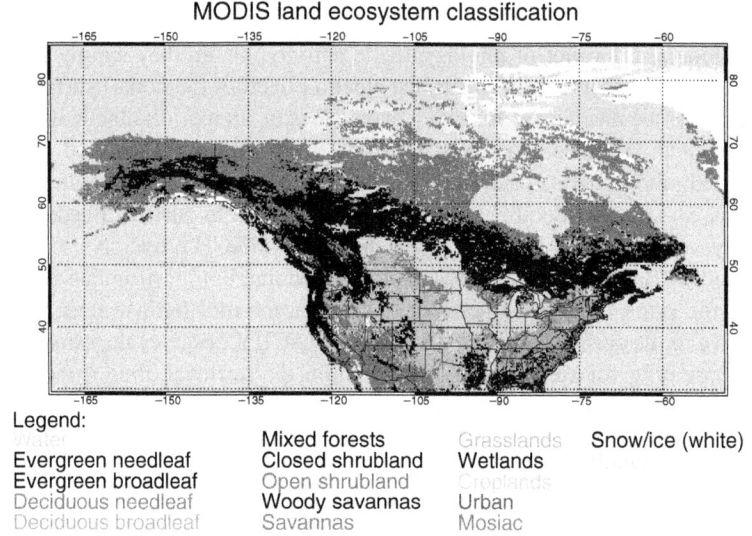

Fig. 11.7 Map of the land vegetation distribution over North America. The vegetation map is taken from the Land Ecosystem Classification Product which is a static map generated from the official MODIS land ecosystem classification data set, MOD12Q1 for year 2000, day 289 data (October 15, 2000) (see http://modis-atmos.gsfc.nasa.gov/ECOSYSTEM/index.html) (*Color version available in Appendix*)

farmed areas and grass plains. The good agreement with surface station data shows that SCIAMACHY can track surface CO_2 and the agreement with AIRS indicates that these large scale features are not linked to the surface albedo, hence they are most likely to be real. Tagged CO_2 model simulations reveal that the biggest contribution to this low swath of CO_2 comes from the North American terrestrial biosphere (Palmer et al. 2008). Intercontinental transport from Asia also influences North America continent, as a significant component of the low CO_2 VMRs over Alaska and northern Canada comes from Asia's boreal forests. Over Siberia, the CO_2 uptake occurs over such an extensive area it is difficult to clearly identify any correlation with vegetation type and cover.

A comparison between SCIAMACHY CO_2 and MODIS Enhanced Vegetation Index (EVI) and the Normalized Difference Vegetation Index (NDVI) at over twenty sites within the US also provides further evidence that SCIAMACHY can detect a biospheric CO_2 signature in the atmosphere (Barkley et al. 2007). At locations where the vegetation activity is high, i.e. where there are large seasonal swings in the magnitude of vegetation indices, the CO_2 variability is also significant. Whereas over locations where the vegetation activity is low, the CO_2 variability is also low. Such findings strongly suggest that SCIAMACHY is able to track changes in lower tropospheric CO_2 that arise from changes in terrestrial vegetation activity, or in other words, SCIAMACHY is able to identify regional terrestrial CO_2 sinks.

Unlike North America, where a great wealth of observational data is exists, few field observations and measurement campaigns (relevant for satellite validation) are available over Siberia, preventing analyses like those discussed above. However, the studies over North America do provide confidence in the SCIAMACHY CO_2 retrievals. Thus, the widespread CO_2 uptake observed during summer over Siberia *is* most likely because of the intense vegetation activity, highlighting the important role of Siberia in the global carbon cycle and the need for close environmental monitoring of this region in the future.

11.7 Summary

SCIAMACHY has the ability to measures CO_2 from space and more importantly is able to detect the signature of the terrestrial biosphere at regional scales, though this is probably at the limit of the instrument's sensitivity. However, the retrievals are still not well characterized enough for inverse modelling purposes. So whilst a CO_2 signal is detectable, the question remains just how useful the data is quantitatively. Qualitatively at least, SCIAMACHY is able to distinguish continental differences, especially between North America and Eurasia, and the retrievals over Siberia indicate that the regional CO_2 exchange between the atmosphere and terrestrial biosphere is considerable.

Probably the most important aspect of the SCIAMACHY CO_2 retrievals however, is the encouragement it should give to the future OCO & GOSAT missions. Both of these missions are dedicated to CO_2 science (unlike SCIAMACHY) and have high specification instruments and more sophisticated retrieval algorithms (e.g. Bösch et al. 2006). Furthermore, much effort has been invested in understanding potential biases and sampling errors associated with future satellite CO_2 measurements (Corbin and Denning 2006; Lin et al. 2004; Miller et al. 2007) and to understand how they will improve model inversions (Chevallier et al. 2007). The data from these instruments should provide the first operational monitoring of CO_2 distributions with measurement precisions suitable for inverse modelling. It is highly anticipated that the forthcoming satellite data will ultimately lead to an improved understanding and quantification of regional carbon fluxes and place tighter constraints on the global carbon cycle budget.

References

Barkley MP (2007) Measuring atmospheric CO_2 from space. Ph.D. thesis, University of Leicester
Barkley MP, Frieß U, Monks PS (2006a) Measuring atmospheric CO_2 from space using Full Spectral Initiation (FSI) WFM-DOAS. Atmos Chem Phys 6:3517–3534
Barkley MP, Monks PS, Frieß U, Mittermeier RL, Fast H, Körner S, Heimann M (2006b) Comparisons between SCIAMACHY atmospheric CO_2 retrieved using (FSI) WFM-DOAS to ground based FTIR data and the TM3 chemistry transport model. Atmos Chem Phys 6:4483–4498
Barkley MP, Monks PS, Engelen RJ (2006c) Comparison of SCIAMACHY and AIRS CO_2 measurements over North America during the summer and autumn of 2003. Geophys Res Lett 33:L20805. doi:10.1029/2006GL026807
Barkley MP, Monks PS, Hewitt AJ, Machida T, Desai A, Vinnichenko N, Nakazawa T, Yu Arshinov M, Fedoseev N, Watai T (2007) Assessing the near surface sensitivity of SCIAMACHY atmospheric CO_2 retrieved using (FSI) WFM-DOAS. Atmos Chem Phys 7:3597–3619
Bösch H et al (2006) Space-based near-infrared CO_2 measurements: Testing the Orbiting Carbon Observatory retrieval algorithm and validation concept using SCIAMACHY observations over Park Falls, Wisconsin. J Geophys Res 111:D23302. doi:10.1029/2006JD007080
Bovensmann H, Burrows JP, Buchwitz M, Frerick J, Nöel S, Rozanov VV, Chance KV, Goede A (1999) SCIAMACHY – mission objectives and measurement modes. J Atmos Sci 56:127–150
Buchwitz M, Rozanov VV, Burrows JP (2000) A near infrared optimized DOAS method for the fast global retrieval of atmospheric CH_4, CO, CO_2, H_2O, and N_2O total column amounts from SCIAMACHY/ENVISAT-1 nadir radiances. J Geophys Res 105:15231–15246
Buchwitz M, de Beek R, Nöel S, Burrows JP, Bovensmann H, Bremer H, Bergamaschi P, Körner S, Heimann M (2005) Carbon monoxide, methane and carbon dioxide columns retrieved from SCIAMACHY by WFM-DOAS: Year 2003 initial data set. Atmos Chem Phys 5:3313–3329
Buchwitz M, de Beek R, Nöel S, Burrows JP, Bovensmann H, Schneising O, Khlystova I, Bruns M, Bremer H, Bergamaschi P, Körner S, Heimann M (2006) Atmospheric carbon gases retrieved from SCIAMACHY by WFM-DOAS: version 0.5 CO and CH_4 and impact of calibration improvements on CO_2 retrieval. Atmos Chem Phys 6:2727–2751
Buchwitz M, Schneising O, Burrows JP, Bovensmann H, Notholt J (2007) First direct observation of the atmospheric CO_2 year-to-year increase from space. Atmos Chem Phys Discuss 7:6719–6735
Chédin A, Serrar S, Scott NA, Crevoisier C, Armante R (2003) First global measurement of midtropospheric CO_2 from NOAA polar satellites: Tropical zone. J Geophys Res 108(D18):4581. doi:0.1029/2003JD003 439
Chevallier F, Bréon F-M, Rayner PJ (2007) Contribution of the Orbiting Carbon Observatory to the estimation of CO_2 sources and sinks. Theoretical study in a variational data assimilation framework. J Geophys Res 112:D09307. doi:10.1029/2006JD007375

Conway TJ, Tans PP, Waterman LS, Thoning KW, Kitzis DR, Masarie KA, Zhang N (1994) Evidence for interannual variability of the carbon cycle from the National Oceanic and Atmospheric Administration/Climate Monitoring and Diagnostic Laboratory Global Air Sampling. J Geophys Res 99:22831–22855

Corbin KD, Denning AS (2006) Using continuous data to estimate clear-sky errors in inversions of satellite CO_2 measurements. Geophys Res Lett 33:L12810. doi:10.1029/2006GL025910

Denning AS, Fueng IY, Randall D (1995) Latitudinal Gradient of atmospheric CO_2 due to seasonal exchange with land biota. Nature 376:240–242

Dils B et al (2006) Comparisons between SCIAMACHY and ground-based FTIR data for total columns of CO, CH_4, CO_2 and N_2O. Atmos Chem Phys 6:1953–1976

Engelen RJ, McNally AP (2005) Estimating atmospheric CO_2 from advanced infrared satellite radiances within an operational four-dimensional variational (4DVar) data assimilation system: Results and validation. J Geophys Res 110(D18):305. doi:10.1029/2005JD005982

Friedlingstein P, Dufresne J-L, Cox PM, Rayner P (2003) How positive is the feedback between climate change and the carbon cycle? Tellus 55B:692–700

Gloudemans AMS, Schrijver H, Kleipool Q, van den Broek MMP, Straume AG, Lichtenberg G, van Hees R, Aben I, Meirink JF (2005) The impact of SCIAMACHY near-infrared instrument calibration on CH_4 and CO total columns. Atmos Chem Phys 5:2369–2383

Gottwald M et al (2006) SCIAMACHY monitoring the earth's changing atmosphere. DLR, Insitut fúr Methodik der Fernerkundung (IMF), Germany

Gurney KR et al (2002) Towards robust regional estimates of sources and sinks using atmospheric transport models. Nature 415:626–630

Houweling S, Bréon F-M, Aben I, Rödenbeck C, Gloor M, Heimann M, Ciais P (2004) Inverse modeling of CO_2 sources and sinks using satellite data: a synthetic inter-comparison of measurement techniques and their performance as a function of space and time. Atmos Chem Phys 4:523–538

Houweling S, Hartmann W, Aben I, Schrijver H, Skidmore J, Roelofs G-J, Bréon F-M (2005) Evidence of systematic errors in SCIAMACHY-observed CO_2 due to aerosols. Atmos Chem Phys 5:3003–3013

Krijger JM, Aben I, Schrijver H (2005) Distinction between clouds and ice/snow covered surfaces in the identification of cloud-free observations using SCIAMACHY PMDs. Atmos Chem Phys 5:2279–2738

Lin JC, Gerbig C, Daube BC, Wofsy SC, Andrews AE, Vay SV, Anderson BE (2004) An empirical analysis of the spatial variability of atmospheric CO_2: Implications for inverse analyses and spaceborne sensors. Geophys Res Lett 31:L23104. doi:10.1029/2004GL020957

Machida T, Nakazawa T, Muksyutov S, Tohjima Y, Takahashi Y, Watai T, Vinnichenko N, Panchenko M, Arshinov M, Fedoseev N, Inoue G (2001) Temporal and spatial variations of atmospheric CO_2 mixing ratio over Siberia. In: Proceedings of the sixth international CO_2 conference, Sendai, Japan

Miller CE et al (2007) Precision requirements for space-based XCO_2 data. J Geophys Res 112:D10314. doi:10.1029/2006JD007659

Olsen SC, Randerson JT (2004) Differences between surface and column atmospheric CO_2 and implications for carbon cycle research. J Geophys Res 109(D02301). doi:10.1029/2003JD003968

Palmer PI, Barkley MP, Monks PS (2008) Interpreting the variability of space-borne CO_2 column-averaged volume mixing ratios over North America using a chemistry transport model. Atmos Chem Phys 8:5855–5868

Patra PK et al (2006) Sensitivity of inverse estimation of annual mean CO_2 sources and sinks to ocean-only sites versus all sites observational networks. Geophys Res Lett 33:L05814. doi:10.1029/2005GL025403

Prentice IC, Farquhar GD, Fasham MJR, Goulden ML, Heimann M, Jaramillo VJ, Kheshgi HS, Le Quéré C, Scholes RJ, Wallace DWR (2001) The Carbon Cycle and Atmospheric Carbon Dioxide. In: Houghton JT, Ding Y, Griggs DJ, Noguer M, van der Linden PJ, Dai X, Maskell K, Johnson CA (eds). Climate Change 2001. The Scientific Basis. Contribution of Working

Group I to the Third Assessment Report of the Intergovernmental Panel on Climate Change Cambridge University Press, Cambridge, United Kingdom and New York, NY, USA, pp 881

Rayner PJ, O'Brien DM (2001) The utility of remotely sensed CO_2 concetration data in surface source inversions. Geophys Res Lett 28:175–178

Remedios JJ, Parker RJ, Panchal M, Leigh RJ, Corlett G (2006) Signatures of atmospheric and surface climate variables through analyses of infrared spectra (SATSCAN-IR). Proceedings of the first EPS/METOP RAO Workshop, ESRIN, Italy

Rödenbeck C, Houweling S, Gloor M, Heimann M (2003) CO_2 flux history 1982–2001 inferred from atmospheric data using a global inversion of atmospheric transport. Atmos Chem Phys 3:1919–1964

Rozanov VV, Buchwitz M, Eichmann KU, de Beek R, Burrows JP (2002) SCIATRAN – a new radiative transfer model for geophysical applications in the 240–2400 nm spectral region: The pseudo-spherical version, presented at COSPAR 2000. Adv Space Res 29(11):1831–1835

Sabine CL, Freely RA, Gruber N, Key RM, Lee K, Bullister JL, Wanninkhof R, Wong CS, Wallace DWR, Tilbrook B, Millero FJ, Peng T-H, Kozyr A, Ono T, Roso AF (2004) The oceanic sink for anthropogenic CO_2. Science 305:367–371

Siegenthaler U, Stocker TF, Monnin E, Lüthi D, Schwander J, Stauffer B, Raynaud D, Barnola J-M, Fischer H, Valérie Masson-Delmotte JJ (2005) Stable carbon cycle–climate relationship during the Late Pleistocene. Science 310:1313–1317

Chapter 12
Climatic and Geographic Patterns of Spatial Distribution of Precipitation in Siberia

A. Onuchin and T. Burenina

Abstract The spatial–temporal distribution of precipitation is a function of atmospheric circulation and the orography of the terrain. Due to these factors the spatial–temporal distribution of precipitation differs on global, regional and local levels. Three vast Siberian ecoregions (Western Siberia, Central Siberia and Eastern Siberia) are differing in their space-temporal patterns of precipitation distribution.

The spatial distribution of precipitation over West Siberia follows a geographical zonation: precipitation changes from 300 mm in the south to 400–500 mm in the forest zone. The areas of Central and East Siberia with extreme continental climates and mountain relief differ in precipitation and moisture characteristics to a great extent. In Central Siberia precipitation varies between 325 and 525 mm, in East Siberia between 250 and 330 mm.

Climatic and geographic patterns of the spatial precipitation distribution in Central Siberia are studied on a regional level. Computer models of spatial precipitation distribution were developed for the Yenisei Mountain Chain, Eastern Sayan, and the South-eastern Trans-Baikal region.

Owing to irregular spatial distribution of precipitation three groups of landscapes were defined: (1) slopes of west, north-west and south-west aspect with orographic precipitation; (2) shadow slopes in mountain regions; (3) plain landscapes. Obtained equations show correlations between the amount of precipitation and altitude, geographical latitude, distance from barrier ridge and other parameters.

Keywords Distribution of precipitation • Ecoregion • Siberia • Spatial precipitation patterns • Geographical zonality

---

A. Onuchin (✉) and T. Burenina
V.N. Sukachev Institute of Forest,
SB RAS, Akademgorodok, 50, Krasnoyarsk 660036, Russia
e-mail: onuchin@ksc.krasn.ru; burenina@ksc.krasn.ru

12.1 Introduction

Atmospheric precipitation is the major moisture circulation component that has a profound influence on terrestrial ecosystem climate, microclimate, and hydrological regimes. This explains the numerous studies undertaken to provide a better understanding of spatial and temporal precipitation patterns under various vegetation and climatic conditions (Arkhangelsky 1960; Berg and Shenrok 1925; Bradley 1966; Burenina et al. 2002; Glebova 1958; Gorec and Younkin 1966; Govsh 1962; Kolomyts 1975; Kopanev 1966; Matasov 1938; Matveyev 1968, 1984; Mellor and Smith 1966; Richter and Petrova 1960; Richter 1963, 1984; Shpak and Bulavskaya 1967; Sosedov 1962, 1967; Tikhomirov 1956; Trifonova 1962; Vinogradov 1964). Spatial precipitation non-uniformity is most readily apparent in mountains, is attributable to trajectories of air masses within the ocean–land system, distance from the ocean, atmospheric processes characteristic of different periods of time, and underlying surface features. According to the current general precipitation scheme, precipitation increases with increasing elevation and in steadily low-pressure belts, one of which lies between 60° and 70° N latitude, and decreases with the distance from the ocean. However, this general scheme contains almost as many exceptions as strong trends.

Precipitation patterns are extremely complicated in mountainous countries. Air flows occurring around mountain systems and single (separate) elevations enhance precipitation on windward slopes, while reducing its amount on downwind slopes (Beyer 1966; Guralnik et al. 1972). The outermost upwind slopes, acting as natural moist air mass breaks, receive more precipitation than those located deeper in mountain systems, even if these in-mountain slopes are higher compared to the outermost ones (Ladeishchikov 1982). In this case, precipitation amount is controlled by a number of factors, such as air mass water content and movement relatively to mountain "barriers", thermal layering of the atmosphere, and underlying surface characteristics. Precipitation is known to vary with elevation above sea level (a.s.l.). In mountains, precipitation usually increases up to a certain elevation called a peak-moisture zone, beyond which precipitation stops to increase and can even decrease with elevation due to decreasing water vapour concentration in the upper atmospheric layers.

These relationships exhibit different behaviour depending on specific orographic and climatic characteristics of mountainous countries. This dependence was supported by research studies conducted high in the Alps (Berg 1938), Pamir-Altai and the Pamir highland (Kotlyakov 1968). Among other factors, these differences are accounted for by foehn, warm and dry wind that occurs in mountains due to downdraughts found in an atmospheric layer not less than 0.5–1 km deep and promotes moisture evaporation. The relationship between precipitation amount and topography was the focus of a number of studies (Burenina 1998; Hartzman 1971; Korytny 1980; Onuchin 1987; Onuchin and Burenina 2002; Shultz 1972). In closed mountain hollows, precipitation depends on the distance from the surrounding barriers (Korytny 1980). Precipitation modelling for upwind slopes breaking the prevailing atmospheric moisture transfer is a sophisticated process, since it has to consider many more influences (Onuchin and Burenina 2002). Where a moisture transfer trajectory coincides with large valley orientation, atmospheric moisture can be brought to a fairly small region and discharged there through heavy snow or rainfalls (Suslov 1954). Although spatial and temporal precipitation patterns are known to be controlled by numerous landscape-specific factors, precipitation modelling is usually reduced, with very few exceptions, to interpolating scarce data provided by a sparse weather station network and assessing the precipitation gradient. The qualitative trends identified so far for spatial precipitation patterns are still under considered in development of the quantitative methods adapted to the local conditions of the territory.

Little is known about spatial precipitation distribution in Siberia, with most of the available publications focusing on spatial precipitation non-uniformity in the southern Siberian Mountains (Grudinin 1979, 1981; Lebedev 1979, 1982). Mountain range direction, elevation a.s.l. and location with respect to winds carrying moisture are responsible for this non-uniformity. Orographic diversity was the reason of dividing mountains of southern Siberia into regions differing in annual precipitation and its vertical gradient (Chebakova 1986; Grudinin et al. 1975). A number of scientists (Afanasyev 1976; Ladeishchikov 1982), in their studies of the precipitation patterns in lake Baikal catchment, noted that the vertical precipitation gradient varied in mountains from 20 to 1,000 mm per 100 m elevation step depending on the distance from Lake Baikal and the angle between slopes and wet winds. These factors introduce considerable ambiguity into interpretations of precipitation changes with increasing elevation a.s.l.

Regional and zonal snowfall matches the general macro-scale precipitation occurrence trends found for Siberia, except that the snow cover distribution is much more of a mosaic due to numerous influences. Snow cover development on mountain slopes, under the forest canopy and in open sites, and spatial snow pack patterns in northern Eurasia were addressed by a number of earlier studies (Onuchin 1984, 1985, 2001; Onuchin and Burenina 1996a, b). Since snow cover development requires an extended separate discussion, we leave it outside the scope of this chapter and focus on general spatial precipitation trends in Siberia.

12.2 Study Area and Methods

Siberia stretches from the Ural Mountains eastward as far as the Lena river and includes the West Siberian Plain, Central Siberian Tableland, the watershed of Lena river, the Altai-Sayan mountain range, and the Lake Baikal region. Our study area covered the territory from 50° to 70° N latitude and from 60° to 160° E longitude.

Our ground truth data were obtained in the Putoran Plateau, Yenisei Mountain Chain, north-eastern and central Siberia respectively (Fig. 12.1), south-western Siberia, Eastern Sayan, and the south-eastern Trans-Baikal region. The study sites thus covered the entire range of Siberian vegetation zones, from tundra to steppe and all altitudinal vegetation zones in mountains.

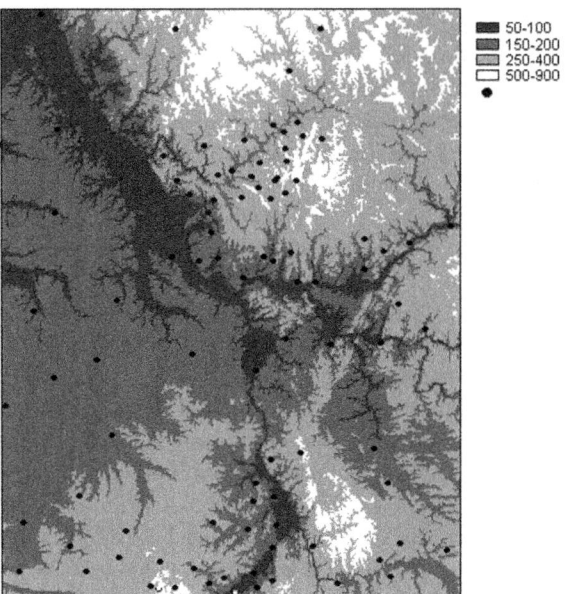

Fig. 12.1 Location of points of snow precipitation measurements in Central Siberia. *Dots* are points of snow precipitation measurement; the scale is altitude (m)

The precipitation distribution was investigated at ecoregional[1] and local levels. In the latter case, spatial precipitation models were developed for different areas within ecoregions accounting for local orography and air mass circulation. Macroscale (i.e. ecoregional) precipitation distribution was analyzed using weather data provided by 130 weather stations situated in the study area (USSR Climate Guide 1956, 1969, 1970). Weighted average annual precipitation was calculated by spatially averaging for each study area (forest provinces or forest districts according to the current Russian forest regionalization by Korotkov (1994)).

Siberia is remarkable for contrasting natural conditions; its weather station network is highly irregular, and particularly sparse in the north. Therefore, one of the most reliable methods to obtain average precipitation for an area involves building average precipitation isoclines and measuring areas between pairs of adjacent isoclines by planometric techniques (Schwer 1984). Average precipitation values were calculated for the ecoregions of interest by the following equation:

$$F_s = \frac{S_k}{S} \times \sum_1^m \frac{F_k + F_{k+1}}{2} \qquad (12.1)$$

where S_k is area between a pair of isolines corresponding to F_k

---

[1]Ecoregion refers to a big area identified based on a landscape approach. Ecoregion examples in Siberia are West-Siberian Plain, Central Siberian table land, and Altai-Sayan mountain system. Ecoregional boundaries coincide with those of forest oblasts as identified by Korotkov (1994).

and F_{k+1} average precipitation values, m is the number of these areas, and S is the entire area for which averaging is done.

Local precipitation distribution was calculated based on weather station information combined with our in situ precipitation measurements. The data obtained were subjected to multiple regression analysis (Lvovsky 1988).

12.3 Results

12.3.1 *Spatial Precipitation Patterns in Siberian Ecoregions*

The interactions of the atmosphere with the underlying land surface, with the former being dynamic in time and the latter having noticeable spatial variability, are responsible for precipitation development and, hence, its spatial and temporal patterns.

Atmospheric precipitation is a key moisture cycling component which controls hydrological regimes of terrestrial ecosystems to a great extent. Precipitation pattern is a major factor accounting for landscape differentiation. The precipitation amount varies, in turn, depending on regional atmospheric circulation and underlying surface characteristics.

Geographical contrasts of Siberia are reflected in both latitudinal and meridian precipitation occurrence in this region. For this reason, it is most appropriate to analyze macro-scale precipitation distribution based on large ecoregions. According to the current Russian forest regionalization (Korotkov 1994), eight forest regions (defined as "ecoregions") found in Siberia contain 32 forest provinces, which are divided into subprovinces (or forest districts) (Fig. 12.2). This forest regionalization considers precipitation amount by latitude and continentality. Table 12.1 shows the spatial distribution of precipitation for forest provinces and forest districts inside of large Siberian ecoregions. The numbers of forest provinces and districts in Table 12.1 are given according to Fig. 12.2.

A latitudinal precipitation change characteristic of the West Siberian Plain is manifest in the decreasing precipitation amounts proceeding south and northward from the taiga forest provinces. Precipitation is fairly high (505–508 mm) in the northern and southern taiga, while it decreases further to the north and averages only 378 mm in the north of the Trans-Ural-Yenisei province, which is actually forest-tundra. The lowest precipitation (296 mm) was found for the Kulunda province situated in the steppe zone.

Precipitation tends to decrease west-eastward in eastern Siberia. The western part of the Central Siberian tableland can be considered as a transition to the dry continental climate of eastern Siberia. While the maximum annual precipitation (617–675 mm) was found to occur in the Putoran and Yenisei forest provinces, the Anabar province and Kotuy-Olenek forest district received the lowest amount of precipitation (325–342 mm). The non-uniform precipitation distribution found within the western part of the Central Siberian tableland is most probably caused by its orography.

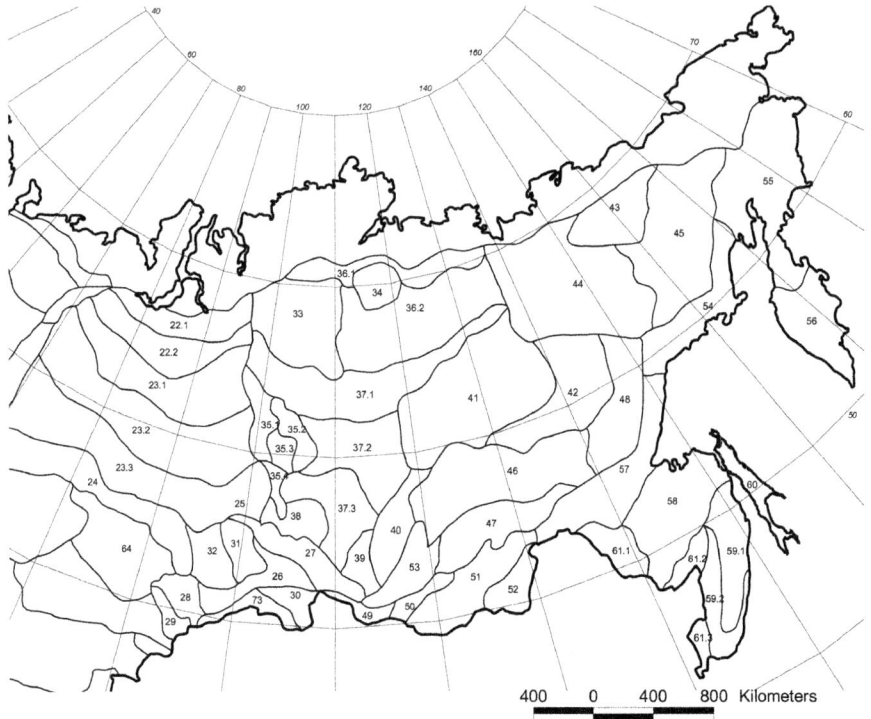

Fig. 12.2 Scheme of forest regionalization of Russia by Korotkov (1994)

The Lena-Veluy and Aldan forest provinces situated in central Yakutia appeared to be fairly dry, with annual precipitation ranging from 298–335 mm. Precipitation exhibited a decrease moving southward, down to the Trans-Baikal region. All study areas were determined to receive precipitation mainly through rainfall, with snowfall accounting for only 25–30% of the annual total.

Since mountains are known to enhance atmospheric processes responsible for precipitation development, precipitation is much heavier in mountains compared to the plains surrounding them. Precipitation changes with elevation depend on the atmospheric circulation over a particular mountain massif location. For any mountain system, there exists a certain upper precipitation threshold depending on elevation controlled by rising air. Besides global air mass transfer, precipitation in mountains is considerably impacted by large-scale thermal circulation between a mountain system and the adjacent plain.

The mountains of southern Siberia have considerable differences in their precipitation distribution associated with mountain range height, morphometric characteristics, and position relative to the prevailing moisture transfer. An annual precipitation of 600–1,048 mm was found to be common in the Altai-Sayan ecoregion, whereas the northern and southern Trans-Baikal ecoregions were determined to receive 450–527 and 373–465 mm per year, respectively. Precipitation amount

Table 12.1 Precipitation patterns in Siberian ecoregions

No	Forest provinces and districts	Precipitation (mm)		
		Warm period (April–October)	Cold period (November–March)	Annual
West Siberian Plain forest oblast (ecoregion)				
22.	Trans-ural forest province: near-tundra forests and open woodlands			
22.1.	Forest-tundra district	378	256	122
22.2.	District of northern taiga open stands and open woodlands	466	330	136
23.	Trans-Ural-Yenisei taiga forest province			
23.1.	Northern taiga forest district forest district	508	375	133
23.2.	Central taiga forest district	505	378	127
23.3.	Southern taiga and subtaiga forest district	471	369	102
24.	Irtish-Ob forest-steppe province	414	368	46
64.	Kulunda steppe province	296	254	42
Altai-Sayan Mountain forest oblast (ecoregion)				
26.	Northern Altai-Sayan forest province	781	651	130
27.	Eastern Sayan forest province	663	542	121
28.	Central Altai forest province	772	430	342
29.	Western Altai forest province	986	594	392
30.	Eastern Tuva (Tojin) forest province	660	514	146
31.	Khakasia-Minusinsk forest province	463	387	76
32.	Salair-Kuznetsk forest province	1,078	668	410
Central Siberian Tableland forest oblast (ecoregion)				
33.	Putoran mountain forest province	617	422	195
34.	Anabar forest province	325	250	75
35.	Yenisei forest province	675	461	214
36.	Kotuy-Olenek forest province	342	268	74
37.	Angara-Tunguska taiga forest province			
37.1.	Lower Tunguska forest district	548	377	171
37.2.	Podkamennaya Tunguska forest district	489	358	131
37.3.	Fore-Angara forest district	465	355	110
38.	Kansk forest-steppe province	435	382	53
39.	Upper Angara forest province	425	360	65
40.	Upper Lena forest province	411	356	55
Central Yakutian plain forest Oblast (ecoregion)				
41.	Lena-Viluy forest province	298	204	94
42.	Aldan forest province	335	230	105

(continued)

Table 12.1 (continued)

		Precipitation (mm)		
		Warm period (April–October)	Cold period (November–March)	Annual
Yan-Kolyma mountain forest Oblast (ecoregion)				
43.	Lower Kolyma forest province	256	171	85
44.	Yan-Indigirka forest province	278	181	97
45.	Kolyma forest province	302	170	132
Northern Trans-Baikal Forest Oblast (ecoregion)				
46.	Baikal-Stanovaya forest province	527	307	220
47.	Upper Vitim-Olekin tableland forest province	496	289	207
48.	Uchura-Maya forest province	450	308	142
Southern Trans-Baikal mountain forest oblast (ecoregion)				
49.	Jidin forest province	465	362	103
50.	Selenga forest province	373	317	56
51.	Chikoy-Ingodin forest province	446	336	110
52.	Dahur forest province	391	304	87
Fore-Baikal mountain forest oblast (ecoregion)				
53.	Fore-Baikal forest province	596	405	191

appeared to vary in southern mountain forests found in each of the ecoregions under study. This difference was determined to be 100–150 mm for the forest provinces of the Southern and Northern Trans-Baikal ecoregions, while it exceeded 500 mm for those in the Altai-Sayan ecoregion Table 12.1). Furthermore, the amount of precipitation presumably varies within provinces with elevation and slope aspect.

Annual precipitation exhibited a considerable variability in the Altai Mountains. West-facing slopes were found to receive over 1,600 mm, while the value appeared to be as low as 300 mm for interior downwind slopes, narrow valleys, and hollows, with Altai and Tarbogatai south-facing slopes looking at Zaisan lake being especially dry (about 250 mm of precipitation). Annual precipitation was found to range from 1,000–1,300 mm on western slopes in the Mountainous Shoria. Places of the Salair mountain range and Kuznetsk Alatau that are exposed to moist western wind appeared to receive twice as much precipitation as the adjacent valley. The Sayan Mountains also showed precipitation contrasts. The boundaries of the Western Sayan regions identified as differing in precipitation (Chebakova 1986; Lebedev 1982) appeared to coincide with those of the forest vegetation regions, which differ in vegetation belt. Precipitation changes with elevation are presented in Table 12.2.

As is clear from Table 12.2, the northern upwind macro slope of Western Sayan receives the greatest amounts of precipitation yearly: up to 1,700 mm for the excessively humid Jebash and Amyl ranges and 1,200–1,450 mm for lower ranges with high precipitation. The Western Sayan foothills and Minusinsk hollow are fairly dry

Table 12.2 Precipitation distribution depending on elevation in Western Sayan (Chebakova 1986)

Altitudinal vegetation belt	Elevation a.s.l. (m)	Annual precipitation (mm)
Excessively humid regions		
Forest-steppe	300–350	550–580
Light-needled forest	350–400	580–950
Dark-needled forest:		
Mixed fir/Siberian pine chern stands	400–900	950–1,400
Mountain taiga mixed Siberian pine/fir stands	800–1,300	1,400–1,500
Subalpine belt (mixed Siberian pine/fir open stands)	1,300–1,800	1,500–1,650
Mountain tundra	1,800–2,100	165–1,700
Highly humid regions		
Light-needled forest	700–1,000	760–950
Dark-needled forest:		
Mountain taiga Siberian pine stands	700–1,500	750–1,200
Subgolets-taiga Siberian pine stands[a]	1,500–1,800	1,200–1,350
Mountain tundra	1,800–2,200	135–1,450
Moderately humid regions		
Steppe	2,509–400	300–350
Forest-steppe (mountain hollows)	400–800	350–550
Light-needled forest:		
Mixed subtaiga larch/Scots pine stands	500–1,200	400–750
Mountain taiga larch stands	800–1,500	500–800
Dark-needled forest:		
Mountain taiga Siberian pine stands	1,100–1,600	700–850
Subgolets-taiga Siberian pine stands	1,600–1,900	850–950
Dry regions		
Steppe	800–1,800	250–450
Light-needled forest:		
Mountain taiga larch stands	1,200–2,000	350–500
Subgolets-taiga larch stands	2,000–2,200	500–600
Dark-needled forest:		
Subgolets-taiga Siberian pine and mixed Siberian pine/larch stands	1,800–2,200	480–600
Mountain tundra	2,200–3,000	600

[a] Subalpine belt for more continental climate conditions

(350–400 mm), since western and south-western air masses precipitate abundantly when going through the high parts of Kuznetsk Alatau, western Sayan, and the Tanu-Ola mountain range. The Alash plateau has a markedly dry climate, with annual precipitation totalling 250–300 mm and steppe being the most widespread vegetation type.

Considerable elevation-induced precipitation changes were observed in each of the above regions. The greatest precipitation gradient (100–200 mm/100 m) was found for low and middle mountains of excessively humid regions (Chebakova 1986). This gradient drops drastically down to 20 mm/100 m in high mountains, since air masses precipitate intensively while moving upslope and thus loose much of their water before they reach high-mountain vegetation. An average annual precipitation gradient of 70 mm/100 m determined for highly humid regions exhibited a decrease in highland. This gradient appeared to vary from 50 to 35 mm per 100 m elevation step. The lowest gradient was calculated for Alash plateau.

The Eastern Sayan watershed mountain range stretching from southeast to northwest is a well pronounced climatic boundary between a highly humid (>1,200 mm of precipitation) south-southwest areas and the dry, highly continental (annual precipitation less than 400 mm) Sayan highland situated east-northeast of this range.

12.3.2 Spatial and Temporal Precipitation Pattern Models

Weather data used in the current methodologies of building generalized precipitation maps are provided by a sparse weather station network. This drawback is aggravated by the fact that these data are not always representative of the area of interest. Atmospheric circulation and underlying surface are the two factors controlling spatial and temporal precipitation patterns. The former factor has temporal dynamics, whereas the latter varies in space.

While atmospheric moisture content and thermal layering change in time and cyclone and anticyclone trajectories often differ from that of the general air mass transfer, there are seasonal location-specific characteristics of the atmospheric circulation that occur every year. For this reason, regional average multi-year precipitation distribution is mainly a function of topography. Our studies of spatial precipitation patterns conducted in the Fore-Baikal region support this.

12.3.2.1 The Central Part of the Krasnoyarsk Region

The relationship between precipitation and topography varies depending on landscape characteristics. We identified three groups of landscapes in the central part of the Krasnoyarsk region. These groups differ in spatial precipitation patterns and in landscape characteristics controlling the precipitation amount. The first group, so-called "barrier landscapes", can be divided into two subgroups of the upwind western and south-western slopes of Yenisei mountain range and the eastern Sayan foothills. Southwest-facing slopes of the Yenisei mountain range located at a right angle to the prevailing wind direction fall into the first subgroup. The second subgroup covers west- and northwest-facing slopes of the Yenisei mountain range and the eastern Sayan foothills located at an acute angle to the prevalent south-westerly winds. Landscapes of the second group are found in

West-Siberian Plain (Ket and Kas riverheads). Intermountain valleys of the above mountain systems that fall within the wind shade and, thus, within precipitation shade under the prevailing regional moisture transfer make up the third landscape group ("shaded" landscapes).

The data obtained were analyzed for the relationship of precipitation distribution with geographical locality characteristics (elevation above sea level, a.s.l., the distance down to the barrier foot, location coordinates, and distance from the locality to the highest barrier point and relief features such as terrain roughness, i.e. the ratio between elevation difference total and a profile length, and a screening coefficient that shows how much of a location is shaded by a barrier). As a result of this analysis, precipitation models were obtained fore the two barrier landscape subgroups (Eqs. 12.2 and 12.3), the plain landscape group (Eq. 12.4), and shaded landscapes (Eq. 12.5).

$$X = -309 + 227 \cdot \ln H + 83 \cdot J - 149 \cdot \ln H \cdot L_p + 801 \cdot L_p - 898 \cdot (h - L_b / 100)^2 \quad (12.2)$$

$$R^2 = 0.61$$
$$\sigma = 93$$
$$F = 15.4$$

where X is the average multiyear precipitation (mm); H is elevation a.s.l. (m); J is the relief form coefficient taken equal to 1, 0, or -1 for mountain peaks, slopes, and valleys, respectively; L_p is the distance to the barrier foot (km); h is the height of the first barrier occurring in the way of the prevailing moisture transfer (m); L_b is the distance to the highest point of the first barrier occurring in the way of the prevailing moisture transfer (km); R^2 is the multiple coefficient of determination; σ is the standard error; and F is the Fisher criterion.

$$X = 484 + 246.6 \cdot \ln H \cdot \ln(10 \cdot L_p + 1) - 228.4 \cdot \ln + (10 L_p + 1) \quad (12.3)$$

$$R^2 = 0.72 \quad \sigma = 53.9 \quad F = 26.9$$

$$X = -4125 + 0.52 \cdot H + 80.4 \cdot S - 173 \cdot U \quad (12.4)$$

$$R^2 = 0.49 \quad \sigma = 40 \quad F = 3.9$$

Where S is latitude (°N); U is relief roughness within a 20 km orographic profile section beginning from the barrier foot along the prevailing moisture transfer axis.

$$X = -746.4 + 0.23 \cdot H + 20.2 \cdot S \quad (12.5)$$

$$R^2 = 0.62 \quad \sigma = 23.7 \quad F = 26.1$$

Equation 12.2 indicates that the elevation of a locality, the distance from it to the barrier foot, and screening of its parts by barriers are the main factors responsible for precipitation amount on the upwind slopes of the Yenisei mountain range.

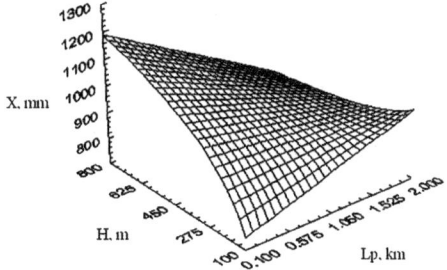

Fig. 12.3 Correlation of precipitation with altitude and distance from foot barrier ridge for the Yenisei chain of hills. X – mean annual precipitation (mm); H – altitude (m); Lp – distance from foot barrier ridge (km)

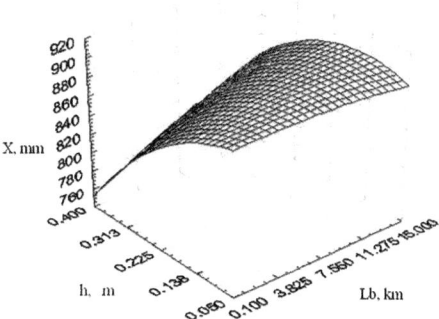

Fig. 12.4 Correlation of precipitation with elevation of the first barrier ridge and distance from first barrier ridge for Yenisei chain of hills. X – mean annual precipitation (mm); h – elevation of the first barrier ridge (m); L_b – distance from first barrier ridge (km)

While precipitation decreases with increasing distance to a barrier foot, is shows an increase with increasing elevation. However, the latter trend is observed only within two longitude degrees away from the barrier foot, beyond which distance the trend is reversed (Fig. 12.3). Single (separate) elevations and orographic barriers lower than the main watershed, but similarly oriented, appeared to reduce precipitation received by barrier-shaded sites. Immediately behind a barrier, precipitation decreases depending on the barrier height. Precipitation was observed to increase with increasing distance from the precipitation sample point to the barrier, at non-shaded precipitation sample points (Fig. 12.4).

Equation 12.3 shows a monotonous, non-linear increase in precipitation with increasing elevation and distance from the barrier foot for west- and northwest-facing slopes of the Eastern Sayan and Yenisei mountain range foothills located at acute angles to the prevailing south-westerly winds (Fig. 12.5).

In plain (flat) landscapes, precipitation amount appeared to depend mainly on latitude, elevation, and relief roughness. As is clear from Eq. 12.4, precipitation increases from south to north and with increasing elevation. Precipitation was found to decrease with increasing relief roughness within a 20 km distance from a precipitation sample point in the direction opposite to that of the prevailing moisture transfer.

In "shaded" landscapes (Eq. 12.5), precipitation amount changes mainly with latitude and elevation, with other factors having much less influence. The vertical precipitation gradients obtained for these landscapes indicate a complicated

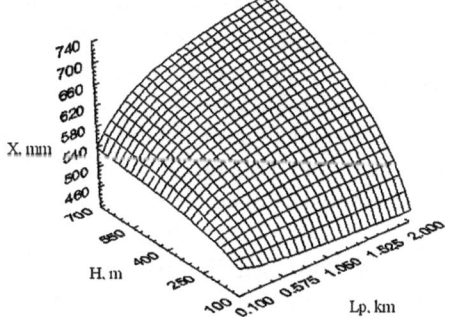

Fig. 12.5 Correlation of precipitation with altitude and distance from foot barrier ridge for East Sayan. X – mean annual precipitation (mm); H – altitude (m); L_p – distance from foot barrier ridge (km)

distribution of its amount. While the model-based annual precipitation gradient appeared to be 23 and 52 mm/100 m for "shaded" and plain landscapes, respectively, it showed a drop from 60 mm/100 m at an elevation of 100–200 m down to 20 mm/100 m at 600–700 m elevation for "barrier" landscapes, where air flows strike on barriers at acute angles. In case of perpendicular barrier location to the prevailing moisture transfer, the precipitation gradient was found to decrease from 80 mm/100 m near a barrier foot (100–200 m a.s.l.) down to as low as zero near the watershed part of the Yenisei mountain range (800–900 m a.s.l.).

The windward slopes of the Yenisei mountain range found in the central part of the Krasnoyarsk region were determined to enjoy the highest precipitation (ca. 800 mm), while 600 mm is precipitated on the upwind slopes of the northern foothills of Eastern Sayan and southern foothills of the Yenisei mountain range. Plain and "shaded" landscapes were calculated to receive about 550 and 450 mm of precipitation, respectively.

Precipitation appeared to have the highest spatial non-uniformity in barrier landscapes, whereas it exhibited a relatively uniform occurrence for "shaded" landscapes. Upwind slopes of the Yenisei mountain range were found to have the most complicated spatial pattern of precipitation (six factors accounted for 61% of the total precipitation variability). Conversely, spatial precipitation pattern appeared to be fairly simple for "shaded" landscapes (two factors accounted for 62% of the total precipitation variability).

12.3.2.2 South-Eastern Fore-Baikal Region

The orographic diversity of this region makes it virtually impossible to build any universal model describing the relationship between precipitation and relief. For this reason, our modelling efforts were initially limited to the northern Khamar-Daban macro-slope. Apart from elevation a.s.l., the distance from a barrier foot to an in-mountain locality, and slope, this model considers the shortest distance to the regional axial line going through "Angara Gate"[2]. This latter parameter was incorporated, because our preliminary analysis of multi-year precipitation data provided by Baikal weather stations revealed a distinct precipitation trend to decrease west and east of Tankhoy weather station situated opposite "Angara Gate".

The precipitation models obtained for the Khamar-Daban regions found west (Eq. 12.6) and east (Eq. 12.7) of the axial line of "Angara Gate" appeared to be qualitatively and quantitatively different:

$$\ln X = 4.89 + 0.32 \ln H - 0.027 L_1 + 0.0147 L_2 - 0.35 \ln L_2 + 0.037 \ln H \cdot \text{lm} L_1 \quad (12.6)$$

$$R^2 = 0.42, \ \sigma = 206, \ F = 82,$$
$$L_1 \geq 0.5 \text{ km}, \ L_2 \geq 0.5 \text{ km}$$

$$\ln X = 1,1 + 0.7 \ln H - 0,0067 \ln L_2 + 0,001 X_t / \ln L_1$$
$$+ 0,0004 X_t / \ln L_2 + 0,0008 \cdot \ln H \cdot \ln L_1 \cdot \ln L_2 \quad (12.7)$$

$$R^2 = 0.75, \ \sigma = 114, \ F = 220,$$
$$L_1 \geq 0.5 \text{ km},$$
$$L_2 \geq 0.5 \text{ km}$$

where H is elevation a.s.l. (m); L_1 is the distance from a barrier foot to an in-mountain locality down the prevailing wind (km); L_2 is the shortest distance to the axial line going through "Angara Gate" (km); Xt is the annual precipitation at Tankhoy (representative) weather station (mm); R^2 is the multiple coefficient of determination; σ is the precipitation mean-square error (mm); and F is the Fisher criterion (factor).

As is clear from Eq. 12.6 derived for western Khamar-Daban, there exists a distinct negative correlation between precipitation amount and the distance from a barrier foot to an in-mountain locality situated down the prevailing wind (L^1), while precipitation shows a positive correlation with elevation a.s.l. The model built for eastern Khamar-Daban (Eq. 12.7) is structurally different from Eq. 12.6. Equation 12.7 reveals a more complicated relationship of annual precipitation (X_t) occurring on the eastern Khamar-Daban slope with the distance from a barrier foot to an in-mountain locality situated down the prevailing wind (L_1) and the distance to the "Angara Gate" axial line (L_2) as compared to the western Khamar-Daban model. According to Eq. 12.7, precipitation decreases with increasing distance from the axial line, presumably due to increasing climate continentality west-eastward. Also, these two equations indicate that L_1 and L_2 are not as important regarding

---

[2] "Angara Gate" is the Angara river valley where this river flows out of Lake Baikal. This valley stretches from south-east to north–west and is the only place on the northern Baikal bank where northwestern moist air masses can pass.

precipitation as elevation above sea level. Variables L_1 and L_2 appear to be positively correlated with annual precipitation (Tankhoy weather station), however, the influence of these two factors decreases proceeding from the barrier foot into the mountain system down the prevailing wind, and with increasing distance from the "Angara Gate" axial line. Our precipitation distribution models obtained for the south-eastern Fore-Baikal region thus describe general dependences of precipitation on locality elevation and situation regarding the prevailing wind direction, as well as its relationship with a representative weather station data.

12.4 Discussion

Our analysis of precipitation patterns in Siberia revealed that the precipitation distribution found for the Siberian plain landscapes exhibits the highest agreement (the best fit) with the global precipitation pattern, according to which precipitation increases with elevation and in a steadily low-pressure zone between 60° and 70° N latitudes and decreases proceeding into continents.

The precipitation pattern found for upwind mountain slopes appeared to differ considerably from that for plains. This difference can be attributed to a complicated transformation of water-laden air flows where they encounter mountain systems. In this case, both snow and rainfall patterns are controlled by a much greater number of factors than in plains, particularly where barriers are located at right angle to the prevailing moisture transfer and induce drastic changes of air mass speed, direction, moisture and heat exchange with the environment (e.g. the Yenisei and Khamar-Daban mountain ranges). Moisture-carrying air flow changes were determined to be less distinct where they meet with barriers at acute angles (e.g. northern and southern foothills of the Eastern Sayan and southern Yenisei mountain range, respectively). Air masses coming to east-facing macro-slopes of the Yenisei mountain range and the eastern Sayan foothills, as well as to inter-mountain hollows, and big river valleys occurring at the right angle to the prevailing air mass transfer direction were determined to have little water. Any influence on precipitation pattern is minimal here and, therefore, precipitation occurs uniformly.

These precipitation trends appeared to be most pronounced at a macro-slope scale, while the impact of the slope aspect on precipitation was found to decrease at meso- and micro-scale. Sosedov (1967), in a precipitation study in Trans-Ily Alatau, found that, with the same moisture availability and slope screening effect, snowfall amount did not depend on either meso-, or micro-slope aspects. In this case, slope aspect had obvious influence only on the precipitation measurement error, since little snow fell into measuring buckets due to high wind speed. Our study showed no significant snowfall variability among slope aspects at meso- or micro-scale. Therefore, it makes sense to discuss only the influence of the macro-slope aspect on the amount of winter precipitation, namely, when precipitation sample plots cover a range of orography and vegetation zone-specific moisture availabilities.

The most complicated precipitation pattern was observed where mountain massifs were dissected by big winding rivers (Kureyka and Khantaika rivers) making the terrain extremely rough. It appeared to be very hard to quantify the influence of any factor on precipitation patterns in these areas.

12.5 Conclusion

Our data analysis showed that spatial precipitation non-uniformity found for Siberia can be attributable to the location of this region in inland Eurasia. In the West-Siberian Plain, precipitation occurrence was determined to depend on latitude and reach its maximum in the central taiga forest zone. A general trend of decreasing precipitation proceeding from southwest to northeast identified for eastern Siberia was often found to be broken due to the orographic diversity of this region. This diversity results in a highly variable precipitation pattern: the upwind slopes of even fairly low elevations were found to receive more precipitation than valley, plateaus, and hollows.

Precipitation appeared to differ greatly between north- and south-facing slopes in the southern Siberian Mountains located right in the centre of Eurasia. Because of orographic variability, southern Siberian Mountains were divided into regions differing in annual precipitation and precipitation gradient.

The regional and local precipitation patterns revealed by our study for Siberia were based upon in developing regional precipitation models that consider relief parameters and orographic barrier location with respect to the prevailing direction of moisture transfer.

Acknowledgment This study was supported by the Russian Foundation for Basic Sciences No.07-05-00016.

References

Afanasyev AN (1976) Baikal catchment water resources and balance. Nauka, Novosibirsk
Arkhangelsky VA (1960) Estimating vertical precipitation gradients in Sikhoteh-Alin region. Far East Hydrometeorol Inst Trans Issue 11:118–129
Berg LS (1938) Climatology fundamentals. Gidrometeoizdat, Leningrad
Berg LS, Shenrok AM (1925) Snow covers depth in the European USSR. Gidrometeoizdat, Leningrad
Beyer VV (1966) Technical meteorology. Gidrometeoizdat, Leningrad
Bradley ChC (1966) The snow resistograph and slab avalanche investigation. Publ/Assoc Intern Hydrol Scient 69
Burenina TA (1998) A precipitation distribution model in south-eastern Fore-Baikal region. J Geogr Nat Resour (Russ) 2:142–145
Burenina TA, Onuchin AA, Stakanov VD (2002) Liquid and solid precipitation patterns. In: Pleshikov FI (ed) Forest ecosystems along Yenisei Meridian. Publications of the Siberian Branch of the Russian Academy of Sciences, Novosibirsk, pp 48–50

Chebakova NM (1986) Climate characteristics in mountainous relief. In: Polikarpov NP, Chebakova NM, Nazimova DI (eds) Climate and mountain forests of Southern Siberia. Nauka, Novosibirsk, pp 44–89
USSR Climate Guide (1956) USSR climatological reference book, part 2, vol 18–24. Gidrometeoizdat, Leningrad
USSR Climate Guide (1969) Data for certain years, parts 2, vol 21. GUGS, Krasnoyarsk
USSR Climate Guide (1970) Krasnoyarsk Region and Tuva, part 4, vol 17–24. Gidrometeoizdat, Leningrad
Glebova MYa (1958) Snow cover in Western Europe. GGO Trans Issue 85:18–37
Gorec PA, Younkin RJ (1966) Sinoptic climatology of heavy snow-fall over the central and eastern Unated States. Month Wealth Rev 94:11
Govsh RK (1962) Variability of snow covers characteristics in trans-Baikal region. GGO Trans Issue 130:43–61
Grudinin GV (1979) Snow cover. In: Bufal VV, Hlebovich IA (eds) Geosystems of Western Sayan Foothills. Nauka, Novosibirsk, pp 117–133
Grudinin GV (1981) Snow cover in the south of Minusinsk hollow. Nauka, Novosibirsk
Grudinin GV, Kovalenko AK, Korytny LM (1975) On elevation gradients of precipitation and air temperature in western Sayan foothills. In: Studying Siberian nature, economics, and population. Publications of the Institute of Geography of Siberia and Far East, Irkutsk, pp 74–75
Guralnik II, Dubinsky GP, Mamikonova SV (1972) Meteorology. Gidrometeoizdat, Leningrad
Hartzman IN (1971) Problems of geographical zoning and discrecity of hydrometeorological fields in mountain conditions of monsoon climate. Far East Hydrometeorol Inst Trans Issue 35:3–31
Kolomyts EG (1975) Snow cover development and distribution in Sosvin part of Fore-Ob region. In: Sosvin part of Fore-Ob region. Publications of the Institute of Geography of Siberia and Far East, Irkutsk, pp 158–214
Kopanev ID (1966) On snow cover distribution. GGO Trans Issue 195:12–29
Korotkov IA (1994) Forest vegetation regionation of Russia and former USSR Republics. In: Alexeev VA, Berdsy RA (eds) Carbon of forest and bog ecosystems. Publications of the Institute of Forest, Krasnoyarsk, pp 29–47
Korytny LM (1980) On precipitation patterns in inter-mountain hollows. In: Climate and waters of Siberia. Nauka, Novosibirsk, pp 128–132
Kotlyakov VM (1968) Earth's snow cover and glaciers. Gidrometeoizdat, Leningrad
Ladeishchikov NP (1982) Big lake climate characteristics. Nauka, Moscow
Lebedev AV (1979) Water and heat balances of natural complexes. In: Protopopov VV (ed) Environmental importance of Baikal catchment forests. Nauka, Novosibirsk, pp 79–136
Lebedev AV (1982) Hydrological role of mountain forest of Siberia. Nauka, Novosibirsk
Lvovsky EN (1988) Statistical methods for building empirical equations. Higher School Publications, Moscow
Matasov MI (1938) Snow cover in Yakutian Autonomous Republic. In: Problems of Arctic Region. pp 134–157 No 5–6
Matveyev PN (1968) Snow cover development in spruce forest of Tyan-Shan. J Forest Sci (Russ) 1:79–83
Matveyev PN (1984) Hydrological and protective role of Kirgizian mountain forests. Ilim, Frunze
Mellor M, Smith IH (1966) Strength studies of snow. U.S. Cold regions. Res. and Eng. Lab., Haniver, Res. Report, No 163
Onuchin AA (1984) Snow accumulation in open site of Khamar-Daban forest ecosystems. In: Environmental changes induced by forest ecosystems. Publications of the Institute of Forest, Krasnoyarsk, pp 75–86
Onuchin AA (1985) Elevation-specific solid precipitation changes caused by Khamar-Daban mountain forests. In: Hydrological studies in USSR mountain forests. Illim, Frunze, pp 109–119

Onuchin AA (1987) Solid precipitation changes caused by Khamar-Daban mountain forests. Ph.D. thesis Abstract, Publications of the Institute of Forest, Krasnoyarsk

Onuchin AA (2001) General snow accumulation trends in boreal forests. Izvestiya Acad Sci USSR (Geogr) 2:80–86

Onuchin AA, Burenina TA (1996a) Climatic and geographic patterns of snow density in northern Eurasia. Arctic Alpine Res 34(1):99–103

Onuchin AA, Burenina TA (1996b) Spatial and temporal patterns of snow cover density in northern Eurasia. Meteorol Hydrol (Russ) 12:101–111

Onuchin AA, Burenina TA (2002) Modelling spatial and temporal precipitation patterns. In: Forest ecosystems of Yenisei Meridian. Publications of the Russian Academy, Siberian Branch, Novosibirsk, pp 50–54

Richter GD (1963) Investigation of snow cover in various parts of the USSR; estimating its necessary amount and proposing its management methodologies. Glaciol Invest Mat; Chronicle Events Discuss 7:3–34

Richter GD (1984) Role of snow cover in geophysical processes. Trans Russ Acad Inst Geogr 40:17–31

Richter GD, Petrova LA (1960) Global terrestrial snow cover pattern. In: Snow cover geography. Publications of the Russian Academy, Moscow, pp 3–46

Schwer ZA (1984) Precipitation patterns on continents. Gidrometeoizdat, Leningrad

Shpak IS, Bulavskaya TN (1967) Snow cover patterns, elevation-caused snow gradients and load in Trans-Carpathian region. Ukrainian NIIGMI Trans 66:15–37

Shultz VL (1972) Some results and further development of methods for estimating runoff in mountainous countries. Central Asian Hydrometeorol Inst Trans 62:44–61

Sosedov IS (1962) Influence of mountain slope aspect on snow cover pattern; a case study in Small Almaatinka river catchment. In: Snow cover: Its pattern and role in the national economy. Publications of the Russian Academy, Moscow, pp 87–97

Sosedov IS (1967) Investigating snow water balance on Trans-Ily Alatau mountain slopes. Alma-Ata, pp 5–6

Suslov SP (1954) Physical geography of the USSR; the Asian part. Uchpedgiz, Moscow

Tikhomirov BA (1956) Characteristics of tundra sow cover and its impact on vegetation. In: Snow melt water. Publications of the Institute of Geography of Siberia and Far East, Irkutsk. Moscow, pp 89–108

Trifonova TS (1962) On spatial snow cover pattern changes. GGO Trans 130:21–36

Vinogradov VN (1964) Snow cover distribution in Kamchatka. In: Kamchatka geography issues, vol 2. Petropavlovsk-Kamchatsky, Publications of the Institute of Geography of Siberia and Far East, Irkutsk. pp 27–44

Part IV
Information Systems

Chapter 13
Interoperability, Data Discovery and Access: The e-Infrastructures for Earth Sciences Resources

S. Nativi, C. Schmullius, L. Bigagli, and R. Gerlach

Abstract The ever-increasing need to integrate knowledge from the diverse disciplines of the Earth System sciences requires to switch from data-centric systems towards service-oriented enabling infrastructures. Important international initiatives and programmes are defining a standard baseline for interoperability of geospatial information, models and technologies, in particular for data discovery and access. We describe the design of an e-infrastructure for Earth Sciences, from the point of view of the data, services and distribution model. This design is implemented in the Siberian Earth System Science Cluster (SIB-ESS-C), an e-infrastructure supporting the generation and distribution of products and information about central Siberia, along with advanced analysis tools for Earth Sciences.

Keywords Interoperability • Geospatial information • Spatial data infrastructure • Siberia • Earth Observation

13.1 Introduction

Scientific and technological advancements in sensors, remote sensing and aerospace industry are set to increase exponentially the availability of geospatial information, in the near future. It is estimated that there are currently around 100,000 in-situ stations and 50 environmental satellites.[1] Likewise, the advancements of research

---

[1] http://ec.europa.eu/research/environment/themes/article_1357_en.htm

S. Nativi (✉) and L. Bigagli
Italian National Research Council – IMAA and University of Florence, Prato, Italy
e-mail: nativi@imaa.cnr.it; lorenzo.bigagli@pin.unifi.it

C. Schmullius and R. Gerlach
Friedrich-Schiller-University, Jena, Italy
e-mail: c.schmullius@uni-jena.de; roman.gerlach@uni-jena.de

in environmental sciences, supported by the increasing capacity of computational platforms and telecommunication infrastructures, will allow to deepen our understanding of natural phenomena. Therefore, there is an ever-increasing need to integrate knowledge from the diverse disciplines engaged in studying the constituent parts of the complex Earth system.

Earth system analysis is a real challenge for scientists as much as it is for information technology. In fact, the scope and complexity of Earth system investigations demand for the formation of distributed, multidisciplinary collaborative teams. This requires the integration of different discipline information systems, characterized by: heterogeneous and distributed data and metadata models, different semantics and knowledge, diverse protocols and interfaces, different data policies and security levels (Foster and Kesselman 2006). Advanced e-infrastructures (aka cyber-infrastructures) will support the formation and operation of a Earth System Science Community, based on multidisciplinary knowledge integration. Developing an advanced enabling infrastructure to facilitate the Earth system analysis implies to scale from specific and monolithic systems (data-centric) towards independent and modular (service-oriented) information systems (Foster and Kesselman 2006).

Advanced e-infrastructures will provide scientists, researchers and decision makers with a persistent set of independent services and information that scientists can integrate into a range of more complex analyses. The importance of the geospatial information to support the decision process and the management of environmental issues at the different scales (i.e. national, international and global) was already recognized and outlined by the United Nations Conference on Environment and Development (Rio de Janeiro, June 1992) and by the General Assembly for the implementation of Agenda 21 (New York, June 1997).

In these years, there were launched important initiatives and programmes by the European and International Communities to design and build such advanced e-infrastructures in order to collect, manage and share geospatial information providing the Society with Earth and environmental information in a handy and near real-time way. These initiatives are resulting decisive to reach out to the different Earth Sciences disciplines and systems. The most relevant to our topic are briefly presented below.

13.1.1 GEOSS

In 2005, member countries of the Group on Earth Observations (GEO) agreed on a 10-year implementation plan for a Global Earth Observation System of Systems (GEOSS).[2] In 2006, GEO has begun implementation of the GEOSS 10-Year Implementation Plan as endorsed by the Third Earth Observation Summit. GEOSS

---

[2] http://earthobservations.org/geoss.shtml

is a worldwide effort to build upon existing national, regional, and international systems to provide comprehensive, coordinated Earth observations from thousands of instruments worldwide, transforming the data they collect into vital information for society. GEOSS will meet the need for timely, quality long-term global information as a basis for sound decision making, and will enhance delivery of benefits to society in nine Societal Benefit Areas (SBAs), identified as key applications of GEOSS, namely: Disasters, Health, Energy, Climate, Water, Weather, Ecosystems, Agriculture, Biodiversity. GEOSS Architecture and interoperability process are investigated by a couple of pilot initiatives: the Architecture Implementation Pilot (AIP) (Percival 2008) and the Interoperability Process Pilot Project (IP3) (Khalsa et al. 2008).

13.1.2 GMES

The European Global Monitoring for Environment and Security (GMES)[3] initiative is a concerted effort promoted by the European Community and the European Space Agency to bring data and information providers together with users, so they can better understand each other. GMES will support the implementation of public policies at European or national level that deal with, for example, agriculture, environment, fisheries, or regional development, external relations, security. GMES is set to be the main European contribution to GEOSS. The main GMES objective is to make environmental and security-related information available to the people who need it through enhanced or new services. The services identified by GMES can be classified in three major categories:

- **Mapping**, including topography or road maps but also land-use and harvest, forestry monitoring, mineral and water resources that do contribute to short and long-term management of territories and natural resources. This service generally requires exhaustive coverage of the Earth surface, archiving and periodic updating of data.
- **Support for emergency management** in case of natural hazards and particularly civil protection institutions responsible for the security of people and property.
- **Forecasting** is applied for marine zones, air quality or crop yields. This service systematically provides data on extended areas permitting the prediction of short, medium or long-term events, including their modeling and evolution.

The widespread and regular availability of technical data within GMES will allow a more efficient use of the infrastructures and human resources. It will help the creation of new models for security and risk management, as well as better management of land and resources.

---

[3] http://www.gmes.info

13.1.3 INSPIRE

The Directive 2007/2/EC of the European Parliament and of the Council of 14 March 2007 establishing an Infrastructure for Spatial Information in the European Community (INSPIRE),[4] as published in the official Journal on the 25 April 2007, establishes a regional Spatial Data Infrastructure (SDI) in Europe, also addressing some aspects of environmental monitoring. INSPIRE is conceived to serve policymakers, planners and managers at European, national and local level and the citizens and their organizations, delivering to the users integrated spatial information services was. The INSPIRE Directive entered into force on the 15 May 2007. Five Drafting Teams have been designing the directive implementation rules, as far as its architecture, data policy and monitoring process are concerned. The first approved regulation concerns metadata.

In the following sections, we elaborate on remote sensing in Siberia and we introduce the current standard baseline for interoperability of geospatial information, presenting the main adopted models and technologies, in particular for data discovery and access. We introduce the Siberian Earth System Science Cluster (SIB-ESS-C), a large database of datasets and value-added products spanning the central Siberian region. SIB-ESS-C realizes a initial SDI (i.e. an e-infrastructure) to generate and distribute products and information about central Siberia, along with advanced analysis support for Earth Sciences. This is a valuable example of how scientific data can be published and accessed under the interoperability paradigm. We present some results concerning the implementation of advanced access, discovery and processing services for SIB-ESS-C. The infrastructure architecture applies relevant international standards and best-practices; its interoperability with the introduced relevant initiatives is argued.

13.2 The Siberian Earth System Science Cluster

The main goal of the Siberian Earth System Science Cluster[5] is to provide an infrastructure for spatial data to facilitate Earth system science studies in Siberia. The region under study covers the entire Asian part of the Russian Federation from 58° E–170° W and 48–80° N. The region comprises a significant part of the Earth's boreal biome, but also includes a large portion of the arctic biome and a small portion of the temperate biome in Northern Eurasia. The watersheds of the rivers Ob, Yenissei and Lena representing the main freshwater source of the Arctic Ocean are located in this region. Figure 13.1 depicts the interested geographic area. With respect to Global Climate Change several studies identified Siberia as one of the hotspots where temperature changes are more pronounced than in other regions of the world (Hansen et al. 1999; Zhaomei et al. 2001; Arctic Climate Impact

---

[4] http://www.ec-gis.org/inspire

[5] http://www.sibessc.uni-jena.de

13 Interoperability, Data Discovery and Access   217

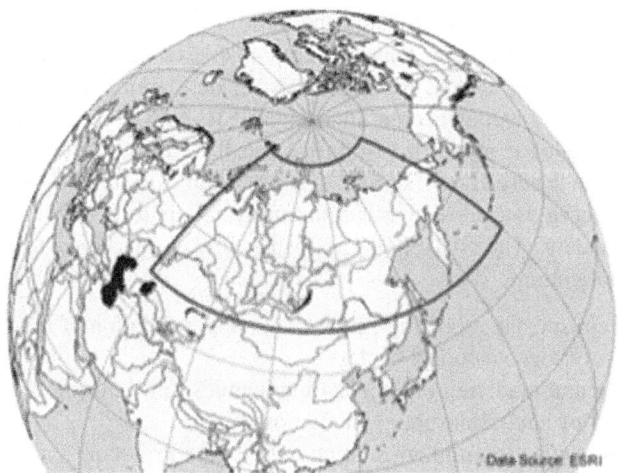

Fig. 13.1 The region of interest of SIBERIA-II (inner areas) and SIB-ESS-C (outer bounding box) (*Color version available in Appendix*)

Assessment 2004). Understanding the system, its underlying processes and their interaction is crucial and requires interdisciplinary research. The availability and access to data and information across discipline boundaries is a prerequisite to any integrated research approach. Within different scientific fields (e.g. biology, geography, oceanography) specific data models, data formats and tools evolved over the years making it difficult to easily share data and information across them.

13.2.1 Objectives

SIB-ESS-C emerged from the need to preserve a collection of Earth observation data products created during previous research projects and make this data accessible to the scientific community as well as the general public. In order to publish data products in a consistent and well documented manner metadata describing the content, history and quality of the data is required. The data discovery process relies heavily on the availability of metadata and its publication using common standards and Internet services. Hence, the first objective of SIB-ESS-C is to create metadata for all data products and publish it through a catalog service allowing users to identify and locate the data resources. This also includes the development of a Web interface to perform queries against the catalog service. Once a user is aware of a data resource, access to the data becomes important. Traditionally, data products have been retrieved by downloading data files from an FTP site. In SDIs like SIB-ESS-C, web services are deployed for direct data access via Web. In addition to data access a user may decide to visualize or analyze the dataset of interest. The SIB-ESS-C system will provide Web-based tools to explore the spatio-temporal characteristics of the published data products. Other SIB-ESS-C goals include generating added-value data products and building up time series. This implies a standard and open processing environment capable of handling vast amounts of data, i.e. a computing cluster, but also tools for data archiving, storage management and automated metadata creation. Indeed, SIB-ESS-C must be considered as one node in a global network of similar Earth Science Clusters. In fact, the integration into a network of similar systems enables SIB-ESS-C to offer data products and services to a broad user community and, in turn, to benefit from other resource providers. The SIB-ESS-C infrastructure is conceived to facilitate the following applications:

- Earth system science modeling (input to models, validation of model results)
- Modeling of biogeochemical cycles
- Monitoring and modeling of vegetation dynamics (e.g. shifting of tree line)
- Assessment of land-atmosphere interaction
- Support to convention implementation (e.g. Kyoto Protocol)
- Assessing the environmental impact of socio-economic development

SIB-ESS-C was designed to be fully interoperable with the infrastructures developed by GEOSS, GMES and INSPIRE. In particular, SIB-ESS-C may represent a valuable testbed to implement the GMES vision and technological solution by providing researchers, decision makers and citizens with Earth System Science information. SIB-ESS-C infrastructure might be a valuable case in point as for the GMES land-monitoring Fast-Track service (EC 2005). SIB-ESS-C infrastructure fits in the GEOSS purpose: to achieve comprehensive, coordinated and sustained observations of the Earth system, in order to improve monitoring of the state of the Earth, increase understanding of Earth processes, and enhance prediction of the behavior of the Earth system (GEO 2007). In keeping with GEOSS view, SIB-ESS-C data products and services are expected to contribute to the societal benefit areas: Ecosystem, Climate, Water, Energy, Heath, Disasters.

13.3 The Interoperability Infrastructure

IEEE defines interoperability as the ability of two or more systems or components to exchange information and to use the information that has been exchanged (IEEE 1990). In the geospatial information area interoperability is mainly achieved through the access to common and open technology and the implementation of standards.

The continuous development of sophisticated Information and Communication Technologies (ICT) solutions provides fundamental tools to tackle interoperability. Technologies developed by consortiums like W3C[6] and OASIS[7], including

---

[6] http://www.w3.org
[7] http://www.oasis-open.org

HTML, XML and Web Services, have been used by a broad range of scientific and business communities to address heterogeneity as far as information and programming interfaces are concerned. On that premises, the international geospatial research community is strongly pursuing the specification and the standardization of frameworks (i.e. data and service models, with related profiles and extensions) of general ICT solutions for geospatial information management, including Earth Observation and Environmental Monitoring, in a coordinated, consensus-driven effort lead by standardization organizations such as: ISO TC211[8], Open Geospatial Consortium (OGC)[9] and World Meteorological Organization (WMO)[10]. According to GEOSS 10-year Implementation Plan (GEO 2005) interoperability is achieved by a Service-Oriented Architecture (SOA), in which contributed components interact by passing structured messages over network communication services. Such interactions will take place according to agreed-to "interoperability arrangements" that should be based on non-proprietary, open standards. Therefore, interoperability is mainly pursued by standardization. Rather than attempting to define new specifications, GEOSS seeks to recognize standard specifications agreed to by consensus, with preference given to formal international standards. However, many Earth science disciplinary communities introduced contracts suited for their specific components; in the GEOSS interoperability framework they are called "special arrangements". GEOSS promoted the Standard and Interoperability Forum SIF[11] to discuss and recognize them.

Therefore, the research and experimentation focuses on the design and implementation of enabling infrastructures that support geospatial resources sharing by means of a minimum set of protocols, standard specifications and best practices. Such facilities, known as SDIs, can be defined as the relevant base collection of technologies, policies and institutional arrangements that facilitate the availability of and access to geospatial data. Different hierarchical levels of SDIs are reckoned, e.g. global, regional, national, local. A SDI provides a basis for spatial data discovery, evaluation, and application for users and providers within all levels of government, the commercial sector, the non-profit sector, academia and by citizens in general (Nebert 2004).

Earth science data infrastructures must consider a couple of other important service categories: observation and measurement, and processing and knowledge extraction services. They are important to interact with sensor and modeling systems, to work out value-added products and serve policy makers. Figure 13.2 provides an overview of a general SDI for Earth science data.

As far as SDI architecture specification is concerned, the international standardization process is based on a couple of well-accepted principles: to follow a Model Driven Architecture (MDA) approach (Miller and Mukerji 2003) and implement it as open distributed system (ISO 19101 2002; ISO/PDTS 19101-2; Nebert 2004).

---

[8] http://www.isotc211.org/
[9] http://www.opengeospatial.org/
[10] http://www.wmo.int
[11] http://seabass.ieee.org/groups/geoss/

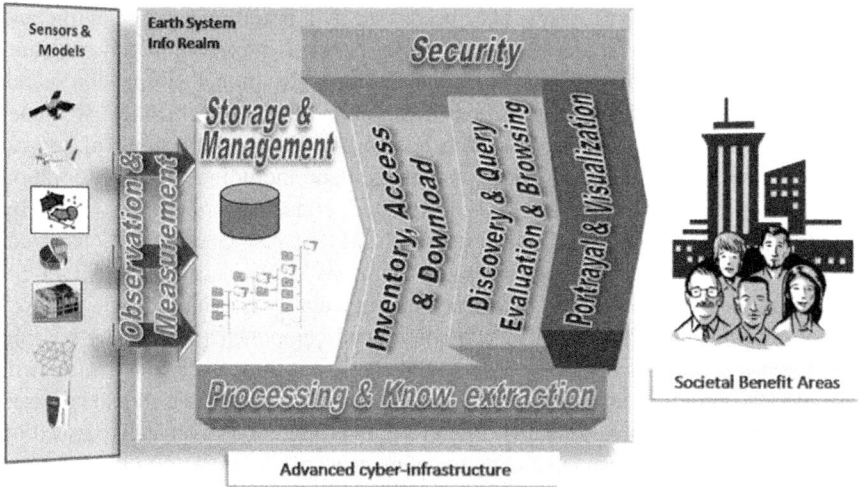

Fig. 13.2 Simplified architectural schema of an advanced infrastructure for Earth sciences information (*Color version available in Appendix*)

ISO TC211 has been developing an MDA for geospatial information e-infrastructures; OGC has developed well-accepted service-oriented specifications to implement open distributed systems. These specifications process follows a standard approach: the ISO Reference Model for Open Distributed Processing (RM-ODP) (ISO/IEC 10746 1998), that uses an object modeling approach to describe distributed systems. In order to simplify the problems of design in large complex systems five *viewpoints* provide different ways of describing the system. Highly simplified RM-ODP *Information, Computational,* and *Engineering* views of the SIB-ESS-C architecture are briefly described in the following sections. The RM-ODP *Enterprise* view has been summarized in section 13.2.

13.4 Information View

This architectural view is concerned with information modelling. Thus the information view defines the semantics of information and of information processing, without having to worry about specific implementation details.

Information classes considered for SIB-ESS-C infrastructure are: discrete features, coverages, observations and maps. According to MDA, the basic class of the information conceptual model is the geospatial feature. In fact, a geospatial feature may be defined as *an abstraction of a real world phenomenon implicitly or explicitly associated with a Earth location* (ISO 19107 2003). A coverage is a feature subtypes which is defined by ISO as: *a feature that associates positions within a bounded space (its domain) to feature attribute values (its range)*. In other words, it is both a feature and a function. Examples include a raster image, a polygon overlay or a digital elevation matrix (ISO/FDIS 19123 2005).

13.4.1 Coverage Versus Features

Indeed the Earth sciences community deals with geospatial phenomena. Earth sciences data capture and represent discrete and continuous real world phenomena. Discrete phenomena are recognizable objects that have relatively well-defined boundaries or spatial extent (e.g. measurement stations). While, continuous phenomena vary over space and have no specific extent (e.g. temperature field); continuous phenomenon value is only meaningful at a particular position in space and time. ISO TC211 introduced two fundamental data types to map real world phenomena: features and coverages. Historically, geospatial information has been treated in terms of two fundamental types called vector data and raster data. Vector data deals with discrete phenomena, each of which is conceived of as a feature. The spatial characteristics of a discrete real-world phenomenon are represented by a set of one or more geometric primitives (e.g. points, curves, surfaces or solids) (ISO/FDIS 19123 2005), whereas the other phenomenon characteristics are treated as feature attributes. Generally, a single feature is associated with a single set of attribute values. ISO 19107 provides a schema for describing features in terms of geometric and topological primitives.

Raster data deal with real-world phenomena that vary continuously over space. They contain a set of values, each associated with one of the elements in a regular array of points or cells. Raster data are a commonly used example of Coverage. In fact, the coverage concept generalizes and extends the raster structure type by referring to any data representation that assigns values directly to spatial position. A coverage associates a position within a spatial/temporal domain to a value of a defined data type. It realizes a function from a spatial/temporal domain to an attribute domain (the co-domain) (ISO/FDIS 19123 2005).

13.4.2 Observations and Measurements

In addition to feature and coverage models, another relevant specification for Earth sciences resources is the OGC Observation and Measurement information model (Cox 2006). An Observation is defined as an event with a result which has a value describing some phenomenon. The observation event is modelled as a feature type within the context of the ISO general feature model (ISO 19101 2002; ISO 19109 2005). An observation results in an estimate of the value of a property of the feature of interest; if the property varies on the feature of interest, then the result is a coverage, whose domain is the feature. According to this best practise, in a physical realisation the result will typically be sampled on the domain, and hence represented as a discrete coverage.

In summary, instruments and sensors observe and measure properties of feature of interest (e.g. shape, position, temperature, height, density, direction, intensity, etc.). Observations and measurements generate datasets; they can be modeled and stored as either feature (i.e. boundary) or coverage data. This mainly depends on the observed property variability over the domain which characterizes the feature of interest.

The SIB-ESS-C information model follows this approach managing and processing both feature (i.e. boundary) and coverage datasets which stem from remote and in-situ observations and measurements. Figure 13.3 depicts the SIB-ESS-C information conceptual model. The model is expressed as a Unified Modeling Language (UML) class diagram (ISO/IEC 19501 2005). UML is a well-accepted paradigmatic modeling language used by domain experts. In fact, UML provides a collection of modeling constructs and an associated graphical notation for modeling a problem domain as a class diagram. This is the reason why ISO TC211 selected UML static structure diagram as part of its Conceptual Schema Language for rigorous representation of geographic information.

With reference to the schema, a coverage acts as a function to return one or more feature attribute values for any direct position within its spatiotemporal domain. A coverage dataset is characterized by a coverage function which associate a domain to range-set

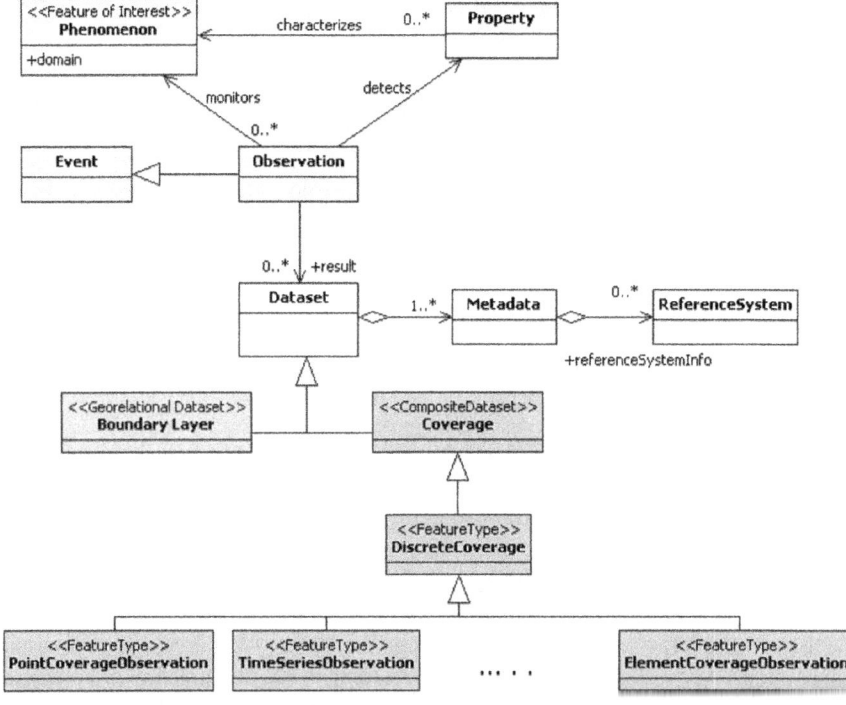

Fig. 13.3 The general dataset conceptual model (*Color version available in Appendix*)

13 Interoperability, Data Discovery and Access

Table 13.1 SIB-ESS-C data products

EO product	Source	Temporal coverage	Spatial resolution	Spatial coverage	Partner responsible
Phenology	SPOT-VGT AVHRR	2000–2003 annual	1 and 10 km	Entire SIBERIA-II Region	Center for the Study of the Biosphere from Space (CESBIO), France
Disturbances	MODIS, AVHRR ATSR-2	1992–2003 on yearly basis	1 km	Entire SIBERIA-II Region	Centre for Ecology and Hydrology Monks Wood, UK
Freeze/thaw	QuikSCAT	2000–2003	10 km	Entire SIBERIA-II Region	TU Wien, Institute of Photogrammetry and Remote Sensing (IPF), Austria
Water bodies	ASAR WS	2003–2004	150 m	Entire SIBERIA-II Region	TU Wien, Institute of Photogrammetry and Remote Sensing (IPF), Austria
Snow depth	SSM/I	2000–2003	25 km	Entire SIBERIA-II Region	Center for the Study of the Biosphere from Space (CESBIO), France
Snow melt	SSM/I	2000–2003	25 km	Entire SIBERIA-II Region	Center for the Study of the Biosphere from Space (CESBIO), France
Land cover	MODIS	2001–2004 annual	500 m	Entire SIBERIA-II Region	University of Wales Swansea, UK
Continuous field land cover	MODIS VCF MODIS LC	2003	500 m	Entire SIBERIA-II Region	Friedrich-Schiller-University Jena, Institut for Geography, Germany
Topography	SRTM/GTOPO	2000	3 arcsec < 60° N 1 km > 60° N	Entire SIBERIA-II Region	Gamma Remote Sensing AG, Switzerland

values. Spatial referencing is based on coordinates; spatial references relate the features represented in the data to positions in the real world (ISO 19111 2003).

13.4.3 Data Products

The data products currently available through the SIB-ESS-C infrastructure were created by a number of research institutions teamed up in the EU-funded SIBERIA-II project (2002–2005). The objective of SIBERIA-II Earth observation was to deliver geo-observational products that aid in improving the modeling approaches and in turn address the key scientific questions of the project: What is the current average greenhouse gas budget of the region and what is its spatial and temporal variability? How will it change under future climatic and anthropogenic impacts? To achieve the goals of the SIBERIA-II project, a diverse set of multi-sensor Earth observation data was used. The definitions of land surface products to be derived from EO data, their spatial and temporal scales were geared towards the project modeling approaches. Table 13.1 summarizes the main properties of the data products available. For a more detailed product descriptions refer to (Delbart et al. 2005; George et al. 2006; Bartsch et al. 2007a; Bartsch et al. 2007b; Grippa et al. 2004; Skinner and Luckman 2004). All data products cover at least a three million square kilometer area in central Siberia defined by the administrative boundaries of the Krasnoyarsk Kray, Irkutsk Oblast, Taymyr and Evenk Okrug. The temporal coverage is usually for four consecutive years (2000–2003). Extending these time series according to the availability of the Earth observation data is intended.

13.4.4 Metadata

Spatial data infrastructures manage and share datasets: they consist of data along with its description: the metadata – data about data. Metadata are crucial to enable data cataloguing allowing its discovery, evaluation and correct use. Metadata allows data localization, extraction, integration and employment. In summary, metadata is important to understand the right data for the right purpose.

To follow a multi-disciplinary geospatial information standard was essential to provide an understanding of data across the different information communities that contribute to SIB-ESS-C. Therefore, SIB-ESS-C adopted ISO 19115 (ISO 19115 2003) as its metadata reference standard. In fact, 19115 provides a structure for describing digital geospatial data and services, defining general-purpose metadata. This standard models information about the identification, the extent, the quality, the spatial and temporal schema, the spatial reference, and the distribution of data. Metadata must be provided for independent datasets, aggregations of datasets, individual geographic features and coverages. GEOSS, GMES and INSPIRE adopts ISO 19115 for their cataloging services.

Table 13.2 Significant SIB-ESS-C metadata (with respective obligation)

Dataset reference date (mandatory)	Spatial representation type (optional)
Abstract describing the dataset (mandatory)	Reference system (optional)
Geographic location of the dataset by four coordinates -the minimum bounding rectangle vertexes (mandatory)	Lineage (optional)
Dataset topic category (mandatory)	On-line resource (optional)
Additional extent information for the dataset (vertical and temporal) (mandatory)	Spatial resolution of the dataset (optional)
Distribution format (mandatory)	Dataset responsible party (optional)
Dataset language (mandatory)	

19115 defined set of metadata elements is quite extensive; thus, only a subset of the full number of elements is generally used, according to the domain and system requirements. For its initial infrastructure, SIB-ESS-C maintains the basic minimum number of metadata elements recommended by the ISO 19115 specification itself, plus few extra elements. These set of minimum metadata elements is called the 19115 "core profile"; it fits to identify a dataset for catalogue purposes. Significant elements of SIB-ESS-C metadata are listed in Table 13.2 along with the respective obligation; they focus on dataset; metadata on metadata are not listed.

13.4.5 Dataset Encoding

At the present development stage data products are provided in GeoTIFF or ESRI Shapefile format. For all data products metadata complying with the ISO 19115 standard is available. Data sets available from the SIB-ESS-C infrastructure are provided free of charge following a simple user registration procedure.

13.5 Computational and Engineering Views

The Computational viewpoint describes the system decomposition in main functional components interacting through well-defined interfaces, according to the SOA. The Engineering viewpoint is concerned with the design of distribution-oriented aspects, that is, the infrastructure required to support distribution. It focuses on the mechanisms and functions required to support distributed interaction between components in the system. In keeping with the advanced infrastructure for Earth sciences information depicted in Fig. 13.2, the present SIB-ESS-C infrastructure services may be organized in the following functional tiers:

- Data Storage & Management
- Data Processing & integration services
- Data Access services
- Data Discovery & Query services
- Data Browsing and Evaluation Services
- Data Portrayal and Visualization services
- Data Download services

The general design strategy pursued for SIB-ESS-C is to implement a service-oriented infrastructure adopting the interface standards published by the OGC, applying the ISO TC211 models and resorting to the W3C solutions in order to achieve interoperability with other information systems. The development has focused on the implementation of free and open source software whenever possible and to utilize existing components that are well established in the Earth sciences, Earth Observation and GIS communities.

Service-Oriented Architecture is based on the notion that it is beneficial to decompose a large problem into a collection of smaller, related pieces: services. This helps to establish a high form of abstraction that encapsulates both application and process logic. For the Earth system domain a service-oriented architecture offers considerable flexibility in aligning information technology functions and processes. SOA is a flexible, extensible architectural framework that enables rapid application delivery and integration across organizations and "siloed" applications (Arsanjani et al. 2007).

In SOA, a service provider publishes a description of the service(s) it offers via a service registry. Service consumer, which may be either a person or process, searches the service registry to find a service that meets a particular need. The goal is total modularization of the distributed computing environment as opposed to recreating the large monolithic solutions of more traditional platforms (Snell et al. 2002). This paradigm is useful for efficiently organizing and utilizing distributed capabilities that may be under the control of different ownership domains, as is the case in GEOSS, GMES and INSPIRE. It provides a uniform means to offer, discover, interact with and use heterogeneous resources. Services interoperability is achieved by applying standard service interfaces based on data and metadata models. We already discussed the data and metadata models in the information view. As for the service interfaces, the SIB-ESS-C infrastructure provides three major service types: resource discovery, access and analysis. They are achieved by implementing standard interfaces like the OGC access protocols. These protocols specify the standard interface functionalities to access and subset feature and coverage-based datasets as well as generic maps (i.e. pictorial images), namely: the Web Feature Service (WFS), the Web Coverage Service (WCS), and the Web Map Service (WMS). However, for specific Earth science disciplinary data well-accepted best practices (i.e. community standards) are implemented, as well. Examples include the CF-netCDF,[12] OPeNDAP,[13] or THREDDS (Nativi et al. 2006)

---

[12] http://badc.nerc.ac.uk/help/formats/netcdf/index_cf.html

[13] http://www.opendap.org/

data and protocol models for the fluid Earth sciences, and the TDWG[14] standards for the Biodiversity and Ecology communities.

13.5.1 Resource Discovery Service

The resource discovery service of SIB-ESS-C utilizes a federated catalogue providing a standard OGC discovery interface: the Catalog Service for Web (CS-W) interface (OGC 07–006r1 2007). It implements the ISO 19115 metadata model described in the system information view. The CS-W is the emerging standard for cataloguing services in geomatics, and its ISO Application Profile (OGC 07–045 2007) is the INSPIRE candidate recommendation for discovery services.

This enables users to perform queries on external catalogues and in turn allows other registries to harvest information about SIB-ESS-C data holdings and services. The implementation of the SIB-ESS-C catalogue is based on an open technology: the GI-cat server (Bigagli et al. 2004; Nativi et al. 2007). An ad-hoc module was added to GI-cat base implementation in order to directly access the database containing the SIB-ESS-C 19115 metadata.

GI-cat is an open solution for developing catalog components which implement distributed discovery, data model mediation and access services. GI-cat provides a consistent interface for querying heterogeneous catalogs and data providers that implement international geospatial standards and special arrangements, making it possible to federate heterogeneous data sources by specifying mediation rules for interoperability. GI-cat is a federated catalog providing a unique and consistent interface that enables the interrogation of heterogeneous data resources. GI-cat exposes an OGC CS-W standard interface and is able to federate heterogeneous catalogs and access servers that implement international geospatial standards, such

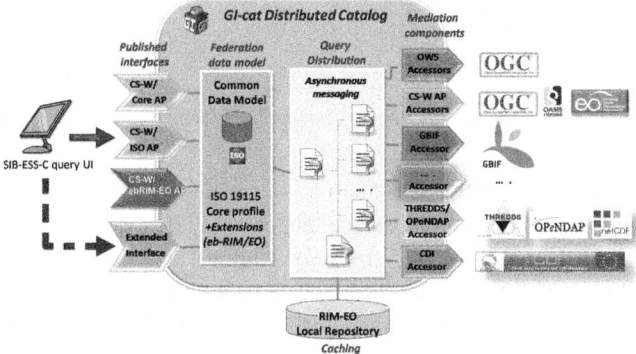

Fig. 13.4 SIB-ESS-C components architecture implementing the discovery and access services (*Color version available in Appendix*)

---

[14] http://wiki.tdwg.org/twiki/bin/view/DarwinCore/DarwinCoreDraftStandard

as the OGC Web services (e.g. WCS, WMS, CS-W). In addition, GI-cat implements a mediation server, making it possible to federate components which implement Community standard services (e.g. THREDDS/OPenDAP and GBIF services). Other functionalities provided by GI-cat are metadata persistency, based on the ISO 19115 data model, and the session and cache management.

The components architecture for the SIB-ESS-C discovery and access services is depicted in Fig. 13.4.

13.5.2 Resource Access Services

Access to SIB-ESS-C data products is provided through OGC Coverage -, Feature- and Map Services allowing users to directly integrate the data as a service into their application or retrieve a file of the requested data product. Access is granted free of charge after a user registration procedure.

13.5.3 Data Visualization Service

A lightweight web interface based on AJAX technologies was developed to directly access SIB-ESS-C service capabilities. From the list of data products returned by the resource discovery service user can select one or more datasets. Resource access services are used to generate a map view (within a Web browser) of these data sets along with auxiliary data supporting orientation and navigation within the view. Moreover, GI-cat publishes a standard catalog interface (i.e. the CS-W ISO interface); thus, any client application which implements such protocol can access the SIB-ESS-C infrastructure being able to discover, query, access, download and visualize the registered datasets.

13.5.4 Data Analysis Services

An advanced feature of SIB-ESS-C will be the online analysis tool to investigate spatial and temporal characteristics (e.g. changes/trends over time) of data products and their relationships (e.g. cross-correlation) or to assess uncertainty of parameters by intercomparing data products from multiple sensors and algorithms. The system provides a Web interface to investigate spatial and temporal characteristics (e.g. changes/trends over time) of data products and to compare data products from multiple sensors and algorithms. A user selects one or two datasets (using the resource discovery service) and specifies the spatial and temporal coverage as well as the analysis method. According to the analysis method selected the system returns a graphical representation of the data set (e.g. time series plot, map). This service will be available for existing SIB-ESS-C data products, but also for external data sets if they are provided through an OGC access service (i.e. WCS and WFS). SIB-ESS-C is investigating the implementation of the standard Web Processing Service (WPS) interface to run data processing, publishing these modules on the Web.

13.5.5 Services Infrastructure Interoperability

The adoption of standard service interfaces allows the SIB-ESS-C infrastructure to contribute to other international efforts, in particular: the Global Change Master Directory (GCMD) and the GEOSS portal.

The GCMD is a comprehensive directory of information about Earth science data and related tools/services, many of which are targeted for the use, analysis, and display of the data. The directory metadata model (i.e. Directory Interchange Format) is compatible with ISO 19115 standard. The GCMD is supported by NASA and contributes to the Committee on Earth Observation Satellites (CEOS).

SIB-ESS-C will register its standard components and services to the respective GEOSS Registries. In fact, the infrastructure implemented services and standards are recognized and supported by GEOSS.

Presently, the SIB-ESS-C infrastructure follows the vision of the implementation rules under specification by the INSPIRE initiative.

13.6 Future Research Activities

A few research aspects to be possibly investigated in the future are: distributed discovery services based on Peer-to-Peer technologies, advanced access services for multidimensional data, and integration of forecasting models via OGC WPS interface. Another possible topic of further activity is the visual presentation of Earth Observation data, typically coverages, that is hard to implement in the general case due to the complexity and heterogeneity of data structures and formats.

References

ISO/IEC 10746 (1998) Open Distributed Processing – Reference Model
ISO/PDTS 19101-2 Geographic information – Reference model – Part 2: Imagery. TC211 working document
OGC 07-045 (2007) Catalogue Services Specification 2.0.2 – ISO Metadata Application Profile, Ver. 1.0.0
OGC 07-006r1 (2007) OpenGIS® Catalog Services Specification, Ver. 2.0.2

Arctic Climate Impact Assessment (2004) Impacts of a warming Arctic: Arctic climate impact assessment. Cambridge University Press, Cambridge

Arsanjani A, Zhang L, Ellis M, Allam A, Channabasavaiah K (2007) S3: A Service-Oriented Reference Architecture. IEEE IT Pro May–June 2007, pp 10–17

Bartsch A, Kidd RA, Wagner W, Bartalis Z (2007a) Temporal and spatial variability of the beginning and end of daily spring freeze/thaw cycles derived from scatterometer data. Remote Sens Environ 106(3):360–374

Bartsch A, Kidd R, Pathe C, Wagner W, Scipal K (2007b) Satellite radar imagery for monitoring inland wetlands in boreal and sub-arctic environments. J Aquat Conserv: Mar Freshwater Ecosyst 17:305–317

Bigagli L, Nativi S, Mazzetti P, Villoresi G (Sept 2004) GI-Cat: a Web Service for Dataset Cataloguing Based on ISO 19115. In: Proceedings of the 15th International Workshop on Database and Expert abase and Expert Systems Systems Applications, IEEE Computer Society Press, Zaragoza (Spain). ISBN 0-7695-2195-9, pp 846–850

Cox S (ed) (2006) Observations and measurements. OGC Best Practices, OGC® 05-087r4

Delbart N, Kergoat L, Le Toan T, L'Hermitte J, Picard G (2005) Determination of phenological dates in boreal regions using normalized difference water index. Remote Sens Environ 97(1):26–38

IEEE (1990) IEEE Standard Computer Dictionary: a Compilation of IEEE Standard Computer Glossaries. New York, NY

Foster I, Kesselman C (Nov 2006) Scaling system-level science: scientific exploration and IT implications. IEEE Comput 39(11)

GEO (2005) In: Battrick B (ed) Global Earth Observation System of Systems (GEOSS) 10-Year Implementation Plan. ESA Publications Division, the Netherlands. ISSN No.: 0250-1589, ISBN No.: 92-9092-495-0

GEO (2007) 2007–2009 Work Plan: Toward Convergence, GEO document

George C, Rowland C, Gerard F, Balzter H (2006) Retrospective mapping of burnt areas in Central Siberia using a modification of the normalised difference water index. Remote Sens Environ 104(3):346–359

EC (2005) GMES: From concept to reality. Communication from the Commission to the Council and the European Parliament, November 11 2005. http://www.gmes.info/library/files/1.%20 GMES%20Reference%20Documents/COM-2005-565-final.pdf.

Grippa M, Mognard NM, Le Toan T, Josberger EG (2004) Siberia snow depth climatology derived from SSM/I data using a combined dynamic and static algorithm. Remote Sens Environ 93:30–41

Hansen J, Ruedy R, Glascoe J, Sato Mki (1999) GISS analysis of surface temperature change. J Geophys Res 104:30997–31022

ISO 19101 (2002) Geographic information – Reference model
ISO 19107 (2003) Geographic information – Spatial schema
ISO 19109 (2005) Geographic information – Rules for application schema
ISO 19111 (2003) Geographic information – Spatial referencing by coordinates
ISO 19115 (2003) Geographic information – Metadata
ISO/FDIS 19123 (2005) Geographic information – Schema for coverage geometry and functions
ISO/IEC 19501 (2005) Information technology – Open distributed processing – Unified Modeling Language (UML) Version 1.4.2

Khalsa SJS, Nativi S, Geller G (2008) The GEOSS Interoperability Process Pilot Project (IP3). Submitted to IEEE TGRS

Miller J, Mukerji J (eds) (June 2003) "MDA Guide Version 1.0.1" OMG Document Num. omg/2003-06-01

Nativi S, Domenico B, Caron J, Bigagli L (June 2006) Extending THREDDS middleware to serve OGC community, Advances in Geosciences. J Eur Geosci Union, 8:57–62. SRef-ID: 1680-7359/adgeo/2006-8-57

Nativi S, Bigagli L, Mazzetti P, Mattia U, Boldrini E (July 2007) Discovery, query and access services for Imagery Gridded and Coverage Data: a clearinghouse solution. In: IEEE International Geoscience and Remote Sensing Symposium, Barcelona, Spain

Nebert D (ed) (2004) Developing spatial data infrastructures: the SDI cookbook ver. 2.0. GSDI publication
Percival G (2008) GEOSS Architecture Implementation Pilot. In: Proceedings of the '08 EGU General Assembly meeting, Vienna, April 2008
Skinner L, Luckman A (2004) Introducing a land cover map of Siberia derived from MERIS and MODIS data. In: Proceedings of IGARSS'04, Anchorage, 20–24 Sept, pp 223–226. G. O
Snell J, Tidwell D, Kulchenko P (Jan 2002) Programming Web Services with SOAP. O'Really edition. ISBN 0-596-00095-2, pp 244
Zhaomei Z, Zhongwei Y, Duzheng Y (2001) The regions with the most significant temperature trends during the last century. Adv Atmos Sci 18(4):481–496

Chapter 14
Development of a Web-Based Information-Computational Infrastructure for the Siberia Integrated Regional Study

E.P. Gordov, A.Z. Fazliev, V.N. Lykosov, I.G. Okladnikov, and A.G. Titov

Abstract To understand dynamics of regional environment properly and perform its assessment on the basis of monitoring and modelling, an information-computational infrastructure is required. Management of multidisciplinary environmental data coming from large regions requires new data management structures and approaches. In this chapter on the basis of an analysis of interrelations between complex (integrated) environment studies in large regions and modern information-computational technologies major general properties of distributed information-computational infrastructure required to support planned investigations of environmental changes in Siberia in the Siberia Integrated Regional Study (SIRS) are discussed. SIRS is a Northern Eurasia Earth Science Partnership Initiative (NEESPI) mega-project co-ordinating national and international activity in the region in line with an Earth System Science Program (ESSP) approach. The infrastructure developed in cooperation of Russian Academy of Science (Siberian Branch) specialists with their European and American partners/counterparts is aimed at supporting multidisciplinary and "distributed" teams of specialists performing cooperative work with tools for exchange and sharing of data, models and knowledge optimizing the usage of information-computational resources, services and applications. Recently developed key elements of the SIRS infrastructure are described in details. Among those are the Climate site of the environmental web portal ATMOS (http://climate.atmos.iao.ru) providing an access to climatic and mesoscale meteorological models and the Climate site of the Enviro-RISKS web portal (http://climate.risks.scert.ru/),

---

E.P. Gordov (✉), I.G. Okladnikov, and A.G. Titov
Siberian Center for Environmental Research and Training and Institute of Monitoring of Climatic and Ecological Systems SB RAS, Akademicheski avenue 10/3, Tomsk 634055, Russia
e-mail: gordov@scert.ru; onuchin@ksc.krasn.ru; titov@scert.ru

A.Z. Fazliev
Institute of Atmospheric Optics SB RAS, Akademicheski avenue 1, Tomsk 634055, Russia
e-mail: faz@iao.ru

V.N. Lykosov
Institute for Numerical Mathematics RAS, Moscow, Russia
e-mail: lykossov@inm.ras.ru

providing an access to interactive web-system for regional climate assessment on the base of standard meteorological data archives. As an example of the system usage recent dynamics of some regional climatic characteristics are analyzed.

Keywords Environmental monitoring and assessment • Climate • Information systems • Meteorology • Regional climate change

14.1 Introduction

The fact that specifics of basic Earth system science and their regional/local environmental applications (Environmental Sciences) make them multidisciplinary and require to involve into studies a number of nationally and internationally distributed research groups is common knowledge nowadays. Really, here multidisciplinary (in virtue of problems treated and in nature of the environmental issues tackled), "distributed" teams of specialists should perform cooperative work, exchange data and knowledge and co-ordinate activities optimizing the usage of information-computational resources, services and applications. Also the community acknowledged that to understand dynamics of regional environment properly and perform its assessment on the basis of monitoring and modelling more strong involvement of information-computational technologies (ICT) is required, which should lead to the development of information-computational infrastructure as an inherent part of such investigations (Gordov 2004a, b). In particular, recently it was stressed that management of multidisciplinary environmental data coming from large regions requires new data management structures and approaches (Parson and Barry 2006). Thus the contemporary challenge is to save efficiency of such efforts via development of a platform/mechanism providing a collaborative working environment for the scientists engaged, as well as giving access to and preservation of scientific information resources, such as environmental data collections, models, results, etc. All these issues are among the priorities within the R&D strategy of major actors in the field now.

It is clear nowadays that a very beneficial synergy effect could be achieved by closely coupling the areas of Environmental Sciences (ES) and Information-Computational Technologies (ICT) that is for an interdisciplinary field concerned with the interaction of processes that shape our natural environment (ecology, geosciences, hydrology, and atmospheric sciences), and the way that these processes are "mapped" into an information system architecture and are dealt with via relevant software tools. Formally the latter belongs to Informatics, which is application of formal and computational methods for analysis, management, interchange, and representation of information and knowledge, while its synergetic usage in ES can be defined as Environmental Sciences Informatics (ESI). Being a subdivision of Informatics, ESI is mainly aimed at a formal representation of the spatial and temporal hierarchical structure of subsystems compounding the regional environment or Earth system as a whole and the relationships between these compounds. At the same time, being a subdivision of Environmental Sciences it is aimed at design, development, and application of tools to acquire, store, analyze, visualize, manage, model, and represent information about the spatiotemporal dynamics of the environment system to interdisciplinary community. In other words ICT or ESI plays a pivotal role in developing the 'underlying mechanics' of the work, leaving the earth scientists to concentrate on their important research as well as providing the environment to make research results available and understandable to everyone. Major efforts here are undertaken either in an attempt to provide GIS platforms with required web accessibility, computing power and data interoperability or to exploit completely the huge potential of web based technologies. In spite of some remarkable achievements (see for details the Open Geospatial Consortium web portal http://www.opengeospatial.org/) we consider attempts to save GIS functionality together with computing power required to support modern models as well as huge data archive sharing not very promising and the approach relying upon web technologies potential was chosen for the development of the information-computational infrastructure required.

There are two key projects strongly employing web technologies' potential nowadays, which mainly determine direction for software tools design in thematic domain of Earth Science, namely, PRISM (Program for Integrated Earth System Modelling, http://prism.enes.org) and ESMF (Earth System Modelling Framework, http://www.esmf.ucar.edu/). PRISM aims at providing the European Earth System Modelling community with a common software infrastructure. A key goal is to help assemble, run, and analyze the results of Earth System Models based on component models (ocean, atmosphere, land surface, etc.) developed in different climate research centres in Europe and elsewhere. It is organized as a distributed network of expertise to help share the development, maintenance and support of standards and state of-the-art software tools. The basic idea behind ESMF is that complicated applications should be broken up into smaller pieces, or components. A component is a unit of software composition that has a coherent function, and a standard calling interface and behaviour. Components can be assembled to create multiple applications, and different implementations of a component may be available. In ESMF, a component may be a physical domain, or a function such as a coupler or I/O system. It should be noted that the announced on-component approach is not yet realized consistently in either of the two projects.

A somewhat different approach is based on the suggestion (De Roure et al. 2001) that each separate computational task (it also might be a data assimilation task as well or a combination of both above) can be represented as an information system, employing the three-level model – data/metadata, computation and knowledge levels. Use of this approach for development of Internet-accessible information-computational systems for the chosen thematic domains and organization of data and knowledge exchange between them looks like a quite perspective way to construct a distributed collaborative information-computational environment to support investigations, especially, in the multidisciplinary area of Earth regional environment studies.

The first step in this direction was done in course of development of the bilingual (Russian and English) scientific web portal ATMOS (Gordov et al. 2004, 2006a) (http://atmos.iao.ru/). ATMOS is designed as an integrated set of distributed but coordinated topical web sites, combining standard multimedia information with research databases, models and analytical tools for on-line use and visualization. The main topics addressed are from the Atmospheric Physics and Chemistry domain. It should be noted that in spite of the fact that the portal middleware employs PHP scripting language (http://www.php.net/), it has quite a flexible and generic nature, which allows one to use it for different applications. Currently on this basis web portals are developed and launched, providing a distributed collaborative information-computational environment to organizations/researchers participating in the execution of EC FP6 projects "Environmental Observations, Modeling and Information Systems" (http://enviromis.scert.ru/) and "EnviroRISKS – Man-induced Environmental Risks: Monitoring, Management and Remediation of Man-made Changes in Siberia" (Baklanov and Gordov 2006) (http://risks.scert.ru/). The portals are also powerful instruments for the dissemination of the project results and open free access to collections of regional environmental data and education resources.

While in the ATMOS portal only the data and computation levels were employed, appearance in 2004 RDF and OWL recommendations and supporting software allowing to get conclusions on the basis specified according to these recommendations knowledge formed a basis to include metadata and knowledge levels into a typical information system, which is especially important for complex environmental tasks and problems solving. In particular, the three-level model was used to implement on the basis of the ATMOS software the distributed information system "Molecular Spectroscopy", employing an elaborated task and domain ontology (Fazliev and Privezentsev 2007). Additional opportunities appeared as a result of Semantic Web development (http://sweet.jpl.nasa.gov/). The Semantic Web would enable a new breed of applications on the basis of knowledge sharing: smart agents instead of search engines. These agents will be able to establish a dialogue with other agents or portals to exchange and request information, determine available resources, settle agreements on operations, cooperate in several tasks and return processed results to their user or forward them to other agents for further processing. In order for this cooperation to work, agents must share information in a common language. Ontologies provide a framework for knowledge expression. The field is still maturing and there is no unified ontology for all knowledge domains (although there are efforts such as Standard Upper Ontology from IEEE and similar activity within the GEOD community – codex web portal for creating and managing personal and community ontologies for scientific research). These new opportunities open additional potential for the information-computational infrastructure under development and also should be made available to the Earth and Environmental Sciences professional community.

This chapter describes first elements of the web based environmental informatics system (with accompanying applications) forming a distributed collaborative information-computational environment to support a multidisciplinary investigation of Siberia as that will form a powerful tool for better understanding of the interactions between ecosystems, atmosphere, and human dynamics in the large

Siberia region under the impact of global climate change. Being generic it should provide researchers with a reference, open platform (portal plus tools) that may be used, adapted, enriched or altered on the basis of the specific needs of particular applications in different regions. In this initial stage major attention was paid to components that are crucial for subsequent applications and aimed at handling/ processing different data sets coming from monitoring and modelling regional meteorology, atmospheric pollution transformation/transport and climate important for a regional environment dynamics assessment under climate change.

Below firstly the Siberia Integrated Regional Study (SIRS) will be described, which forms a test bed and major user community for the system under development. Then the following yet developed elements will be discussed in more details: the ATMOS web portal Climate site current version (http://climate.atmos.iao.ru) providing access to climatic and mesoscale meteorological models; the Enviro-RISKS web portal Climate site (http://climate.risks.scert.ru/) providing access to an interactive web-system for regional climate assessment on the basis of standard meteorological data archives; and a web system for visualization and analysis of air quality data for the city of Tomsk (http://air.risks.scert.ru/tomsk-mkg/). To illustrate the system potential, dynamics of some regional climatic characteristics will be analyzed and discussed with their usage.

14.2 Siberia Integrated Regional Study

The regional (region here is a large geographical area, which functions as a biophysical, biogeochemical and socio-economical entity) aspect of science for sustainability and of international global change research is becoming even more important nowadays. It is clear now that regional components of the Earth System may manifest significantly different Earth System dynamics and changes in regional biophysical, biogeochemical and anthropogenic components may produce considerably different consequences for the Earth System at the global scale. Regions are "open systems" and the interconnection between regional and global processes plays a key role. Some regions may function as choke or switch points (in both biophysical and socio-economic senses) and small changes in regional systems may lead to profound changes in the ways in which the Earth System operates. A few years ago IGBP suggested (IGBP Newsletter 2002, 2003) to develop integrated regional studies of environment in selected regions, which would represent a complex approach to reconstruct the Earth System dynamics from its components behaviour. It considered as a complementary effort to the thematic project approach employed so far in the international global change programs. Nowadays the Integrated Regional Study (IRS) approach is developed by the Earth System Science Partnership (http://www.essp.org/), joining four major Programs on global change research. The IGBP initiative is aimed at development of IRS in the most important regions of the planet puts a set of prerequisites for such studies:

- The concept should be developed in the context of the Earth System as a whole.
- Scientific findings should support sustainable development of the region.
- Qualitative and quantitative understanding of global–regional interconnections and the consequences of changes in these interconnections should be achieved.

The word 'integrated' in IRS refers specifically to two types of integration: (i) 'horizontal integration', involving the integration of elements and processes within and across a region; and (ii) 'vertical integration', involving the two-way linkages between the region and the global system. There are two examples of existing IRS – a matured large biosphere-atmosphere experiment in Amazonia (LBA, http://lba.cptec.inpe.br/lba/indexi.html) and a recently started ESSP Monsoon Asia Integrated Regional Study (MAIRS, http://www.mairs-essp.org/).

Siberia is one of the promising regions for the development of such basic and applied regional study of environmental dynamics (Bulletin of the Russian National Committee for the IGBP 2005). Regional consequences of global warming (e.g. anomalous increase of winter temperatures (Ippolitov et al. 2004)) are strongly pronounced in Siberia. This tendency is supported by the results of climate modeling for twentieth–twenty-second centuries (Volodin and Dianskii 2003). The climate warming not only threatens Siberia with destruction of most extractive and traffic infrastructure caused by the shift of permafrost borders northwards but can also change the dynamics of the natural-climatic system as a whole. Although many projects supported by national (SB RAS, RAS) and international (EC, ISTC, NASA, NIES, IIASA, etc.) organizations are devoted to study the modern dynamics of Siberian environment, scientists know little about the behavior of the main components of the regional climatic system as well as about responses and feedbacks of terrestrial and aquatic ecosystems. A regional budget of the most important greenhouse gases CO_2 and CH_4 is still making first steps with respect to individual land classes. Measurements in situ are limited and still lacking any systematic basis. Responses of boreal forests and Siberian wetlands to climate change and the emerging feedback influencing climate dynamics through exchange of momentum, energy, water, greenhouse gases and aerosols are poorly understood and almost not yet identified. A change of climatic characteristics creates the prerequisites for large and significant biological, climatic and socio-economically coupled land use variations throughout this region. Science issues for region are growing in global importance not only in relation to climate change and carbon, but also for condition and stability of aquatic, arid, and agricultural systems, snow and ice dynamics. IGBP reported recently that the circumboreal region including Northern Eurasia is one of the critical "Switch and Choke" points in the Earth system, which may generate small changes in regional systems potentially leading to profound changes in the ways in which the Earth System operates.

The rather short-term SIRS history has been started in 2002, when Will Steffen (IGBP) at the conference on boreal forests in Krasnoyarsk had suggested launching with assistance of SB RAS and on the basis of its research infrastructure the SIRS project as a part of the implementation of IGBP's and ESSP's regional strategy to develop one of the Integrated Regional Studies here. This idea was supported by a group of Russian scientists and their partners abroad and specific activity begun under the overall coordination of the Siberian Centre for Environmental Research and Training (SCERT) in 2003. The approach adopted was examined and endorsed by the Siberian Branch of the Russian National Committee for IGBP in 2005, which decided that during the first stage of SIRS development it is necessary to focus on four lines of investigation:

- Quantification of the terrestrial biota full greenhouse gas budget, in particular exchange of major biophilic elements between biota and atmosphere
- Monitoring and modelling of regional climate change impact
- Development of SIRS information-computational infrastructure
- Development of an anticipatory regional strategy of adaptation to and mitigation of the negative consequences of global change

The SIRS (http://sirs.scert.ru) current state of the art (Gordov and Begni 2005; Gordov et al. 2006b) is characterized by the appearance of a number of large-scale projects on the Siberian environment in line with the SIRS objectives and the very beginning of their clustering. Among those are thematically relevant SB RAS Integrated projects (2006–2008), RAS Programs projects (2006–2008), as well as EC, ISTC and NASA funded projects and clustering is giving them substantial added value. A key role in these projects is played by the institutes of SB RAS. At the same time, new and larger national and international initiatives are emerging to develop a study of that kind on the territory of the whole Northern Eurasia. Appeared a few years ago as a joint program of RAS and NASA, the "Northern Eurasia Earth Science Partnership Initiative" (NEESPI) now has transformed into the international Program and quite recently was adopted as one of the external projects of IGBP. The list of projects that are currently under the umbrella of NEESPI (http://www.neespi.org/) is quite impressive. After a series of discussions it was agreed that SIRS will be a NEESPI mega-project co-ordinating national and international activity in the region in line with the ESSP approach. Among planned joint steps to consolidate cooperation are the organization of Distributed Centres to support NEESPI activity in the region based in Krasnoyarsk (Forestry and Remote Sensing) and Tomsk (Data and Modelling) as well as co-ordination of training and educational activity aimed at young scientists involvement in this scientific theme.

All the above shows that SIRS as a large-scale multidisciplinary investigation of environmental dynamics in this huge region, badly in need of an information-computational infrastructure supporting this activity. Due to efforts of SB RAS that provided its scientific centres distributed in the region with stable high-speed communication channels (Shokin and Fedotov 2003) SIRS has all prerequisites required to become the test bed for the information-computational system under development.

14.3 Developed Elements of the Infrastructure

14.3.1 ATMOS Climate Site

The bilingual (Russian and English) scientific web portal ATMOS (Gordov et al. 2004, 2006a) developed under an INTAS project in 2002 can be used as an example of such an approach (http://atmos.scert.ru/ and http://atmos.iao.ru). It comprises an integrated set of nine distributed but coordinated scientific information systems (IS) operating in the Internet. These IS were designed by means of web technologies, which allowed to form information resources by a content management system using an administrative console and create workflows projected on the base of Petri nets (Van der Aalst 2004). Two groups of IS are included into the portal. The first group comprises the information systems "West Siberia" (http://west-sib.atmos.scert.ru), "Baikal" (http://baikal.atmos.scert.ru) and "Air Quality Assessment" (http://air.atmos.scert.ru). The second group is formed from information-computational systems and now covers five thematic areas: atmospheric aerosols (http://aerosol.atmos.iao.ru), the atmospheric radiation (http://atrad.atmos.iao.ru), gas-phase chemical reactions in the atmosphere (http://atchem.atmos.iao.ru), molecular spectroscopy (http://saga.atmos.iao.ru) and West Siberian climatic and meteorological characteristics (http://climatel.atmos.iao.ru). This separation reflects a sort of hierarchy appearing on a way to the most complicated computational system of the portal, which is modelling climate. It should be noted that this system requires powerful computational support. Currently it is done on a 20-processor cluster, however soon the 572-processor cluster of Tomsk State University will be used as well.

In the process of the information-computational system "Climate" design integration of three scales of models was organized. Those are global models, regional models and models at city level. The global scale models chosen is a GCM (Alekseev et al. 1998) of the RAS Institute for Numerical Mathematics (INM), regional scale models are presented by the Mesoscale Model 5 (MM5) (Dudhia 1993) and the Weather Research and Forecasting Model (WRF) (Weather Research and Forecasting Model 2007), while the final selection of a city level model is not finalized yet. An access to computational resources of the climate system is free; however the user should register and be authorized first to get access to resources. To diminish the amount of the traffic the user gets results of the modelling as plots. An on-line regime is used only for choice of input parameters and to look through the results obtained. Results of calculations are saved on the database server and available for the user at any time. Technically, the process of task execution is determined by the four-level architecture of the system (client, web server, database server and computational cluster). When input data of the problem are transmitted to the server, the relevant process is created on the cluster and the Torque task manager (TORQUE Resource Manager 2007) controls them. The user has to reserve the server resources for access to domain applications (tasks). For this aim one may form catalogues and tasks, where the results of calculation will be stored.

The INM climate model is equipped with a comprehensive physical package that includes advanced parameterizations of the boundary-layer turbulence and air–sea/air–land interaction, solar radiation transfer, land surface and soil hydrology, and other climatic processes. The detailed model documentation is presented at the AMIP (Atmospheric Model Intercomparison Project) site (http://www-pcmdi.llnl.gov/projects/modeldoc/amip2/dnm_98a/index.html). The space resolution of the model is 5° in longitude, 4° in latitude and 21 levels in vertical from the Earth surface to the height of about 30 km. The web-interfaces for this model implemented during INTAS project and can be found at http://climate.atmos.iao.ru.

Interfaces to work with the mesoscale meteorological models MM5 and WRF are accessible via the Internet at the ATMOS portal (http://climate.atmos.iao.ru/star/mm5, and http://climate.atmos.iao.ru/star/wrf). These models are aimed at numerical weather forecast mainly, but can also be used to study convective systems, city heat islands, etc. These interfaces allow the user to change input data, to initiate calculations on the supporting system calculations cluster and to get results in graphical forms. Below those are described for an example of MM5 model usage. It should be noted that the absence of some types of data forced us to restrict the possibility of weather characteristics calculations presented by the MM5 model.

Figure 14.1 shows a scheme of the implemented part (Lavrentiev et al 2006) of MM5, which reflects the sequence of application executions, data flows and briefly characterizes its principal functions. Surface and isobaric meteorological data are interpolated by the TERRAIN and REGRID modules on the longitude–latitude grid into the high definition variable region, which is situated on the Lambert straight angle projection, polar stereographic projection or Mercator projection. INTERPF application interpolates values of physical variables (initial and boundary conditions) from isobaric levels MM5 model's sigma system. Sigma levels near the Earth surface coincide with the relief and as the distance to surface becomes greater they come near the isobaric surfaces.

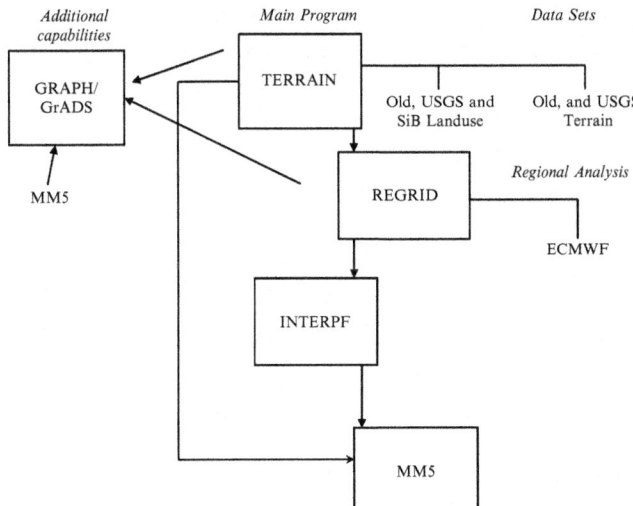

Fig. 14.1 Implemented part of MM5 system

The user of the MM5 model has to present data about the relief, land use and vegetation for the whole territory of interest. Topographical datasets of different resolution are used for geographical location of the MM5 model. The reanalysis data by ECMWF for the period 1991–2002 is used as input data. The sequence of data preparation by the user includes three stages. In the first stage the user defines the coordinates of the region and the number of embedded regions, the type of cartographic projection, the scale of the horizontal grid, on which the user the initializes the relief, the distribution of land use categories and the types of vegetation. The user also defines the resolution of the vertical grid. During the second stage the user assigns the time period of the modelling, the value of temporal digitization of the meteorological data (as a rule 6 or 12 h are stored in DB) and a list of altitudes where calculations have to be done. During the final stage for every embedded region the defined by the user, one makes a choice of the parameterization scheme for microphysics of humidity, cloudiness, boundary layer and radiation in the atmosphere.

Input data files, formed in conversation mode are delivered from the web server to the cluster. The application Torque controls the querying of the task and controls the task line. Computational modules of the MM5 application depend on the number of embedded regions, that's why this application is compiled every time before the execution. Obtained output is stored and provided with unique identifications, which allows the system to connect them with the user and to solve his problems. The results are represented in three groups of physical parameters separated by their distribution in space: 3D (24 parameters) and 2D (31 parameters) variables and combined ones (7 parameters). One can find at the site the full list of the parameters including temperature, pressure, humidity, wind direction and so on. The user can get plots of results in the form of maps with contour lines, colour maps or maps with vector fields. Graphical depiction of the results is done using the GRADS package.

The same approach is applied to design the interface for the WRF model. There are the same four steps, which user has to do for input data assignment. The final list of resulting physical parameters is slightly different from the MM5 list. Figure 14.2 shows a visualization of some results of WRF model runs for the Tomsk region.

14.3.2 Enviro-RISKS Portal and Climate Site

The next element of the infrastructure is formed by the Enviro-RISKS project webportal (http://risks.scert.ru/). This bilingual (Russian and English) resource is aimed at dissemination of information on general environment issues adjusted also for usage in the education process and giving access to regional environmental data and tools to process them online. The portal is organized as a set of interrelated scientific sites, which are opened for external access. The portal engine employs the AIMUS portal software. The portal provides easy access to structured information

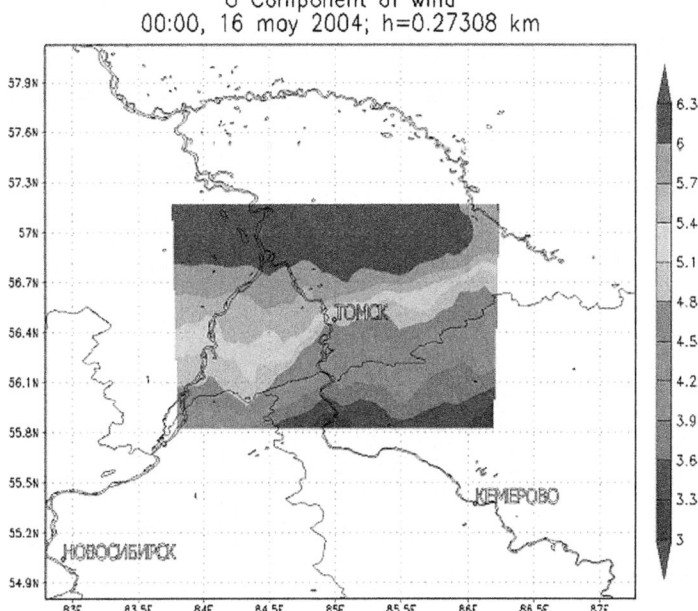

Fig. 14.2 Representation of the zonal component of the wind velocity calculated by the WRF model. The *central part* of the plot shows the area of the West Siberian city of Tomsk

resources on the Siberian environment and to results of project expert group studies devoted to regional environmental management under anthropogenic environmental risks. A built-in Intranet is used as an instrument for project management as well as for exchange and dissemination of information between the project partners. The portal also gives access to gathered and analyzed detailed information on all coordinated projects, gathered and systemized results and findings obtained including relevant observational data and information resources, a distributed database, which will provide access to data on the characteristics of the Siberian environment for the project partners and to relevant metadata for the wider interested professional community. The basic thematic sites currently integrated into the Enviro-RISKS web-portal are the Climate site aimed at access to specially designed analytical tools allowing to retrieve the spatial pattern of selected Siberian climatic characteristics from measured or simulated datasets and the Air Quality Assessment site, which compiles basic and applied aspects of air pollution and environmental impact assessment. A special site is devoted to project management. It comprises information on the project partners, project management, projects/program coordination and gives access to educational recourses gathered by the partners.

The Climate site of the portal (http://climate.risks.scert.ru/) upon a qualified user request gives access to an interactive web-system for regional climate change assessment on the basis of standard meteorological data archives. The system is a

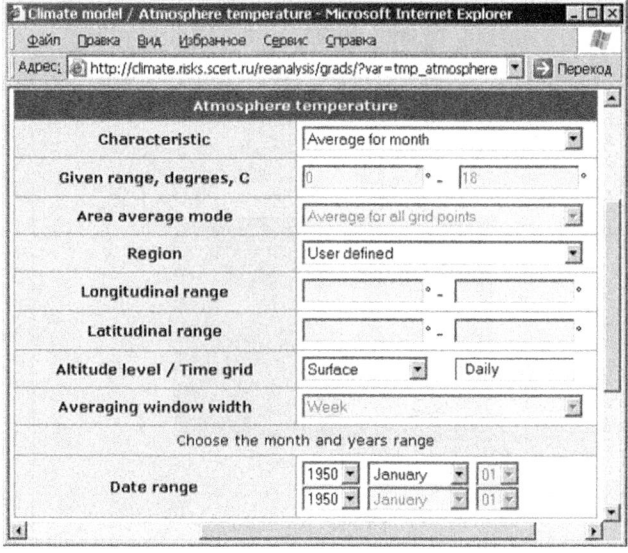

Fig. 14.3 Graphic user interface of the system

specialized web-application aimed at mathematical and statistical processing of huge arrays of meteorological and climatic data as well as on the visualization of results. The data of the first and second NCAR/NCEP Reanalysis editions are currently used for processing and analysis. The Grid Analysis and Display System (GrADS, http://www.iges.org/grads/) and the Interactive Data Language (IDL, http://www.ittvis.com/idl/) are employed for visualization of the results obtained.

The system consists of a graphic user interface, a set of software modules written in the script languages of GRADS or IDL and structured meteorological datasets. The graphic user interface is a dynamic web form to choose parameters for calculation and visualization and designed using HTML, PHP and JavaScript languages (Fig. 14.3).

The set of software modules consists of independent modules implementing analytical algorithms required for meteorological data processing, switched on with assistance of PHP and executed by the GRADS/IDL system, which generates a graphical file containing the results of calculations performed. The latter along with corresponding metadata is passed to the system kernel for subsequent visualization at the web site page. The structured meteorological data are stored on the specialized server and are available only for processing by the system so that the user cannot access data files directly while he can freely get results of their analysis. The screenshot in Fig. 14.4 shows a window of a resulting graphical output.

At present the following climatic characteristics have been chosen for subsequent analysis: temperature, pressure, air and soil humidity, precipitation level and geo potential height. Due to the middleware used the system structure is rather flexible

14 Development of a Web-Based Information-Computational Infrastructure

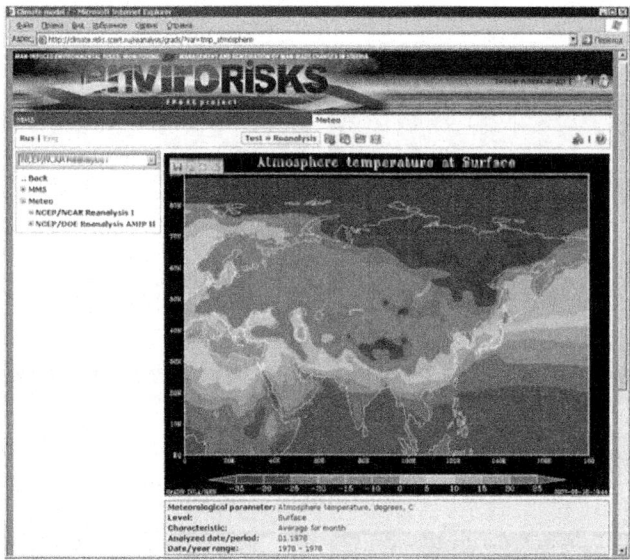

Fig. 14.4 Surface temperature in January, 1978 for chosen geographical domain (40–90° of the Northern latitude and 0–180° of the Eastern longitude)

and this set can be easily expanded by adding new features to the interface as well as to the executable software.

Currently the system allows performing for chosen spatial and time ranges the following mathematical and statistical operations, which are key for a climate assessment: calculation of maximum, minimum, average, variance and standard deviation values, number of days with the value of the chosen meteorological parameter within given range, as well as time smoothing of parameter values using a moving average window. It is also possible to calculate the correlation coefficient for an arbitrary pair of parameters, linear regression coefficients between some pair of characteristics and to determine the first (last) cold (warm) period (such as day, week, month) of the year. The user interface allows one to choose the geographic domain, time interval, characteristic of interest and visualization parameters. For example, a pulldown menu "Region" includes the following options: Siberia, Europe, Asia, Eurasia, the whole Earth and that is defined by the user. The last option choice leads to appearance of "Longitudinal range" and "Latitudinal range" fields in which the user should enter coordinates of the geographical domain of interest. The user can also choose the type of statistical parameter, the interval for averaging, the altitude level, and so on. While calculating averages with a moving window, whose width can be specified as week, month, 3 months, half a year and a whole year, one gets the smoothed sequence of spatial distributions for the parameter of interest. This set is represented as an animation, which can be viewed either in an automatic or a controlled regime. Figure 14.5 shows several shots from such a sequence including the control bar specifying the viewing mode.

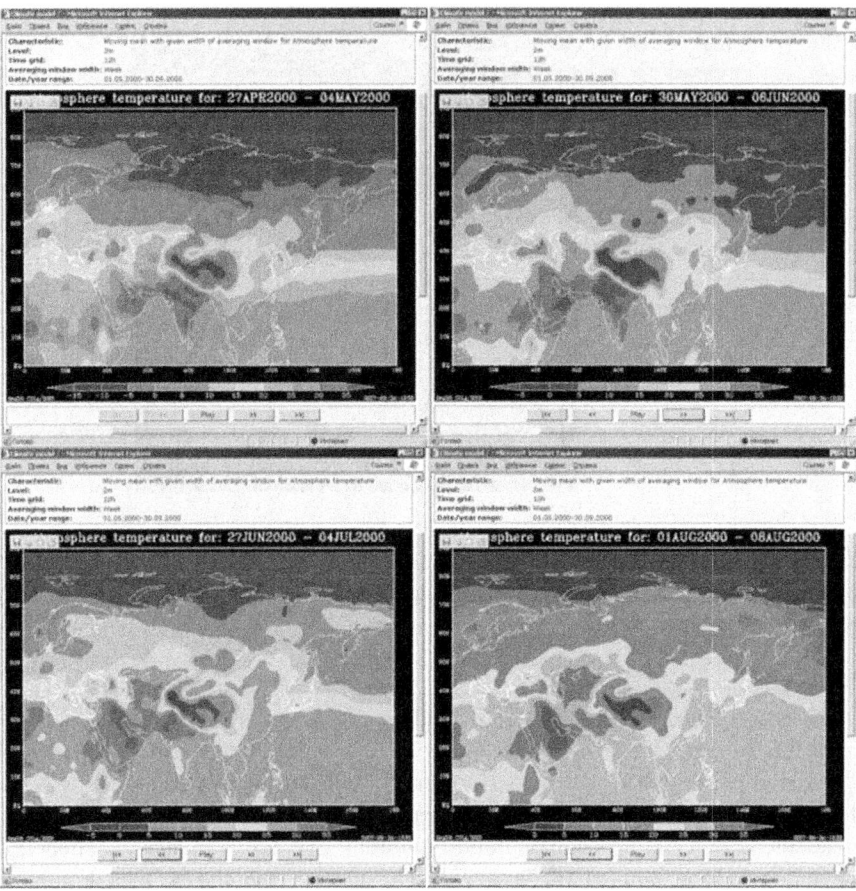

Fig. 14.5 Results of averaging with moving window

The system will be useful for regional meteorological and climatic investigations aimed at determination of trends of the processes taking place. Also it will simplify the work with huge archives of spatially distributed data. It should allow scientific researchers to concentrate on solving their particular tasks without being overloaded by routine work and to guarantee the reliability and compatibility of the results obtained.

14.3.3 Air Quality Assessment Site

The next element deployed at the portal is a web-system for Tomsk air quality assessment based on mathematical modelling of pollution transport and transformations. It is aimed at effective air chemical composition assessment and forecast in the conditions of the industrial city area and its suburbs. The technique applied is based on the meteorological observations, taking into account atmospheric emissions of industrial enterprises and traffic, measurements of the concentration of atmospheric pollutants as well as on the numerical modelling of the gaseous substances' transformation and transport processes in the atmosphere. This system is also based on the ATMOS portal middleware. Currently air quality data sets for an assessment of the Tomsk area air pollution used by the system are obtained with the help of a mathematical model of pollution transport employing meteorological fields calculated with prognostic meteorological models of pollution transport for selected periods within the 2000–2005 interval. The model takes into account transport, dispersion and dry deposition of the pollutants as well as their photochemical transformations.

The system comprises three following parts: generated by the model and converted into binary files by a specially developed Java utility, stored on the server data archives containing calculated fields of pollutant concentrations; graphic user interface; and a set of PHP-scenarios to perform calculations with subsequent visualization. Currently it allows a registered user to get visualized results of such characteristics as average monthly and seasonal pollution along with their annual dynamics as well as daily dynamics for various pollutants. The GrADS open source software has been used for the visualization of results for it has strong capabilities for the table graphic data representation. The graphical user interface is implemented using HTML, PHP and JavaScript languages and represents a dynamic HTML form (Okladnikov and Titov 2006) for choice of input parameters and visualization characteristics (Fig. 14.6).

The form allows user to set the following parameters:

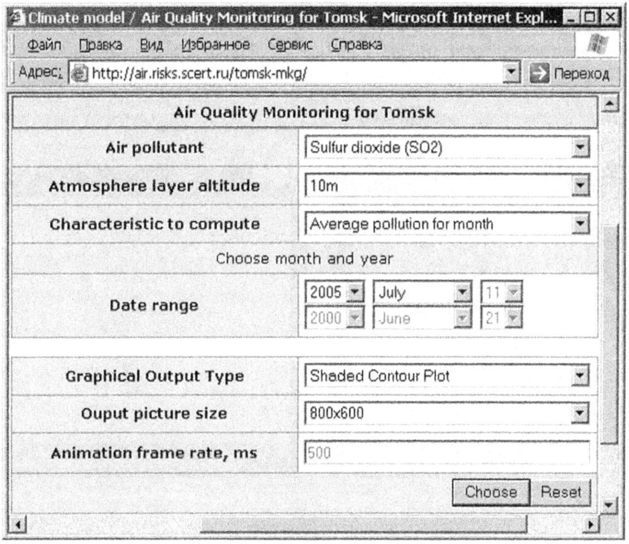

Fig. 14.6 Graphic user interface for air quality assessment system

1. "Air pollutant" with such options as airborne particulate matter, sulphur dioxide, nitrogen dioxide, carbon oxide and ozone.
2. "Atmospheric layer altitude" ranging from 10 to180 m.
3. "Characteristics to compute". At present it is possible to calculate such characteristics as average pollution for month, season and their dynamics within the time interval, and hourly dynamics during the selected day.
4. Date range, graphical output type and picture size. There is also a possibility to choose an animation frame rate to see dynamics of the concentrations. The screen shots in Fig. 14.7 demonstrate an instance of hourly dynamics of sulphur dioxide during July 11, 2005.

The system might be used by regional ecologists and decision makers to determine the characteristics of pollution distribution above the territory and their dynamics under different weather conditions, to estimate input of selected pollution sources (industry enterprises, transport, etc.) into the pollution fields, as well as to estimate

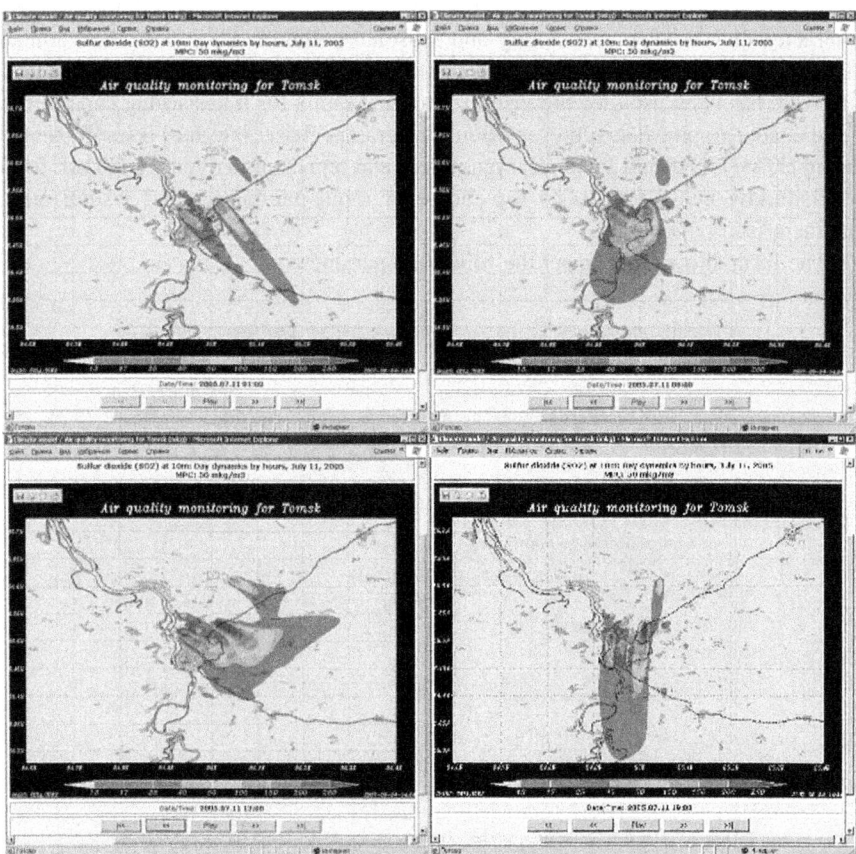

Fig. 14.7 Sulphur dioxide concentration dynamics during 11 July 2005

consequences of possible accidents leading to additional pollutant emissions. Also it might be used to understand the degree of anthropogenic influence on the regional environment and climate. It should be added that the system has generic character and being provided with characteristics of industrial and transport pollution sources, local meteorology data, surface properties and generated by the photo-chemical transport and transformations model pollution data sets it can be easily adjusted for conditions of an arbitrary city.

14.4 Conclusions

As seen from above only a few elements of the SIRS information-computational infrastructure have been elaborated so far. However, it has proved its efficiency already. To illustrate it, a recently reported (Melnikova et al. 2007) analysis of recent dynamics of regional climatic characteristics performed with the use of the "Climate" site described in 14.3.2 is discussed. It should be noted, that since the system is currently processing only fields of meteorological characteristics obtained from the NCEP/NCAR Reanalysis and the NCEP/DOE Reanalysis AMIP II projects, conclusions derived will characterize these datasets mainly. The latter are not very well correlated with the reality in this region due to the poor observational network.

Due to the multifactor formation of atmospheric processes, statistical methods are the most reliable approaches to a quantitative assessment of the characteristics of climatic dynamics and the reported study deals namely with those. In particular, statistical properties of the recent variability of precipitation and near-surface (2 m) temperature in West Siberia were analyzed. To support this analysis a set of functions built into the web system was employed. Among those are the determination of the first warm/cold day, week and month in a year (variability of the warm season duration); determination of the number of days with daily mean precipitation amount in the chosen interval (variability of precipitation amount); determination of the number of days with daily mean temperature in the chosen interval (variability of warming); and calculation of correlation coefficients for different pairs of meteorological parameters (degree of their linear dependence). Both parametric and non-parametric statistical criteria were used for the analysis. The analysis performed shows that the precipitation amount is slightly decreasing during warm seasons and slightly increasing during cold seasons. However, the homogeneity of precipitation datasets allows one to consider these differences during the last 50 years as insignificant. An analysis of daily temperature dynamics for each season reveals their insignificant difference within the obtained series. However, mean temperatures for larger intervals (weeks, 2 weeks and month) for each season form inhomogeneous series. Thus one can state that according to NCEP/NCAR Reanalysis and NCEP/DOE Reanalysis-2 data in West Siberia a significant increase in mean temperatures takes place during spring and summer. For the winter season the results obtained reveal more complicate dynamics. Here, weekly mean temperatures are decreasing, while for 2 weeks and monthly mean temperatures are increasing.

It should be noted that this chapter is aimed to describe only one of the approaches adopted in the region. We do not dwell upon extensive information resources on biodiversity (Fedotov et al. 1998) organized as an e-library (http://www-sbras.nsc.ru/win/elbib/bio/). One more element of the infrastructure is based on the GOFC-GOLD Northern Eurasia Regional Information Network (NERIN) database (http://www.fao.org/gtos/gofc-gold/net-NERIN.html). Its Russian language mirror (http://nerin.scert.ru/) provides access to data and metadata describing different features of the regional environment.

As for further development of the infrastructure for SIRS, firstly we plan to add new meteorological and other environmental datasets into the collected and stored data archive as well as to add new functionality to the web systems under development. Among those a functionality to work with maps, satellite images and remote sensing data should be added. At the second stage we plan to inter-relate the developed element thus arriving at the Distributed Information-Computation System (DICS) based on the approaches described in this Chapter. It will accumulate data also by means of developed applications in thematic domains, and generate thematic data and knowledge, including those ready for computer processing. Sources of this knowledge are tasks and problems of the thematic domain integrated into information systems as applications. The system will be provided with tools, which allow multidisciplinary users to perform calculations correctly and to avoid redundant activity. In the process of user work in the system thematic domain objects as well as processes, in which they are participating will be followed by data specified by task ontology and domain ontology. Software tools, which allow programmers and knowledge engineers to integrate thematic applications into the system and provide DICS users with conditions for correct work will be also developed and employed by the system. This system must add value to existing data, by providing archive databases with relevant APIs and descriptions, in order to enable independent agents to discover, query, retrieve results in usable/understandable formats, and transfer them to other agents for further processing.

As a whole this activity will lead to the appearance of a powerful instrument for a better understanding of the interactions between the ecosystems, atmosphere, and human dynamics in this large region under the impact of global climate change. It will also provide national and international teams carrying out regional environment research with powerful tools to perform multidisciplinary investigations of the Siberian environment under Global Change. In particular, it will provide researchers with a reference open platform (portal plus tools) that may be used, adapted, enriched or altered on the basis of the specific needs of each application. Extended information resources on the state of the Siberian environment available to regional environment managers and decision makers via the Internet is expected to increase their concern for regional environmental issues.

Acknowledgments A support of a number of national and international projects coordinated by the authors is appreciated, especially acknowledged is partial support of SB RAS Basic Research Program 4.5.2 (Project 2) and Integrated Projects 34 and 86, FP6 EC Projects ENVIROMIS 2 (INCO-CT-2006- 031303) and Enviro-RISKS (INCO-CT-2005- 013427) as well as RFBR grants 05-05-98010, 04-07-90219, 06-07-89201 and 07-05-00200.

References

Alekseev VA, Volodin EM, VYa G, Dymnikov VP, Lykosov VN (1998) Modern climate modelling on the base of INM RAS atmospheric model. VINITI, Moscow

Baklanov A, Gordov EP (2006) Man-induced environmental risks: Monitoring, management and remediation of man-made changes in Siberia. Comput Technol 11:162–171

Belikov DA, Starchenko AV (2005) An investigation of secondary pollutants formation (ozone) in the atmosphere of Tomsk. J Atmos Oceanic Optics 18(05–06):391–398

Bulletin of the Russian National Committee for the International Geosphere Biosphere Programme (2005), In: Vaganov et al. First steps on development of Integrated Regional Study of Siberian environment. 4:4–10

De Roure D, Jennings N, Shadbolt N (2001) A Future e-Science Infrastructure. Report commissioned for EPSRC/DTI Core e-Science Programme

Dudhia J (1993) A nonhydrostatic version of the Penn State/NCAR mesoscale model: Validation tests and simulation of an Atlantic cyclone and cold front. Mon Weather Rev 121(5):1493–1513

Fazliev AZ, Privezentsev AI (2007) Applied task ontology for the systematization of informational resources in molecular spectroscopy. In: Proceedings of the Russian Conference on digital libraries: Perspective methods and technologies, Pereyaslavl, Russia, 15–18 Oct 2007

Fedotov AM, Artyomov IA, Ermakov NB, Krasnikov AA, Potyomkin ON, Ryabko BY, Fedotov AA, Khorev AG (1998) Electronic atlas 'Biological variety of the Siberian flora'. Comput Technol 3(5):68–78

Gordov EP (2004a) Computational and information technologies for environmental sciences. Comput Technol 9(1):3–10

Gordov EP (2004b) Modern tendencies in regional environmental studies. Geogr Nat Resour Special Issue:11–18

Gordov EP, Begni G (2005) Siberia integrated regional study development. Comput Technol 10(2):149–155

Gordov EP, De Rudder A, Lykosov VN, Fazliev AZ, Fedra K (2004) Web-portal ATMOS as basis for integrated investigations of Siberia environment. Comput Technol 9(2):3–13

Gordov EP, Lykosov VN, Fazliev AZ (2006a) Web portal on environmental sciences "ATMOS". Adv Geosci 8:33–38

Gordov EP, Begni G, Heiman M, Kabanov MV, Lykossov VN, Shvidenko AZ, Vaganov EA (2006b) Siberia integrated regional study as a basis for international scientific cooperation. Comput Technol 11:16–28

IGBP Newsletter No. 50 (June 2002) IGBP II – Special edition issue, Global Change Newsl 50:1–3

IGBP Newsletter No. 55 (October 2003), In: Brasseur G. 3rd IGBP Congress overview. Global Change Newsl 55:2–4

Ippolitov II, Kabanov MV, Komarov AI, Kuskov AI (2004) Patterns of modern natural-climatic changes in Siberia: Observed changes of annual temperature and pressure. Geogr Nat Resour 3:90–96

Lavrentiev NA, Starchenko AV, Fazliev AZ (2006) Informatsionno-vychislitelnaya sistema dlya kollektivnogo issledovaniya problem atmosfernogo prigranichnogo sloya (Information-computational system for collective research of atmospheric boundary layer). Computat Technol 11:73–79

Melnikova VN, Shulgina TM, Titov AG (2007) Statistical analysis of meteorological data of NCEP/NCAR Reanalysis 1 and Reanalysis 2 for West Siberia. Program and abstracts of international conference CITES-2007. Publishing house of Tomsk CSTI, Tomsk, p 21

Okladnikov IG, Titov AG (2006) Web-system for processing and visualization of meteorological data. In: Gordov EP (ed) Environmental observations, modeling and information systems. Publishing house of Tomsk CSTI, Tomsk, p 42

Parson M, Barry R (2006) Interdisciplinary data management in support of the international polar year. EOS 87(30):295

Shokin YI, Fedotov AM (2003) Integration of informational and telecommunicational resources of Siberian Branch of RAS. Comput Technol 8(Special issue):161–171

TORQUE Resource Manager. http://www.clusterresources.com/pages/products/torque-resource-manager.php. Cited 19 Sept 2007

Van der Aalst (2004) Business process management demystified: A tutorial on models, systems and standards for workflow management. In: Desel J, Reizig W, Rozenberg G (eds) Lectures on concurrency and Petri nets. Springer, Berlin, Heidelberg, New York, pp 1–65

Volodin EM, Dianskii NA (2003) Response of a coupled atmosphere–ocean general circulation model to increased carbon dioxide. Izvestiya, Atmos Ocean Phys 239:170–186

Weather Research and Forecasting Model. http://www.wrf-model.org. Cited 19 Sept 2007

Chapter 15
Conclusions

H. Balzter

The contributions presented in this book have highlighted a multitude of processes leading to environmental change in Siberia. Many different ecosystem types, regions, species, biogeochemical cycles and administrative districts are subjected to various types of change.

Many of the findings presented in this book have relied on satellite-derived data to give new insights into large-scale changes. Other chapters utilise complex dynamic vegetation models to make predictions of what the future might hold. Many chapters would not have been possible without access to invaluable field data and forest survey information. One interesting observation I wish to make is that only by a combined analysis of field observation, model results and satellite imagery is it possible to gain a holistic view of the multiple scales at which environmental change processes operate today.

Siberia will remain getting public attention by the global change community over the next decades in anticipation that some major changes are likely to be observed here. Let us hope that Siberia is not heading for an environmental catastrophe and global climate mitigation efforts will come to fruition in time to put the brakes on climate change.

H. Balzter (✉)
Centre for Environmental Research, Department of Geography,
University of Leicester, University Road,
Leicester, UK
e-mail: hb91@le.ac.uk

Appendix

Fig. 4.7 Siberian pine seedlings can grow while roots are within a thick (>40 cm, insert) moss cover. Mid of WE transect

Fig. 4.8 Siberian pine regeneration forming a second layer under the larch canopy

Fig. 6.2 Larch stand killed by stand-replacing ground fire

Fig. 6.10 Larch regeneration after a ground fire happened in the late summer – begin of fall 1993. Fire event coincides with the year of high cone production. Number of saplings is about 700,000 thousand/ha (2001 year)

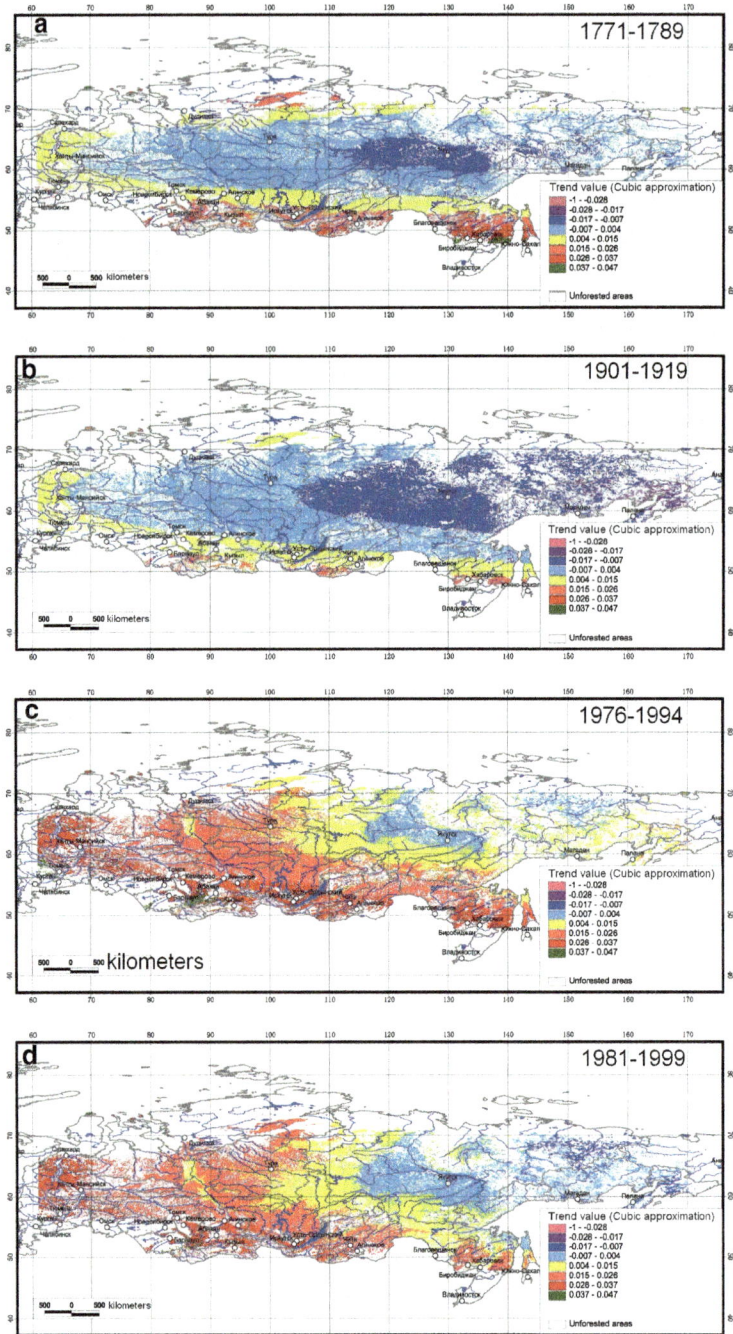

Fig. 7.4 Maps of tree ring growth trends obtained for different time periods. (**a**) 1771–1789 (centre year is 1780); (**b**) 1901–1919 (1910); (**c**) 1976–1994 (1985); (**d**) 1981–1999 (1990). There is common tendency in trend values which decreases from West to East

Appendix 259

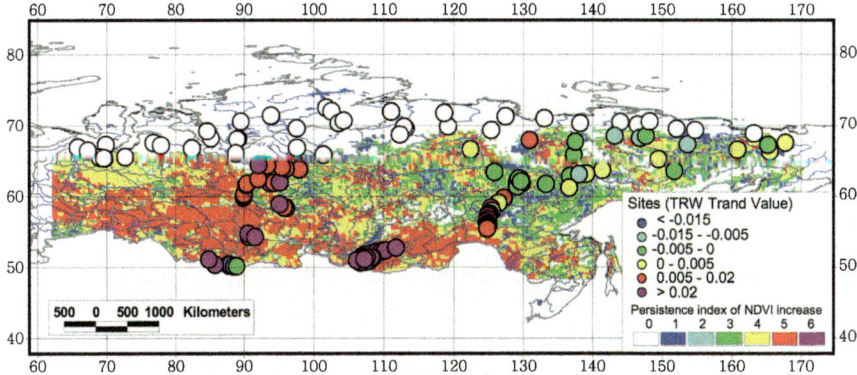

Fig. 7.6 The map of NDVI trends superposed with spatial values of tree-growth trends (*coloured circles*) for Siberia and Far East

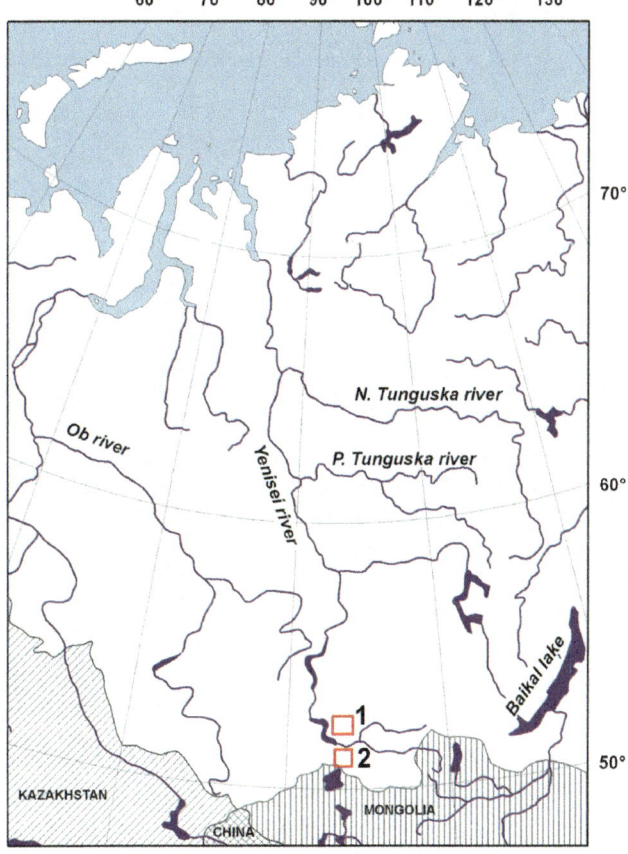

Fig. 8.1 The location of the study area. 1 – North Sayan, and 2 – Tannu – Ola sites

Fig. 8.5 Regeneration is often sheltered by felled predecessors. *Inset*: current year sapling with nut remains on the needles

Fig. 8.6 Siberian pine and larch regeneration reached and surpass the former tee line; also the height is still lower than its predecessors

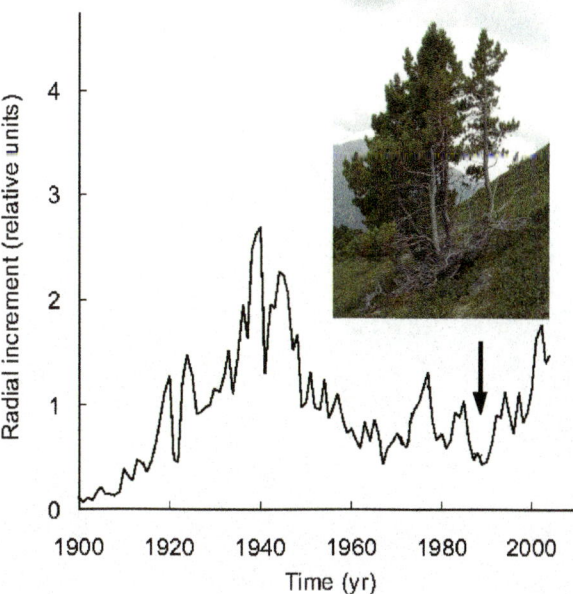

Fig. 8.9 Radial increment dynamics of "post-prostrate" form of Siberian pine. An arrow indicates beginning of prostates transforming into arboreal forms. *Inset*: Siberian pines transformed from krummholz to arboreal form (elevation is about ~2,000 m)

Fig. 8.10 Larch surpasses Siberian pine in winter desiccation resistance, and growing arboreal where pine is prostrate

Fig. 8.11 "Tree-in-skirt" Siberian pine. *Lower brunches* corresponds snow level

Fig. 8.12 Snow accumulation behind propagating trees and regeneration caused a zone of tree vegetation depression. Picture was taken 04 July 2006

Fig. 8.13 Tree clusters are oriented according prevailing winds. The *cluster's shape* is reducing wind impact

Fig. 8.14 Chlorophyll-contain tissues in Siberian pine bark

Fig. 8.15 Willow (Salix spp.) is vegetating in the upper tree limit without leaves. Picture was taken 03 July 2006

Appendix 265

Fig. 8.16 Trees are "diffusing" along elevation gradient. Peripheral trees have a krummholz form

Fig. 8.17 Trees are "diffusing" into tundra zone

Fig. 8.18 "Mother-larch" during centuries is providing seeds for new regeneration waves

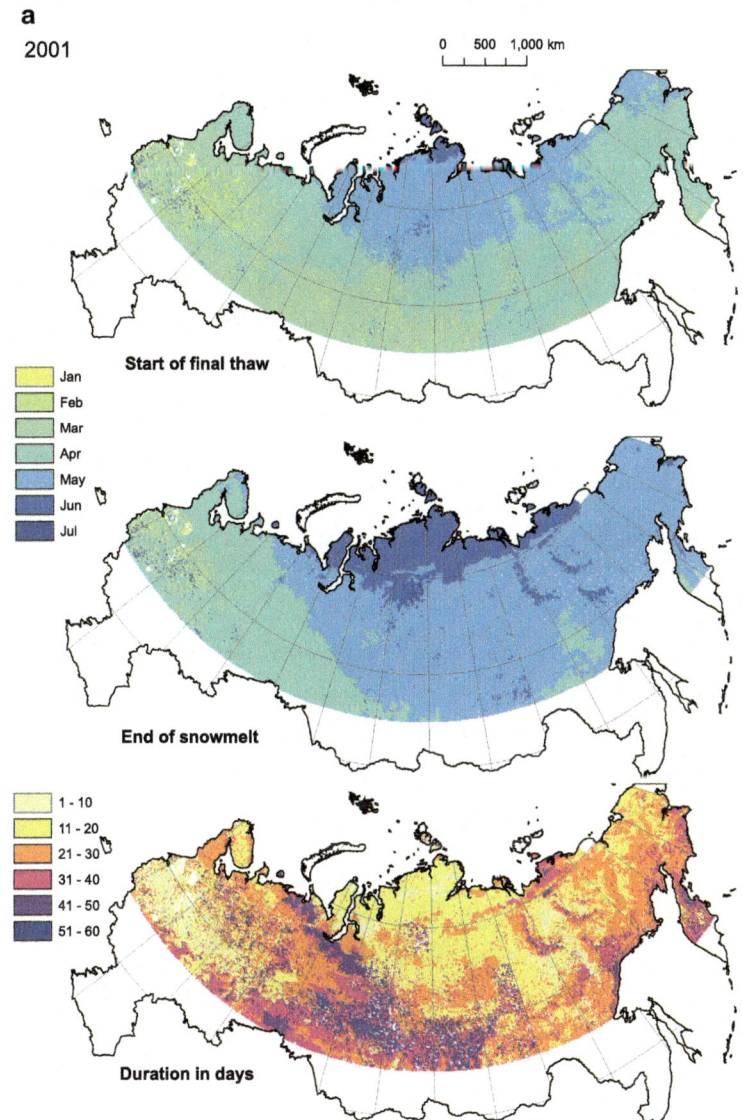

Fig. 9.4 (a) Spring freeze/thaw start, end and duration for Russia 2001 north from 55°N

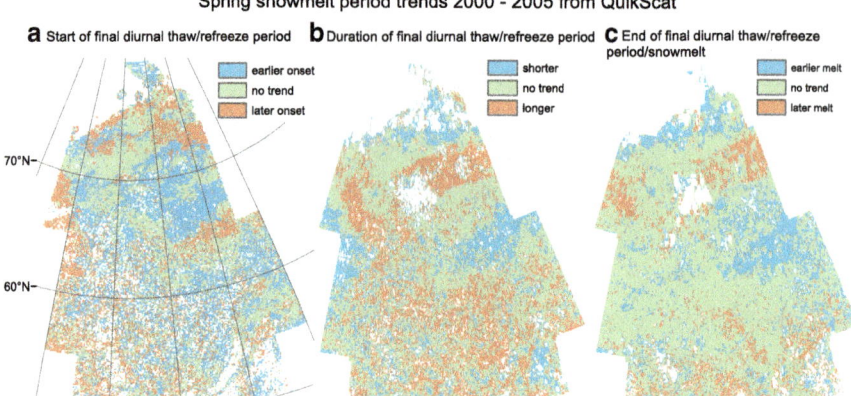

Fig. 9.5 Trends for the snowmelt period 2000–2005 from Seawinds QuikScat in central Siberia: (**a**) start, (**b**) duration, and (**c**) end of diurnal thaw/refreeze period/snowmelt with landscape groups (source: IIASA) within the Siberia II project area; 1 – arctic, 2 – sub-arctic severe, 3 – sub-arctic moderate, 4 – boreal severe, 5 – boreal continental, 6 – sub-boreal (steppe)

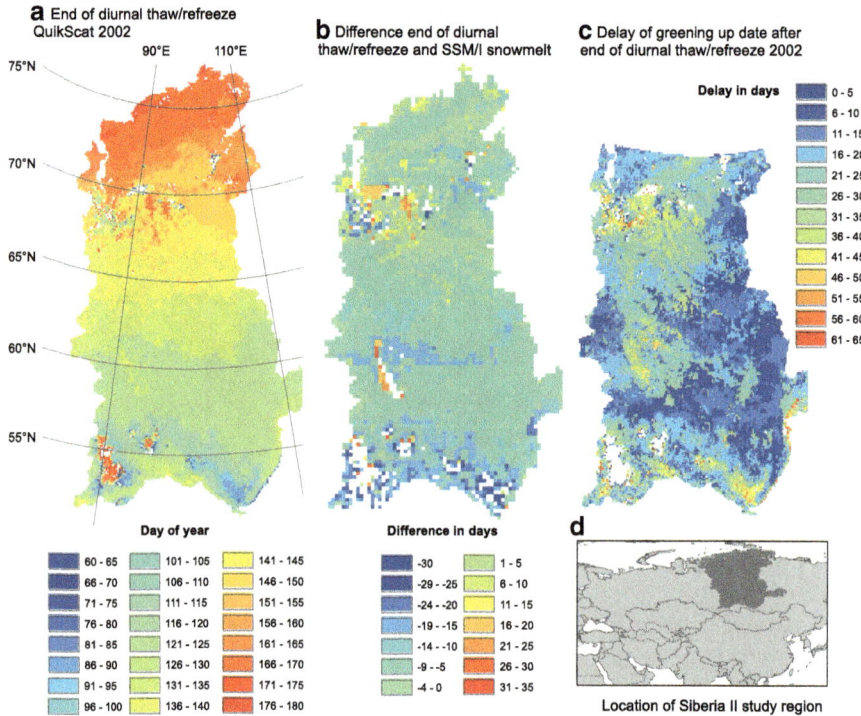

Fig. 9.6 Comparison of spring snowmelt and phenology products from the Siberia II project for 2002: (**a**) end of diurnal thaw/refreeze period from QuikScat (Bartsch et al. 2007a); (**b**) difference between end of diurnal thaw/refreeze period and end of snowmelt from SSM/I (Grippa et al. 2005a, b); (**c**) delay of greening up (Delbart et al. 2005) after end of diurnal thaw/refreeze period; (**d**) location map

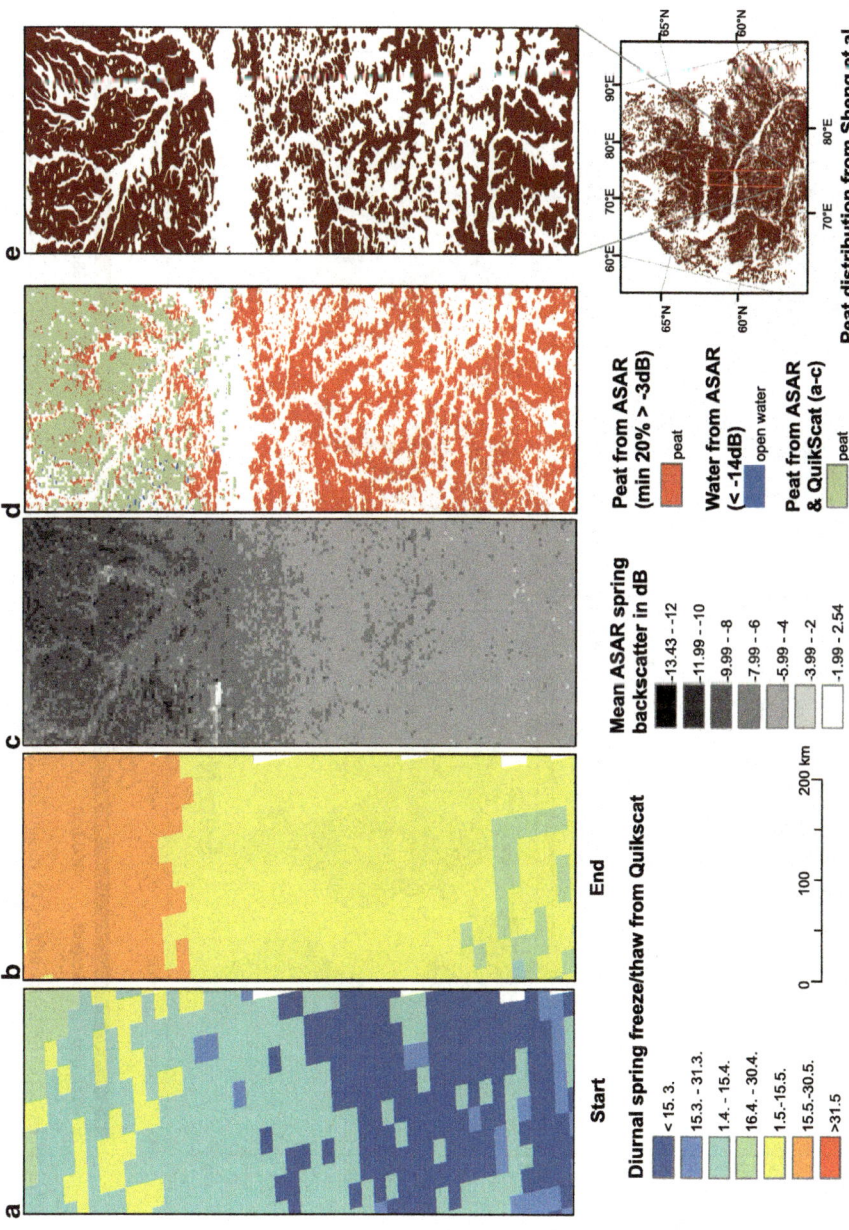

Fig. 9.9 Comparison of QuikScat freeze/thaw dates (2006) (**a**) onset, (**b**) end and (**c**) ENVISAT ASAR GM mean melt period backscatter (2006), (**d**) ASAR GM derived open peat land (July–Sept 2006, threshold method) and (**e**) peat distribution from the West Siberian Lowland database (Sheng et al. 2004). For location see inset map

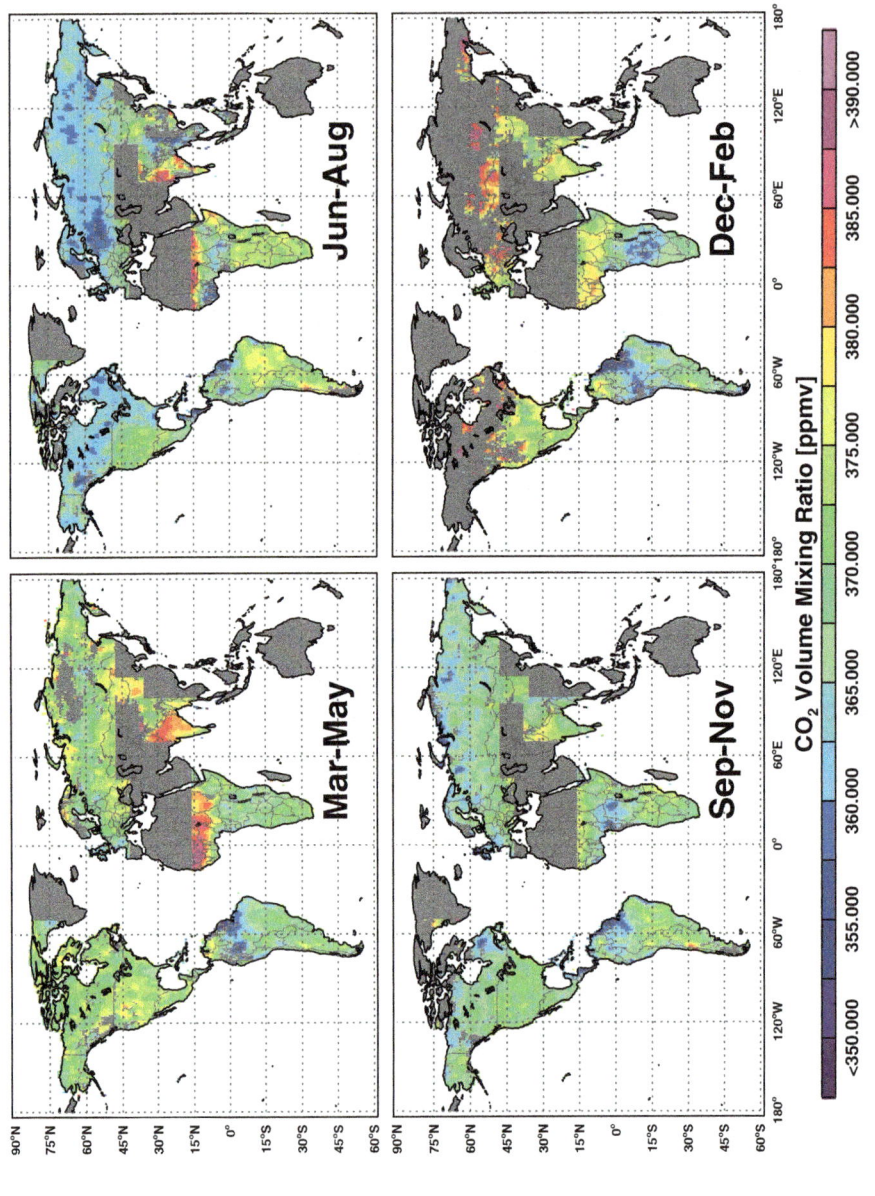

Fig. 11.1 Global SCIAMACHY tri-monthly CO_2 means for 2003 averaged on a $1° \times 1°$ grid and smoothed with a $3° \times 3°$ box car filter

Appendix

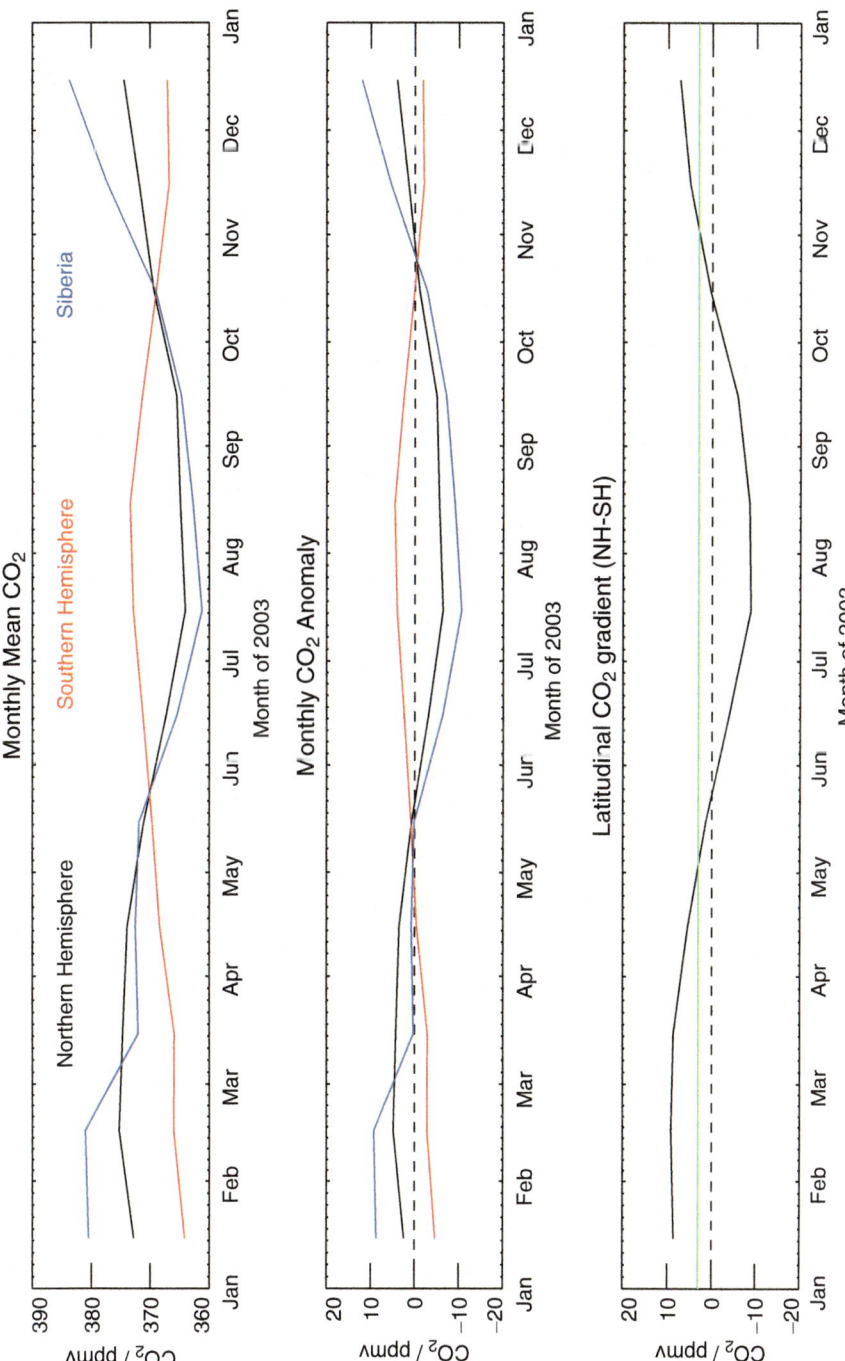

Fig. 11.2 *Top panel*: Time series for 2003 of the monthly means for the northern (*black*) and southern (*red*) hemispheres. The time series for Siberia (*blue*) is also shown. *Middle panel*: Time series of the CO_2 anomalies (i.e. each monthly mean minus the yearly mean). *Bottom panel*: The latitudinal inter-hemispheric gradient (north–south), the 3 ppmv gradient that is expected from surface observations (Conway et al. 1994) is shown as the *green dashed line*

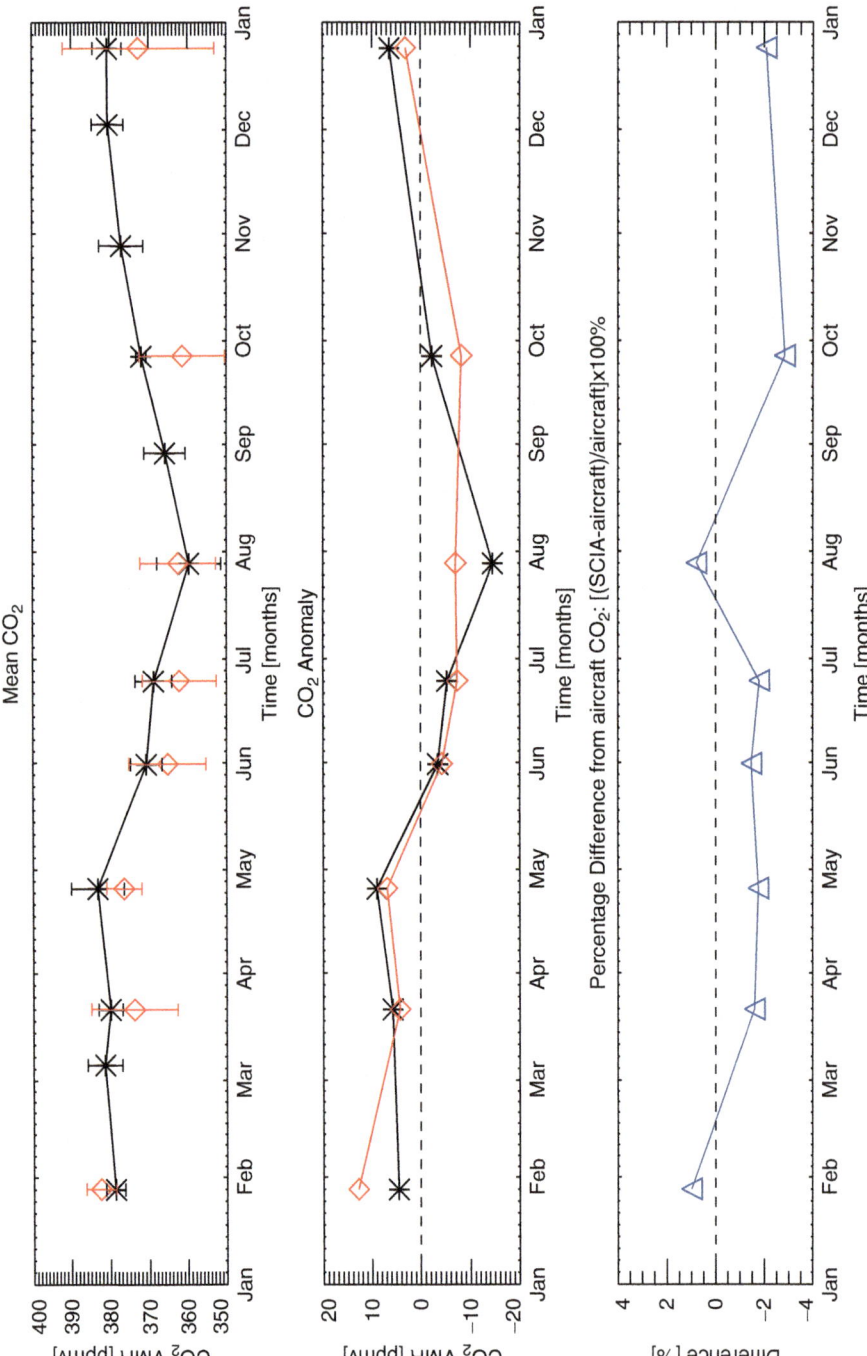

Fig. 11.3 SCIAMACHY (*red*) and aircraft (*black*) measurements over Novosibirsk. *Top panel*: The mean aircraft mixing ratio (over all altitudes) and the SCIAMACHY column VMRs. The *error bars* represent the 1 standard deviation uncertainty. *Middle panel*: The CO_2 anomaly (using only coincidental observations). *Bottom Panel*: The percentage difference between SCIAMACHY and the mean aircraft mixing ratio

Appendix 273

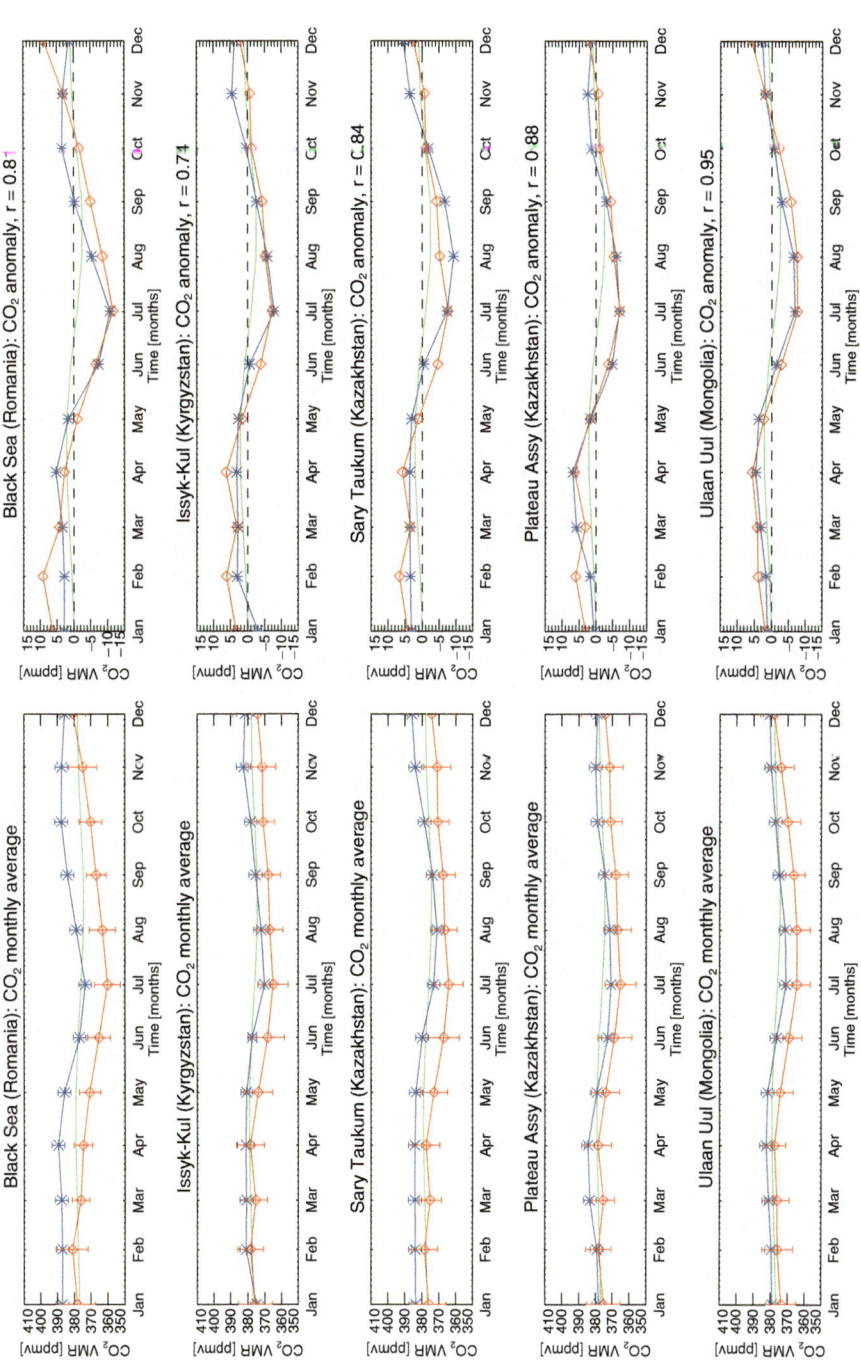

Fig. 11.4 Time series of ground based in situ observations (*blue*) versus SCIAMACHY CO_2 (*red*). *Left panels*: The CO_2 monthly averages with the 1 standard deviation error bars. *Right Panels*: The corresponding CO_2 monthly anomaly. The a priori CO_2 column VMR is shown in *green*. The in situ data is from the World Data Centre for Greenhouse Gases (WDCGG) (http://gaw.kishou.go.jp/wdcgg/)

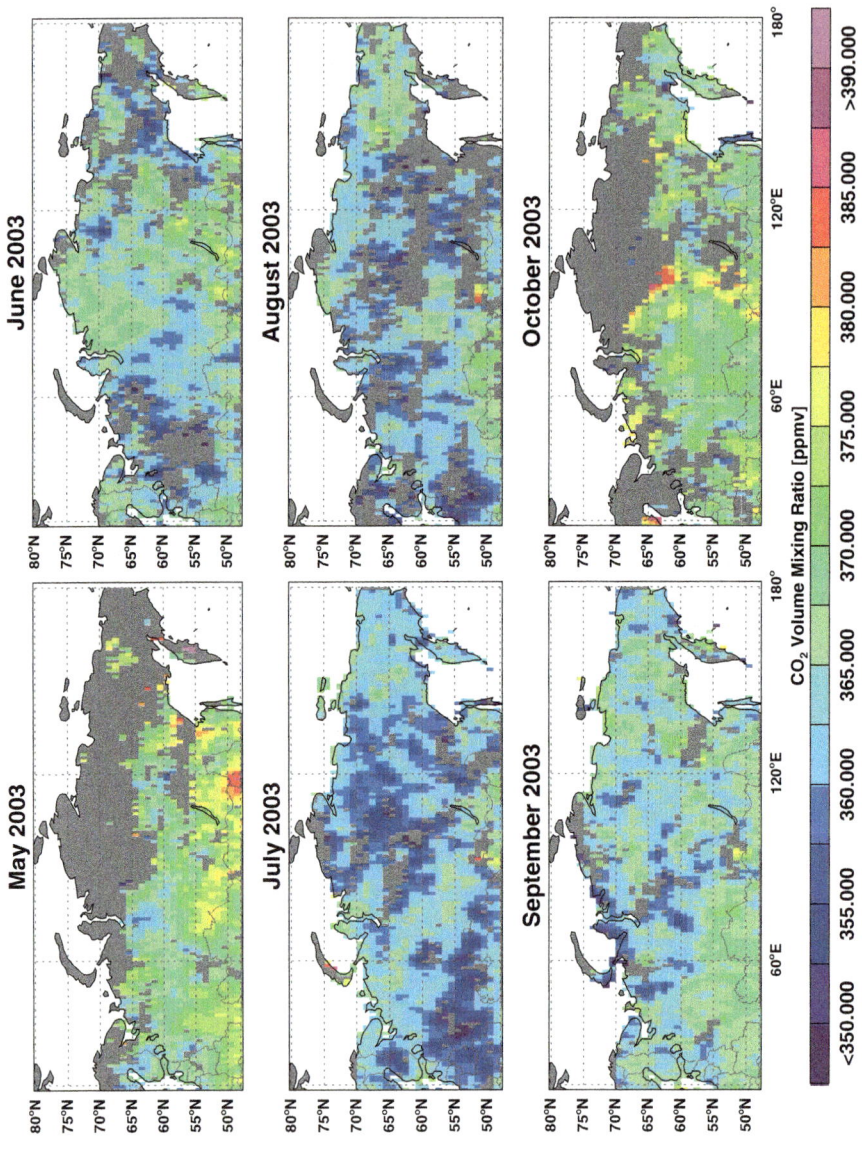

Fig. 11.5 SCIAMACHY monthly CO_2 distributions over Siberia for May–September 2003 averaged on a 1° × 1° grid and smoothed with a 3° × 3° box car filter

Appendix

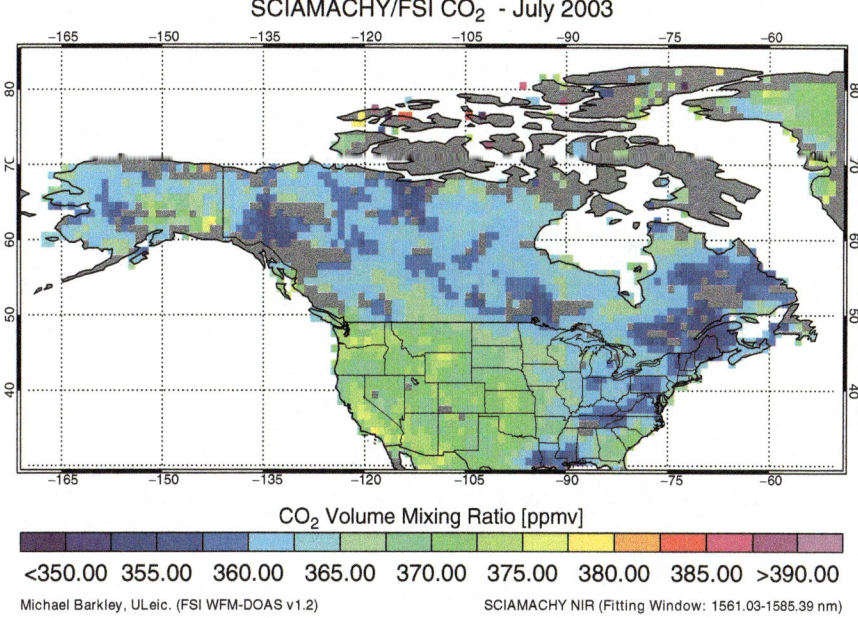

Fig. 11.6 SCIAMACHY CO_2 over North America for July 2003

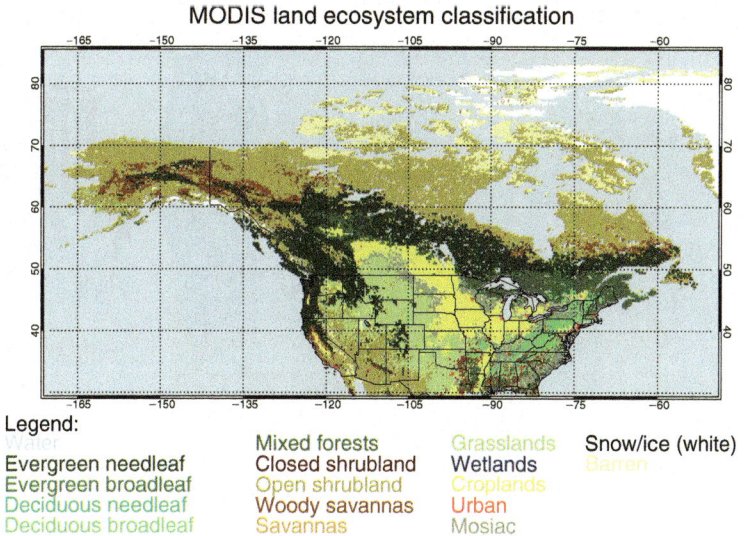

Fig. 11.7 Map of the land vegetation distribution over North America. The vegetation map is taken from the Land Ecosystem Classification Product which is a static map generated from the official MODIS land ecosystem classification data set, MOD12Q1 for year 2000, day 289 data (October 15, 2000) (see http://modis-atmos.gsfc.nasa.gov/ECOSYSTEM/index.html)

Fig. 13.1 The region of interest of SIBERIA-II (green-border areas) and SIB-ESS-C (*red bounding box*)

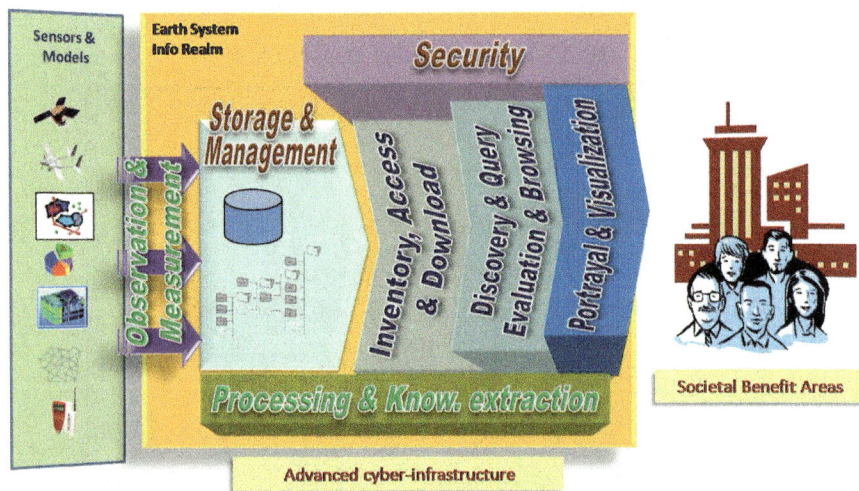

Fig. 13.2 Simplified architectural schema of an advanced infrastructure for Earth sciences information

Appendix

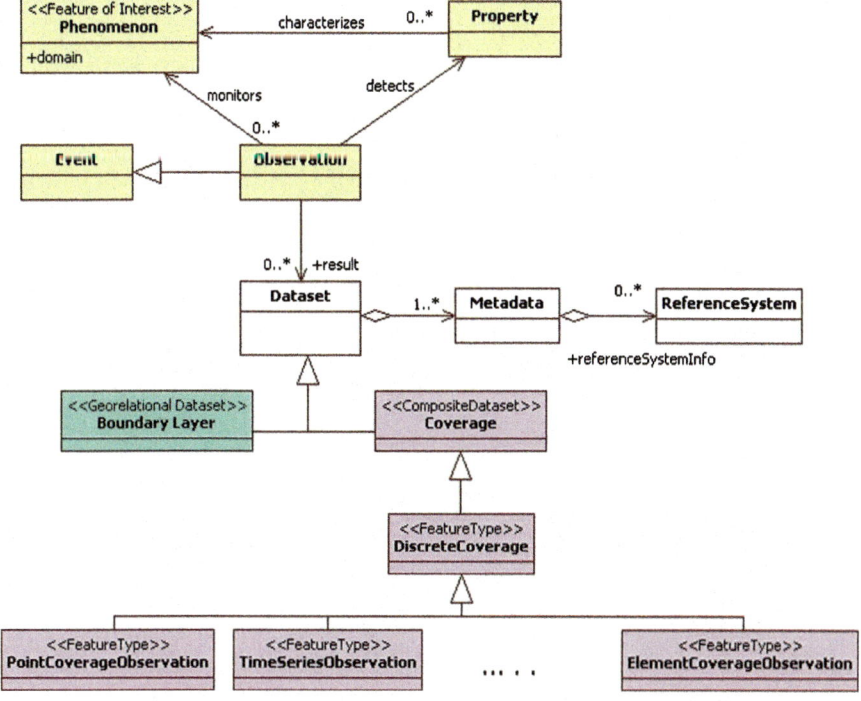

Fig. 13.3 The general dataset conceptual model

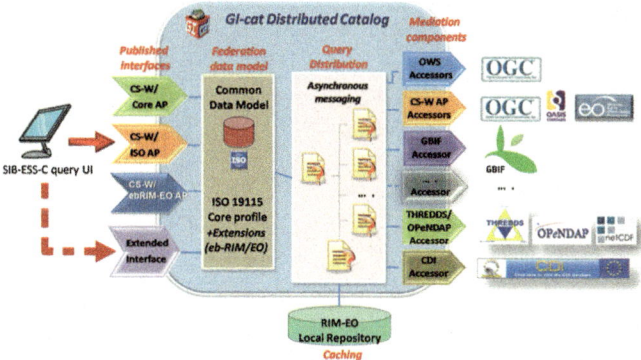

Fig. 13.4 SIB-ESS-C components architecture implementing the discovery and access services

Index

A

Abies sibirica, 38, 39, 41, 43, 45, 46, 54, 69–71, 84, 117
Absorbed photosynthetically active radiation (fAPAR), 34
Active fire, 26, 27, 98
Active layer, 72, 98, 158–161, 163, 164
Active microwave, 136, 138, 141, 146, 153
Advanced Microwave Scanning Radiometer (AMSR-E), 138, 140
Air quality, 215, 237, 240, 243, 246–249
AIRS, 174, 187, 189
Albedo, 33, 34, 64, 131, 177, 187, 189
Along Track Scanning Radiometer (ATSR), 187
Alpine tundra, 118
Altai–Sayan, 37–50, 129, 195, 196, 198, 200
Altitude, 104, 105, 116, 117, 182–184, 196, 204, 205, 242, 245, 248
AMSR-E. *See* Advanced Microwave Scanning Radiometer
Anemone baicalensis, 39, 41, 45, 47
Angara, 5, 31, 32, 55, 206, 207
Aqua, 138
Arboreal, 123–125, 127, 131
Arctic oscillation (AO), 22, 23, 30
ASAR, 139, 151, 152
Aspen, 6, 37–50, 54, 56, 70, 84, 94
Athyrium filix-femina, 39, 44, 45
Atmosphere, 22, 33, 34, 147, 158, 160, 164, 174–177, 179, 189, 194, 197, 218, 235, 236, 238–240, 242, 247, 250
Atmospheric CO_2 concentrations, 102, 173–190

B

Bark, 54, 63, 96, 97, 127, 128
Barrier, 63, 124, 194, 202–208
Betula pendula, 41, 43, 46, 54, 70, 84, 85

Betula pubescens, 54, 70, 84
Biomass burning, 22, 174, 187
Birch, 5, 41, 44, 47, 49, 50, 54, 57, 58, 60, 61, 64, 70, 84, 85, 94, 98
Bogs, 5, 15, 18, 80, 89, 93, 151
Burned area, 5–8, 17, 23–27, 29–30, 32–34, 93, 97, 98

C

Canadian Forest Fire Weather System, 31
Canopy, 7, 8, 17, 18, 26, 63, 94, 98, 121, 195
Carbon cycle, 22, 34, 136, 174, 175, 177, 189
Carbon dioxide (CO_2), 22, 33, 34, 102, 111, 136, 140, 141, 144, 147, 149–150, 173–190, 238
Central Siberia, 5, 22, 29–31, 54, 55, 68, 70, 73–76, 78–80, 84, 85, 129, 141, 142, 145–147, 149, 187, 195–197, 216, 224
Chern, 38–40, 42, 45, 47, 49, 50, 69, 71, 77
Chlorophyll, 4, 5, 7, 8, 11, 12, 17, 18, 127, 128
Climate, 5, 21–34, 39, 42, 46, 47, 50, 54–56, 58–60, 64, 67–81, 86, 94, 97, 98, 101–111, 115–131, 136, 146, 153, 157–168, 174, 194, 196, 197, 201, 206, 215, 216, 218, 235, 237–246, 249, 250, 253
Climate change, 21–23, 33, 46, 68–70, 72–79, 94, 97, 101–111, 131, 136, 153, 164, 174, 216, 237–239, 243, 250, 253
Climate change scenarios, 68–70, 72–74, 77–79
Climate data, 30, 162
Climate model, 21, 30, 111, 158, 159, 240
Climatic anomalies, 69, 72
Cluster analysis, 105
Computing cluster, 218, 240, 241
Crown closure, 43, 54, 64, 84
Cryolithic zone, 157–168

279

D

Data visualization, 217, 228, 229, 237, 244
Desert, 54, 68, 71, 75, 77, 79, 127
Dissimilarity coefficient, 41, 47, 48
Disturbance, 3–18, 22, 26, 27, 38, 49, 55, 98
Duff Moisture Code (DMC), 31, 32
Duschekia fruticosa, 61, 94

E

Earth Observation, 4, 214, 215, 217, 219, 224, 226, 229
Eastern Siberia, 84, 157–168, 197, 208
East Siberia, 89
Ecoregion, 37–50, 196–202
Ecotone, 54, 84, 115–131
Elevation, 77, 78, 84, 86, 89, 97, 116–119, 121, 123, 124, 129, 146, 194, 195, 198, 200–208, 221
Emission, 5, 8–18, 22, 33, 34, 69, 85, 97, 138, 141, 174, 177, 179, 187, 249
Environmental monitoring, 189, 215, 216, 219, 236
ENVISAT, 4, 8, 12, 14, 16, 18, 138, 139, 151–153, 175
ERS-1/2, 138
The European Global Monitoring for Environment and Security (GMES), 215, 218, 224, 226
Evenkia, 78
Experimental plot, 39, 43, 46–48

F

Far East, 22, 103, 111
Feedback, 22, 23, 33–34, 64, 131, 174, 238
Fir, 37–50, 54, 56, 57, 62, 64, 84, 94, 97, 117, 123
Fire, 4–7, 21–34, 38, 46, 54, 55, 60–61, 68, 80, 81, 83–98
Fire danger, 31–33, 88
Fire prevention, 31
Fire regime, 21–23, 30, 34, 68
Fire return interval (FRI), 60, 87–94, 96–98
Floristic composition, 47, 49
Foehn, 194
Forest fire, 18, 26–34, 81, 97
Forest Fire Intensity Dynamics (FFID), 23, 26–30, 32, 33
Fuel, 22, 30, 31, 84, 86, 97, 98, 174

G

Geospatial information, 213, 214, 216, 218–221, 224
Germination, 129, 131
GIS, 15, 18, 108, 151, 226, 235
Global circulation model, 157
Global Land Cover 2000, 23
Global warming, 22, 80, 131, 174, 238
GMES. *See* The European Global Monitoring for Environment and Security
Graphical user interface, 244, 245, 247
Greenhouse gas, 22, 33, 34, 54, 84, 85, 97, 98, 111, 145, 158, 174, 175, 185, 224, 238, 239
Groundwater, 159, 161–163
Growing season, 7, 11, 14, 22, 26, 80, 102, 121, 140, 144

H

Herbaceous layer, 38, 41, 44–45, 47
Heterotrophic respiration, 33, 136, 147, 149
Holocene, 71, 77

I

Infiltration, 33, 159, 161
Information system, 214, 226, 234–236, 240, 250
Infrastructure for Spatial Information in the European Community (INSPIRE), 216, 218, 224, 226, 227, 229
INSPIRE. *See* Infrastructure for Spatial Information in the European Community
Interoperability, 213–229, 235
Invasion, 53–64, 94, 116

K

Krummholz, 123, 124, 127, 129, 131

L

Lake Baikal, 54, 84, 145, 195
Land cover, 23, 27, 145, 187
Larch, 5, 30, 31, 53–64, 70, 76, 83–98, 115–131
Larix cajanderii, 70, 72
Larix dahurica, 72
Larix gmelinii, 70, 76, 78, 85
Larix sibirica, 70, 71, 76, 118
Lena, 5, 147, 148, 157–168, 195, 198, 216

Index 281

Lichen, 16, 84, 96, 97
Lightning, 22, 96, 97
L3JRC, 23–26, 29, 30

M
Matteuccia struthiopteris, 39, 41, 44, 45, 48
Medium Resolution Imaging Spectrometer (MERIS), 4–8, 11–14, 16, 18, 138, 153
MERIS. *See* Medium Resolution Imaging Spectrometer
MERIS Terrestrial Chlorophyll Index (MTCI), 5, 7–14, 16–18
Metadata, 214, 216–218, 224–229, 235, 236, 243, 244, 250
Meteor MSU-E, 7
Meteorology, 61, 237, 249
METOP, 138
Migration, 64, 80, 94–95, 97, 116, 126, 131
Model, 21, 30, 38, 48, 68–73, 76, 77, 111, 158–164, 174, 176, 177, 179, 187, 189, 190, 196, 202, 203, 205–208, 214–219, 221, 222, 224, 226–229, 234–237, 240–243, 247, 249, 253
MODIS, 4, 26, 27, 34, 84, 87, 138, 139, 188, 189
MODIS Enhanced Vegetation Index (EVI), 189
Moisture content, 22, 30, 31, 202
Mortality, 6–9, 11, 17, 18, 94, 97, 118, 119
Moss, 16, 38, 62, 84, 95–97
Mountain, 5, 11, 15, 17, 38–40, 42, 46, 47, 49, 50, 68, 69, 71–74, 76–80, 97, 115–131, 145, 187, 194–196, 198, 200–208
Mount Pinatubo, 30

N
Nimbus-7, 138
Norilsk, 5, 8, 11–14, 17
Normalised Difference Short-Wave Infrared Index (NDSWIR), 26
Normalized Difference Snow Index (NDSI), 140
Normalized differentiated vegetation index (NDVI), 4, 7–16, 102–104, 108, 110, 111, 139, 140, 144, 189
Novosibirsk, 182, 183

O
Ob, 147, 148, 150, 151, 158, 216
Orographic, 11, 194, 195, 203–205, 208

P
Peat land, 136, 147, 149–152
Permafrost, 54, 62, 71–73, 76, 78, 80, 84, 85, 96–98, 136, 148–151, 158–161, 163, 238
Phenology, 136, 144, 146, 147, 149–150
Photosynthetic activity, 4, 34, 187
Picea obovata, 54, 69, 71, 76, 78, 84, 117
Pine, 5, 38, 39, 43, 46–50, 54, 56–58, 60–64, 70, 80, 84, 94, 97, 115–131, 140
Pinus sibirica, 38, 39, 41, 43, 44–47, 50, 54, 69, 71, 76, 78, 84, 117, 118
Pinus silvestris, 38, 39, 41, 43–47, 54, 69, 71, 76, 78, 84, 117, 118
Pollution, 4, 5, 8, 12, 15, 17, 18, 181, 237, 243, 246–249
Population density, 96, 158
Populus tremula, 38, 39, 41, 43, 46, 54, 70, 71, 84
Precipitation, 22, 39, 54–56, 58–61, 63, 64, 68, 69, 72–74, 77, 84, 86, 93, 117, 118, 119, 121, 123, 124, 126, 127, 129, 131, 142, 158–160, 162–167, 193–208, 244, 249
Putorana Plateau, 54, 78, 144, 145, 147

Q
Quickscat, 136, 138, 140–142, 144–149, 151–153

R
Red edge, 4
Reflectance, 4, 5, 7, 8, 13, 23, 26, 187
Regeneration, 45–49, 54–64, 87, 94, 95, 98, 116, 118–124, 126, 127, 129–131
Regional climate change, 237, 239, 243
Regional environmental studies, 235, 237–239, 243, 250
Remote sensing, 23, 30, 34, 40, 135–153, 213, 216, 239, 250
Respiration, 33, 136, 147, 149
River runoff, 136, 140, 141, 147, 148, 157–168
Runoff, 84, 136, 140, 141, 144, 147–149, 157–168
Russian fire danger index (FD), 31, 32
Rybnaya river, 11, 13, 17

S
Sap flow, 140
Saplings, 58, 60–63, 94, 95, 120, 121

Satellite, 3–18, 23, 26, 34, 108, 136, 138, 151, 174, 175, 179, 181, 184, 187, 189, 190, 213, 229, 250, 253
Scanning Multichannel Microwave Radiometer (SMMR), 138
SCIAMACHY, 173–190
Scots pine, 5
Seawinds, 138, 140, 141, 145, 151, 153
Seedlings, 62, 63, 118, 119, 123, 124, 126, 129, 131
Shannon index, 41, 48
Siberian larch, 83–98
Siberian pine, 38, 39, 43, 47–49, 50, 52, 56–58, 60–64, 80, 84, 94, 97, 115–131
Slope, 38, 40, 47, 55, 56, 64, 78, 84, 87, 89, 91, 97, 103, 116–118, 194, 195, 200, 202–208
Snow abrasion, 54, 124, 126, 127
Snowmelt, 135–153, 161
Snow water equivalent (SWE), 136, 140, 162
Soil, 17, 33, 39, 46, 49, 55, 60, 63, 76, 78, 84, 87, 93, 94, 96, 97, 136, 141, 144, 145, 147, 149, 151, 158–163, 241, 244
Soil moisture, 46, 84, 141, 160–163
Southern Siberia, 38, 64, 68, 115–131, 195, 198, 208
Spatial data infrastructure (SDI), 216, 217, 219, 224
Special Spectral Microwave Imager (SSM/I), 138–140, 144–147
Species migration, 94–95, 97, 116, 131
SPOT, 4, 8, 11, 13–16, 23, 84, 87, 138, 139, 144, 145, 147, 153
Spruce, 54, 56, 57, 61–64, 84, 94, 97, 117
Stand density, 116, 131
Steppe, 40, 71, 76, 77, 80, 83, 117, 118, 145, 147, 195, 197, 201
Succession, 38, 39, 41, 43, 46–49, 97
Synthetic Aperture Radar (SAR), 138, 151

T

Taiga, 5, 39, 40, 45–48, 56, 64, 69, 70, 76, 78, 96, 97, 144–147, 149, 151, 197, 208
Temperature, 21–23, 31, 39, 46, 54–56, 58–61, 63, 64, 68–70, 72–74, 78, 84–86, 89–93, 97, 102, 104, 105, 111, 116–124, 126, 127, 131, 138, 142, 149, 150, 158–166, 176, 177, 216, 221, 222, 238, 242, 245, 249
Temperature anomalies, 21, 59, 60, 68, 70, 73, 74

Terra, 4, 26, 84, 87, 138
Terra-MODIS, 26, 84, 87
Thaw, 60, 62, 84, 85, 94–98, 136, 138, 140–153, 158–160, 162
Thermal anomalies, 27
Tomsk, 237, 239, 240, 242, 243, 246, 247
Topography, 5, 84, 85, 87, 89, 97, 148, 149, 202, 215
Transect, 54–58, 61, 62, 118
Transition zone, 118, 136, 151
Tree growth, 42, 96, 98, 102–105, 108–111, 116, 126
Tree height, 41, 42, 87, 126
Tree line, 78, 80, 81, 116, 124, 127, 131, 218
Tree-ring, 54, 88, 89, 102–104, 108, 109, 111
Tundra, 5, 15, 18, 54, 64, 68, 69, 71, 75–80, 83, 84, 97, 115–131, 141, 144, 145, 195, 197
Tunguska, 85

V

Vegetation, 4–18, 22, 23, 26, 27, 33, 34, 41, 55, 67–81, 84, 87, 102, 108, 110, 111, 116, 118, 124, 126, 131, 136, 138, 139, 145, 174, 179, 188, 189, 194, 195, 200–202, 207, 218, 242, 253
Vegetation cover, 4, 7, 15, 16, 18, 41, 55, 76–78, 87
Vegetation model, 69–72, 76, 253
Volcanic eruptions, 23, 30

W

Warming, 21, 22, 34, 50, 68, 69, 71, 73, 76, 78, 80, 102, 104, 110, 111, 115–131, 157–168, 174, 238, 249
Water balance, 33, 159–161
Water content, 26, 145, 151, 194
West Siberia, 5, 22, 68, 71, 77, 150, 152, 187, 195, 197, 203, 208, 240, 243, 249
Wildfire, 31, 38, 54, 55, 60–61, 83–98
Winter desiccation, 54, 63, 124–127

Y

Yakutsk, 182
Yenisei, 68, 76, 84, 142, 146, 148, 150, 158, 195, 197, 202–205, 207
Yenisey ridge, 54–56

CPSIA information can be obtained
at www.ICGtesting.com
Printed in the USA
LVOW02*0557230617
539032LV00002BA/61/P